Listening In

RAF Electronic Intelligence Gathering Since 1945

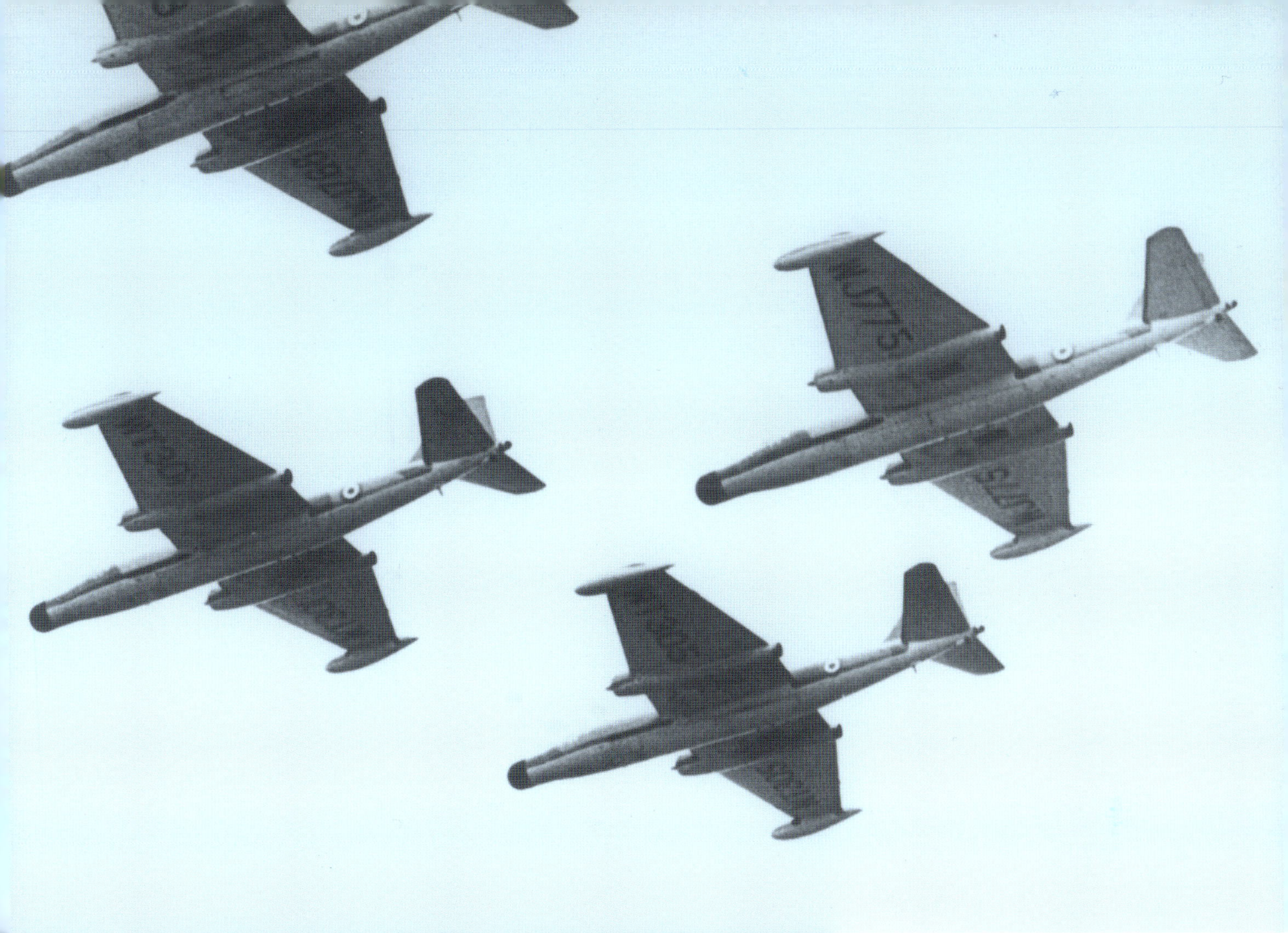

Listening In

RAF Electronic Intelligence Gathering Since 1945

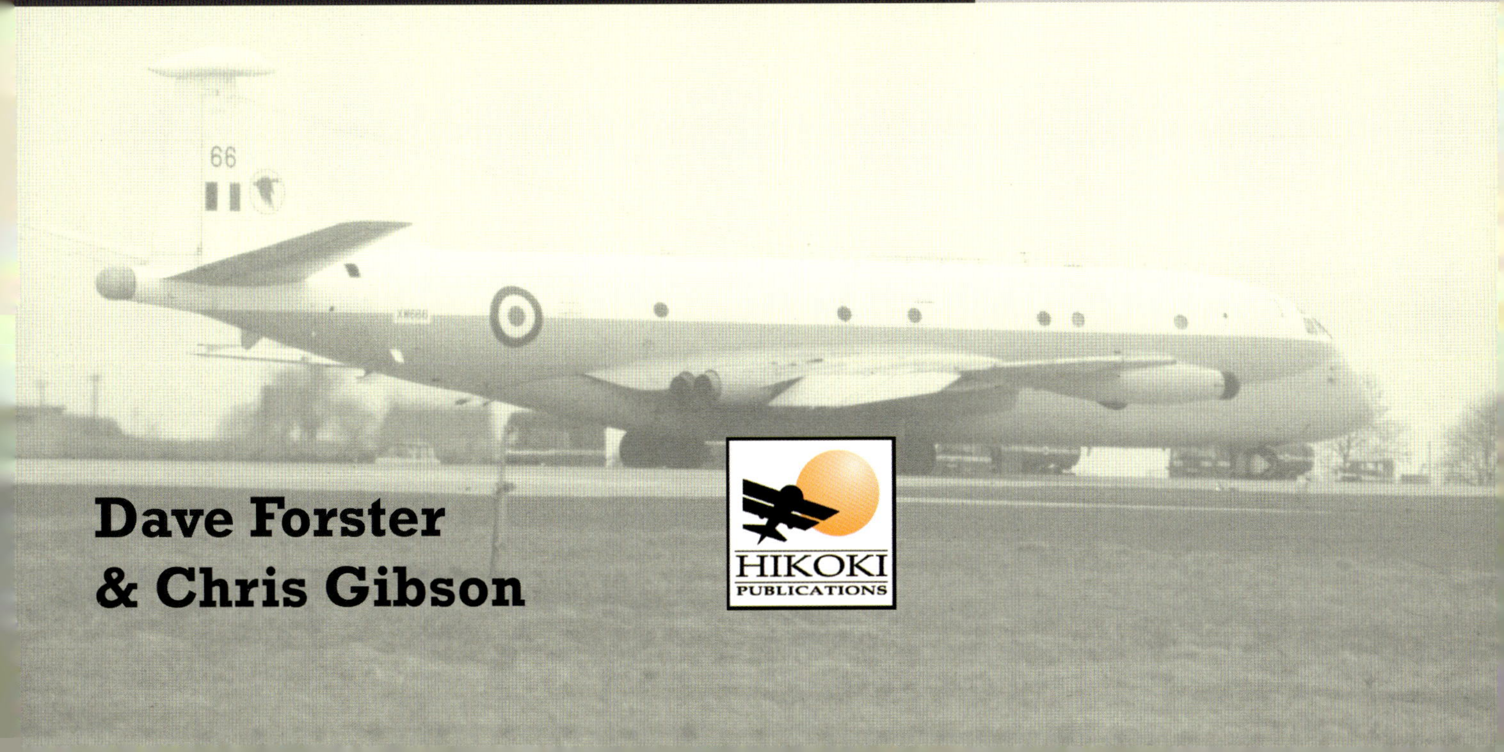

Dave Forster
& Chris Gibson

'Radio counter-measures are a kind of counter-attack and depend on detailed knowledge of the enemy's practice e.g. of the frequencies he will use. …We know something of the present rather backward Russian radar organisation, but we know little of their research, and nothing of their staff requirements for the future. So we can only plan on the supposition … that they will be equipped to our standards. But we cannot assume that they will be equipped like us; they will probably use quite different frequencies and methods'.
Air Marshal Sir Arthur Sanders, Vice-Chief of the Air Staff, December 1949

First published in 2014 by
Hikoki Publications Ltd
1a Ringway Trading Est
Shadowmoss Rd
Manchester
M22 5LH
England

Email: enquiries@crecy.co.uk
www.crecy.co.uk

© Dave Forster

Colour drawings: © Adrian Mann
Line drawings: © Dave Forster

Layout by Russell Strong

ISBN 9 781902 109381

All rights reserved. No part of this book may be reproduced or transmitted in any form or by any means electronic or mechanical, including photocopying, recording or by any information storage without permission from the Publisher in writing. All enquiries should be directed to the Publisher.

Printed in Malta by Melita Press

Front cover illustrations:

Top: Comet 2R XK695 landing at Luqa in Malta. *via Terry Panopalis*

Bottom: Canberra WT305 with turret behind the cockpit housing a radiometer to acquire information on infrared signatures. *Adrian Mann*

Back cover illustrations:

Top: Nimrod R.1 XV249 landing at Waddington, home of the RAF's ISTAR (intelligence, surveillance, target acquisition, and reconnaissance) assets. *Stuart Freer/Touchdown Aviation*

Centre left: B.6(BS) WJ775 in late 1960s configuration. *RAF Wyton via Peter Green*

Centre right: Washington WZ966 in flight showing the fuselage radomes and lack of gun turrets/sighting blisters. *via Paul Stancliffe*

Bottom: The proposed VC10R of 51 Squadron monitoring a Soviet Northern Fleet exercise. *Adrian Mann*

Half-title page: Hawker Siddeley Nimrod R.1 XV249 of 51 Squadron RAF climbs out from RAF Waddington on the start of an Elint sortie. *Stuart Freer/Touchdown Aviation*

Title page, top: Four Canberra B.6(BS) of 51 Squadron perform a formation flypast on 26th June 1971. *Author's collection*

Title page, bottom: Hawker Siddeley Nimrod R.1 XW666 at Wyton in original white over grey colour scheme of the maritime Nimrods and the original Elint fit with antennae in the tail cone, front and undersides of the slipper tanks being the most visible clues. *T Panopalis Collection*

Contents

Introduction 6
Acknowledgements 7
Foreword 8
  1 Noise Listening 9
  2 The Cold War Begins 19
  3 Upping the Stakes 30
  4 Washingtons 44
  5 Jet Elint and the Canberra 50
  6 Probing the PVO Strany 59
  7 Transfer of Control 71
  8 The Comet 91
  9 Elint in the 1960s and 1970s 105
10 World-wide Operations 126
11 New Directions 152
12 Replacing the Comet 158
13 The Nimrod in Service 176
Postscript: *Airseeker* 189
Conclusion 189
Glossary of Terms / Selected Bibliography 190
Index 191

The beginning of a new era for the 'Flying Goose' and 51 Squadron. The RAF's first Boeing RC-135W *Airseeker* (ZZ664) touches down at Waddington in November 2014. The crews had been training on USAF *Rivet Joint* aircraft since 2011. *MoD*

Introduction

'All the business of war… is to endeavour to find out what you don't know by what you do; that's what I called "guessing what was at the other side of the hill".'
Arthur Wellesley, 1st Duke of Wellington

During the Cold War, 1946-90, anyone with an interest in the Royal Air Force was aware that the RAF operated a number of what were officially described as 'Radar Calibration' aircraft. These could often be seen in the circuit at RAF Watton, RAF Wyton and later, RAF Waddington, at the beginning and end of their missions, but where they plied their trade and what that trade actually was, remained veiled in the utmost secrecy. These Washington, Canberra, Comet and Nimrod aircraft were the true spy planes of the Cold War and would have been described as such had their role been revealed to all and sundry. Certainly radar calibration was part of their role, but the lumps, bumps, antennae and dielectric panels that adorned these aircraft hinted at their having a special role, one that involved not an interest in the radars of the RAF, but those of the Soviet Union and its satellite states.

Air Vice-Marshal Sanders in 1949 was well aware that the RAF's wartime edge in radio countermeasures (RCM) and its expertise in what were described as the 'Instruments of Darkness' could provide the United Kingdom with an edge in any forthcoming war against the Soviet Union. The year 1949 saw the emergence of the North Atlantic Treaty Organisation (NATO) and with it a resolve to defend the Western Democracies from what was perceived as a serious threat from beyond the Iron Curtain. Nuclear retaliation was the weapon of choice against a Soviet invasion, given that only the USA possessed these devices, but on 29th August 1949 the balance of power changed when the Soviets carried out operation *Pervaya Molniya* (First Lightning), known in the West as Joe 1, the first Soviet atomic bomb.

From that point the Cold War became an arms race, each side developing and building its way to dominance through ever more technical and costly weapons, with the delivery of nuclear weapons being the primary means of attack. As time passed the destructive force of these weapons became such that their use by any side became unthinkable and the strategy of deterrence came to the fore. Deterrence only works if the other side believes that their opponent has the capability and the intention of using the weapons. Until the deployment of Polaris in 1969 the UK relied on the V-Force to carry the nuclear deterrent and the Valiant, Vulcan and Victor, even when armed with Blue Steel stand-off weapons were required to penetrate Soviet air defences to complete their missions. Therefore knowledge of the Soviets' radar and ground control, their electronic order of battle, was a necessary part of nuclear deterrence and as such was a critical aspect of the United Kingdom's war-fighting ability.

This task was known as electronic intelligence, Elint, and has a somewhat shadowy role in the first rate air forces of the world that involved collection of electronic signals from an opponent's radar and communications systems. The UK, USA and France were particularly active in this field, with the Soviets being just as active in their operations against the West with their activities being more obvious to the layman thanks to oft-presented images of Tupolev Tu-95 *Bears* being intercepted by RAF fighters. The Elint operations and capabilities of the RAF have until now been shrouded in secrecy and

Typifying the RAF's Elint operations during the Cold War, the only clue to a long high-altitude mission is the condensation under the wings of this Nimrod R.1, inbound to RAF Wyton in late 1988. Also of note are the white undersides to the wing pods and lack of AAR probe.

deception, attributes normally more associated with the target of the RAF's operations, the Soviet Union.

The Soviet Union by its very nature was a secretive nation, a closed society that did not lend itself to examination from within, so any attempt to determine what was going on had to be carried out from beyond its borders. Radio and radar had by 1945 become a necessary tool in fighting a war and, since their signals tended to travel in straight lines and were no respecters of borders, could, if detected, located, measured, recorded and analysed, provide a great deal of information on the Soviet Union's weapons development programmes and deployment. Such information also provided a better idea of how Soviet forces operated, their efficacy and, more importantly, any weaknesses or gaps in the defences. In short Elint provided what the Duke of Wellington needed to know – what was on the other side of the hill – albeit an electronic hill.

Listening In lifts the veil on the RAF's Elint activities, explains why such operations were conducted, how the Elint equipment, techniques and platforms evolved throughout the Cold War and examines its continuing role in the United Kingdom's operations in the post-Cold War era. The subject matter is, by its very nature, sensitive and the story can only be told since the opening of a large number of files in the National Archives allowed access to the background and detail of the RAF's Elint work. All the material herein comes from open sources and as such there may be a number of gaps or omissions, particularly with photographs, where information was not available for security reasons. Elint has been perceived as a mysterious, furtive and perhaps underhand activity carried out under a mantle of secrecy. *Listening In* shows that the RAF's Elint activities were not only an interesting side-line of the Cold War but a cornerstone of the political and military decision making process and a major contributor to the defence of the United Kingdom.

Note on Units

Nautical miles (nm) have been used throughout this book. The RAF used statute miles until the 1960s and where these are included in quoted text from original documents, the conversions to nautical miles and kilometres are appended. In the main text, nautical miles are used to allow comparison with post 1960 data. Speeds are quoted in knots (kt) and kilometres per hour (km/hr), mass/weight is quoted in long tons (2,240 lb) and metric tonnes (mt) or pounds (lb) and kilograms (kg). Metric equivalents of all units are shown for clarity. Frequencies are quoted in cycles or its SI equivalent, Hertz. Both can be used interchangeably.

Acknowledgements

The authors would like to thank:
Peter Biggadike, Phil Butler, Shaun Connor, Stuart Freer, Colin Ferrier, Peter Green, Yefim Gordon, Jerry Gunner, Scott Hastings, Ron Henry, Neil Lewis, Adrian Mann, Dr Christopher Morris, Terry Panapolis, Victor Pisani, Alan Powell, Glyn Ramsden, Paul Stancliffe, Russell Strong, Rob Swanson, Robin Walker, Les Whitehouse.

Foreword: Wartime

During the Second World War the UK built up a sophisticated Signals Intelligence (Sigint) organisation, using a network of ground stations and aircraft to intercept, analyse and record Axis radio and radar signals. In the main the ground stations, known as 'Y Stations' ('Y' being a phonetic abbreviation for 'Wireless Intercept') focussed on Communications Intelligence (Comint), intercepting strategic High-Frequency (HF) communications traffic. Aircraft were principally employed collecting tactical Electronic Intelligence (Elint), initially on German blind-bombing systems, later on the German air defence system. The data collected by the airborne Elint programme was used to develop radio counter measures (RCM) against German radio bombing aids; and to develop RCM equipment and tactics to support the bomber offensive against Germany.

By 1944 the main RAF Elint unit was 192 Squadron, part of 100 Group, operating a number of Vickers Wellington Mk.Xs, Handley Page Halifax Mk.IIIs and de Havilland Mosquito Mk.IVs from RAF Foulsham in Norfolk. All three aircraft types were modified for the Elint role and carried various aerials, receivers and ancillary equipment. In the Wellington and Halifax, the Elint equipment was controlled by a dedicated 'Special Operator' (SO), seated at a specially-constructed station on the starboard side of the mid-fuselage, aft of the main crew compartment. Receivers, signal analysers and recorders were installed at the station in wooden equipment racks. The Elint installation in the smaller Mosquito was, due to space constraints, a much more limited affair and normally comprised a single receiver and signal analyser; the equipment being shoe-horned into the cramped cabin, around and behind the Navigator, who doubled-up as Special Operator.

192 Squadron aircraft operated both singly, and in conjunction with the Bomber Command Main Force during Elint investigations, often accompanying bombers to their targets in Germany. Each sortie (known as a 'special duty' or 'Y' flight) had a specific goal and was tasked by the intelligence authorities. By 1944-1945 the principal targets were the various elements of the German air defence system (ground radar, airborne radar, air-to-ground communications and so on) although a large number of sorties were also flown looking for signals in connection with the launching of the V1 flying bomb, and looking for evidence of radio-control of V2 rockets. During a sortie the Special Operator searched his briefed frequencies for signals of interest, making audio and photographic recordings of the more important intercepts. After landing the SO's and Navigator's logs, along with photographic and audio records, would be passed to the Special Signals section for preliminary analysis. A report on the sortie would then be compiled and forwarded to Air Ministry intelligence, Assistant Director of Intelligence ADI (Science), the Telecommunications Research Establishment (responsible for radar, RCM and Elint development) and various other interested parties.

During 1944 the Wellingtons were slowly phased out in favour of the more-capable Halifax Mk.III, and during 1945 the Mosquito Mk.IVs were replaced by Mosquito PR.XVIs. Thus by the end of the war in, May 1945, 192 Squadron was equipped with a number of Halifax Mk.IIIs and Mosquito PR.XVIs.

An Avro Lancaster carrying an *Airborne Cigar* VHF communications jammer delivers a stick of incendiary bombs and a 4,000lb (1,814kg) Cookie. Intelligence gathered by 192 Squadron was vital to the design of Bomber Command RCM systems.
via Tony Buttler

1 Noise Listening

'...Y Section to continue the techniques built up during the last war for the exploitation, both for Intelligence and Countermeasures purposes, of enemy tactical signals, radio aids of all kinds, and other radio "noises"'. Radio Warfare Establishment programme of work, November 1945

The months following the end of the Second World War saw a rapid contraction in the RAF with many units being disbanded. However, special arrangements were made to preserve at least some of Britain's hard-won capability in Sigint. Plans were made to form a Radio Warfare Establishment at RAF Watton in Norfolk to provide a nucleus of expertise in the fields of RCM and Sigint. A number of ground signals intelligence listening stations ('Y' stations) were also retained in the UK, and another was set up in Germany.

The Radio Warfare Establishment (RWE) was formed in August 1945 from the recently-disbanded 100 Group Bomber Support Development Unit (BSDU) and the 100 Group Radio Engineering Unit. As its name suggests, the RWE was intended to maintain an expertise in all forms of radio warfare. The Establishment included a Y Wing, formed from the rump of the recently-disbanded 192 Squadron, tasked with maintaining a tactical Elint capability. The RWE also inherited a large fleet of various specially-fitted radio warfare aircraft from a number of disbanded wartime 100 Group units.

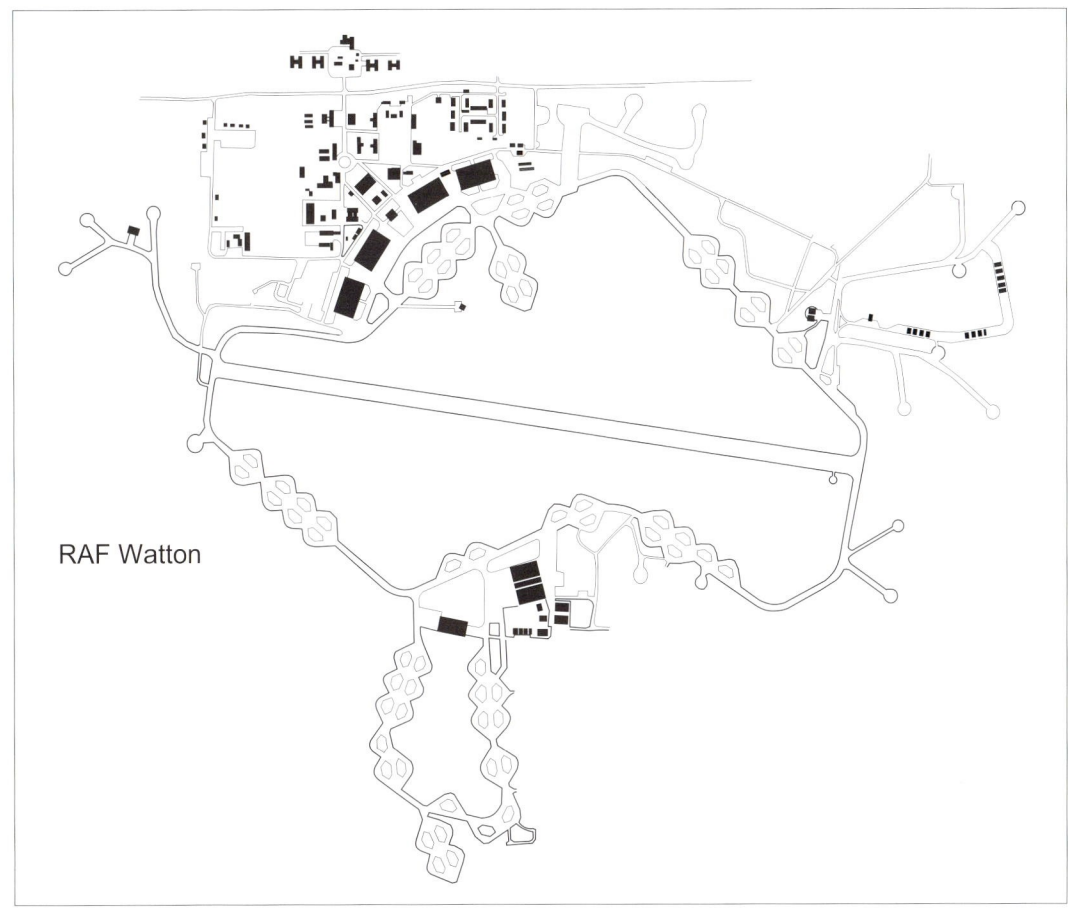

RAF Watton late 1940s. The main site was top left. The CSE research labs were located in the small building between the main hangars. The single runway restricted flying when wind conditions were unfavourable.

9

These comprised 25 Handley Page Halifax B.IIIs, some eight Boeing B-17 Fortress B.IIIs, and twelve de Havilland Mosquitos, including four Mosquito PR.XVI ex-192 Squadron (NS784 '4S-V', NS799 '4S-W', NS809 '4S-X', NS797 '4S-U') for signals intelligence duties.

The intention was to base the Establishment at Watton but that station was still occupied by the USAAF. As an interim measure the main body of the RWE (comprising the HQ Wing, Technical Wing and Y Wing) was initially established at RAF Swanton Morley, while the rest of the Establishment, comprising the Flying and Servicing Wings were temporarily located at RAF Foulsham, formerly home of the BSDU. The USAAF finally vacated Watton in September 1945 and the various Wings of the CSE moved in shortly afterwards.

The RWE 'Programme of Work' was issued at the end of November 1945, specifying the Establishment's main tasks. On the Y side the primary objectives were the completion of a number of half-finished wartime Elint equipment projects and the development of aircraft Y installations for service use. At the end of the war the Telecommunications Research Establishment (TRE), previously responsible for the development of Elint equipment, relinquished all interest in the field, leaving a number of projects uncompleted. The RWE was tasked with continuing the development of the more important of these projects, principally the *Blonde* and *Bagful* automatic search-receivers, airborne direction-finding (D/F) equipment, and airborne sound-recording systems. The aircraft installation tasks were to develop an unattended 'automatic' Y installation, a service Y installation with a single manual search receiver for use in Bomber Command, a service Y installation for communications intercept (also for use in Bomber Command) and a specialist Y installation with manually-operated receivers and analysis equipment. The Y tasks required the establishment of a 'Y Section to continue the techniques built up during the last war for the exploitation, both for Intelligence and Countermeasures purposes, of enemy tactical signals, radio aids of all kinds, and other radio "noises"'.

Y Training Begins

Responsibility for Y work at the CSE rested with Y Wing, initially under the command of Sqn Ldr J F Mazdon (ex-192 Squadron). The first task of the Wing was to establish a signals intelligence training programme to keep their small pool of Special Operators (SOs) in practice. This got under way in November 1945 with the construction of a Ground Trainer and a programme of revision and tests. Flying training was initially limited to navigation practice due to a lack of a suitably-fitted aircraft, but towards the end of the month sufficient search receivers were obtained to fit Anson NK718 as a Y training aircraft and airborne training commenced in mid-December.

Work was also started during November on the re-installation of Elint equipment into two of the Mosquito PR.XVIs (NS799 and NS784) to provide an operational capability. The initial fit of the Mosquitos was almost certainly a single US-built search receiver (AN/APR-4 or AN/ARR-5) and either an oscilloscope and oscillator combination for signal analysis, or a signal recorder. The Mosquito installations were completed by the end of 1945 but, in the end, the aircraft were never flown. The primary reason was that Y Wing were still waiting for an operational programme from the Air Ministry and thus there was no work for the Mosquitos. The other problem was that the Mosquito was not ideal for general Elint investigation work due to its small size. It could only take a single search receiver, had limited navigational facilities, and only carried a crew of two. What was really required was a long-range aircraft, equipped with a good navigation suite and enough space to accommodate a dedicated SO's station complete with a range of receivers and aerials. 192 Squadron had previously used the Halifax B.III in the Elint role but by 1946 that aircraft was obsolete and arrangements were being made to replace the RWE Halifax B.IIIs with Lancasters. Thus by the beginning of 1946, work had started at the RWE on the design of a specialist Y Lancaster, fitted with a single SO station and equipped with a number of manual search receivers. Unfortunately work on the other aircraft Y installation projects called for in the November 1945 'Programme of Work' seems to have been deferred due to a lack of resources.

Control Of Airborne Y

At the end of the war Signals Intelligence in the UK was effectively controlled by the London Signals Intelligence Centre (LSIC) via the inter-service London Signals Intelligence Board (LSIB). The LSIC, previously known as the Government Code and Cypher School(GC&CS), was later renamed the Government Communications Headquarters (GCHQ).

During 1946 there were discussions regarding the control of the CSE Y Wing. The intended

NOISE LISTENING

peacetime Y task of the CSE was mainly to gather Tactical Signals Intelligence on 'foreign' air forces, principally to aid the design of RCM equipment. One view was that, since the task would require access to other intelligence sources, it made sense for the LSIC/LSIB to retain responsibility for Tactical Sigint, with the CSE Y Wing merely collecting intelligence on their behalf. However, it was also recognised that in wartime the RAF would need to take control of the Y Wing via its own Tactical Sigint organisation and there were concerns regarding the disruption that transfer of control might engender. There were also suggestions that having a common authority controlling both Tactical Elint and RCM would have a beneficial 'synergistic' value, and, as a result, a proposal emerged to vest control of Tactical Elint in the CSE, leaving the LSIC/LSIB to concentrate on Strategic Sigint.

In the end a compromise seems to have been reached whereby the LSIC retained overall responsibility for the investigation of foreign radio and radar systems, but devolved control of RCM-related Elint to the Air Ministry. The agreement was that the CSE's primary tasking would come from the LSIC/LSIB via the Foreign Office Research & Development Establishment (FORDE), while follow-up investigations of signals deemed 'RCM subjects of particular interest to the RAF' would fall under the control the Air Ministry, Deputy Director of Signals(B). Located on the North Downs near Sevenoaks, FORDE's main wartime task had been the interception of encrypted German teleprinter traffic for Bletchley Park. Post-war the unit concentrated on long-range technical search and analysis, investigating foreign telecommunications systems.

Although the agreements notionally gave control of airborne Elint to the LSIC/LSIB, the reality was somewhat different. In practice the LSIB actually had little interest in non-communications Sigint (termed 'Noise Listening') and in 1947 the Board decided to confine itself to strategic communications intelligence. As a result of this decision, control of airborne Y operations effectively passed to the Air Ministry Assistant Directorate of Intelligence(Science), with operational control of airborne Y flights vested in DD of S(B).

The Y Lancasters

In April 1946 the Air Ministry appears to have woken up to their neglect of the field and began to take an active interest in airborne Elint for the first time since the end of the war. Work was begun on the preparation of an Elint programme for the RWE and instructions were issued to the Establishment to accelerate work on the Y Lancaster project. As a consequence of the increased priority, the RWE Technical Wing at Watton completed the design work on the Lancaster aerial system and equipment layout by mid-month, and began work on the conversion of a newly-arrived aircraft (Lancaster B.I PD381), shortly afterwards. The Lancaster Elint conversion was termed the 'Prototype Y Lancaster', the intent being that the design would serve as a pattern for further conversions by either industry or by service Maintenance Units, should they be required during a crisis or in time of war. As such the modification work was documented to a sufficiently high standard to allow the work to be easily replicated.

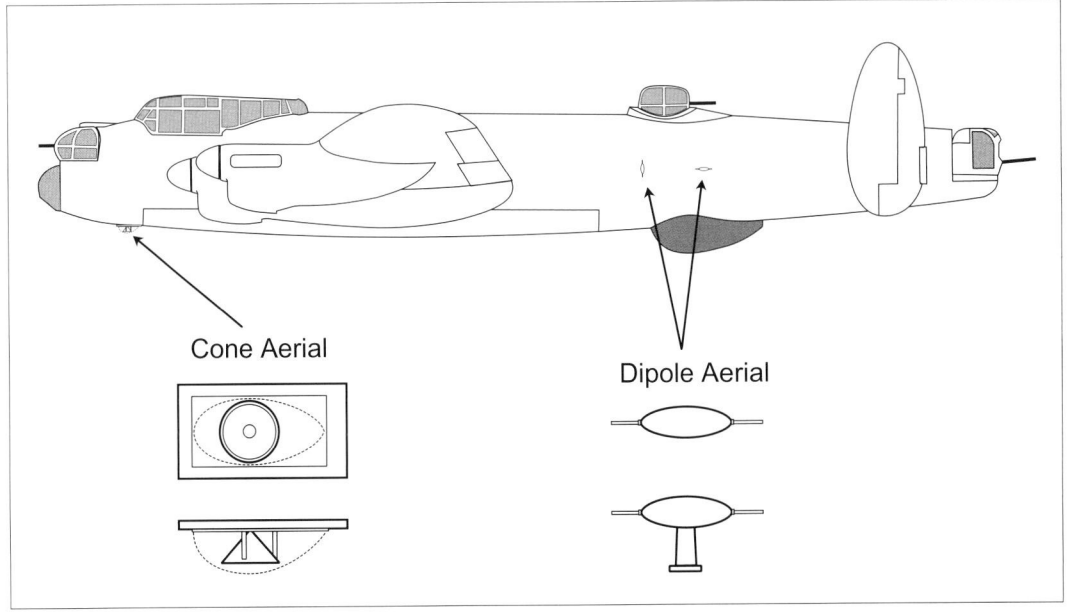

The first 'Y' Lancaster, showing dipole and cone aerials. The dipole aerials were installed in pairs, arranged in the vertical and horizontal planes.

LISTENING IN

Conversion of PD381 did not take long and work was completed in the third week of May. Perhaps not surprisingly the installation was very similar to that previously applied to the Halifax B.IIIs of 192 Squadron during wartime. It comprised a number of external aerials (quarter-wave dipoles either side of the fuselage, a half-wave dipole and cone aerial under the fuselage), a single SO's station in the mid-fuselage with a seat, equipment racking and a table, and at least two receivers: ARR-5, covering the 28-143MHz VHF band, and APR-4 covering the 40MHz -1GHz metric radar band. An Oscilloscope Type 10 and an associated audio oscillator were installed for pulse analysis, and a Boosey & Hawkes magnetic wire recorder (forerunner of the magnetic tape recorder) provided to record signals. The Lancaster Elint fit would allow an SO to intercept radar and other signals in the metric frequency band and establish their basic properties of frequency, polarisation, PRF, and pulse width.

While the Lancaster was under conversion at Watton, the RWE had been struggling to meet their heavy flying programme using the airfield's single, poorly-aligned, runway. The arrival of the Signals Flying Unit from RAF Honiley in April had only made things worse, with the result that by May it had been decided to split the RWE between two locations, leaving the main body (including the scientific and technical elements) at Watton and relocating flying operations to RAF Shepherds Grove in Suffolk, approximately 17 miles (27km) to the south. The advance party of the Flying Wing proceeded to Shepherds Grove in early June 1946 and the move was completed by the end of the month.

Following the completion of post-conversion acceptance tests, Lancaster PD381 (coded 4S-B) made the short hop from Watton to Shepherds Grove on 24th June. After the successful completion of further ground and air-tests at Shepherds Grove, Lancaster PD381 was declared ready for operations by mid-July 1946. The long-term plan was to eventually convert three Lancasters as Y aircraft and work immediately started at Watton on the modification of a second Lancaster, B.III RE121 (coded 4S-B2), to the same standard as PD381.

The decision to proceed with an Elint Lancaster rendered the RWE's Mosquito PR.XVIs largely redundant and in May three of the four aircraft were disposed of, leaving only a single example (NS809) on charge for Special Operator training purposes.

In parallel with the conversion of the Lancaster, the Elint training programme was stepped up and arrangements made to train new SOs. Nine candidates for the Special Air Operator training course were interviewed at the beginning of April, of which five were selected, and the first post-war Special Operators course commenced the same month. An additional Elint training aircraft, in the shape of Anson NK293 was made available from the beginning of April, but unfortunately the

Above: The AN/ARR-5 was the standard airborne HF communications intercept receiver of the 1940s and 1950s. *Author's collection*

Below: The AN/APR-4 was the most widely-used metric-band radar receiver of the 1940s and 1950s. The receiver covered a wide range of frequencies using a number of interchangeable tuning units. *Author's collection*

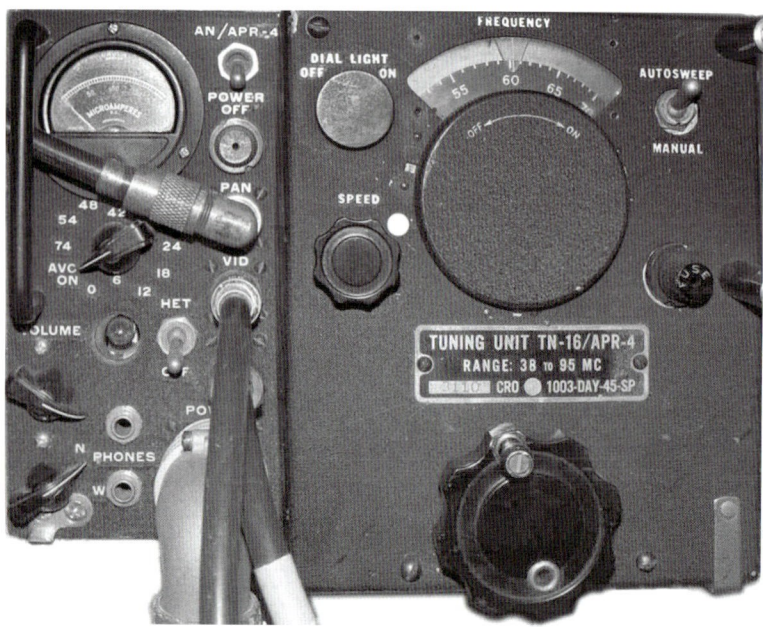

NOISE LISTENING

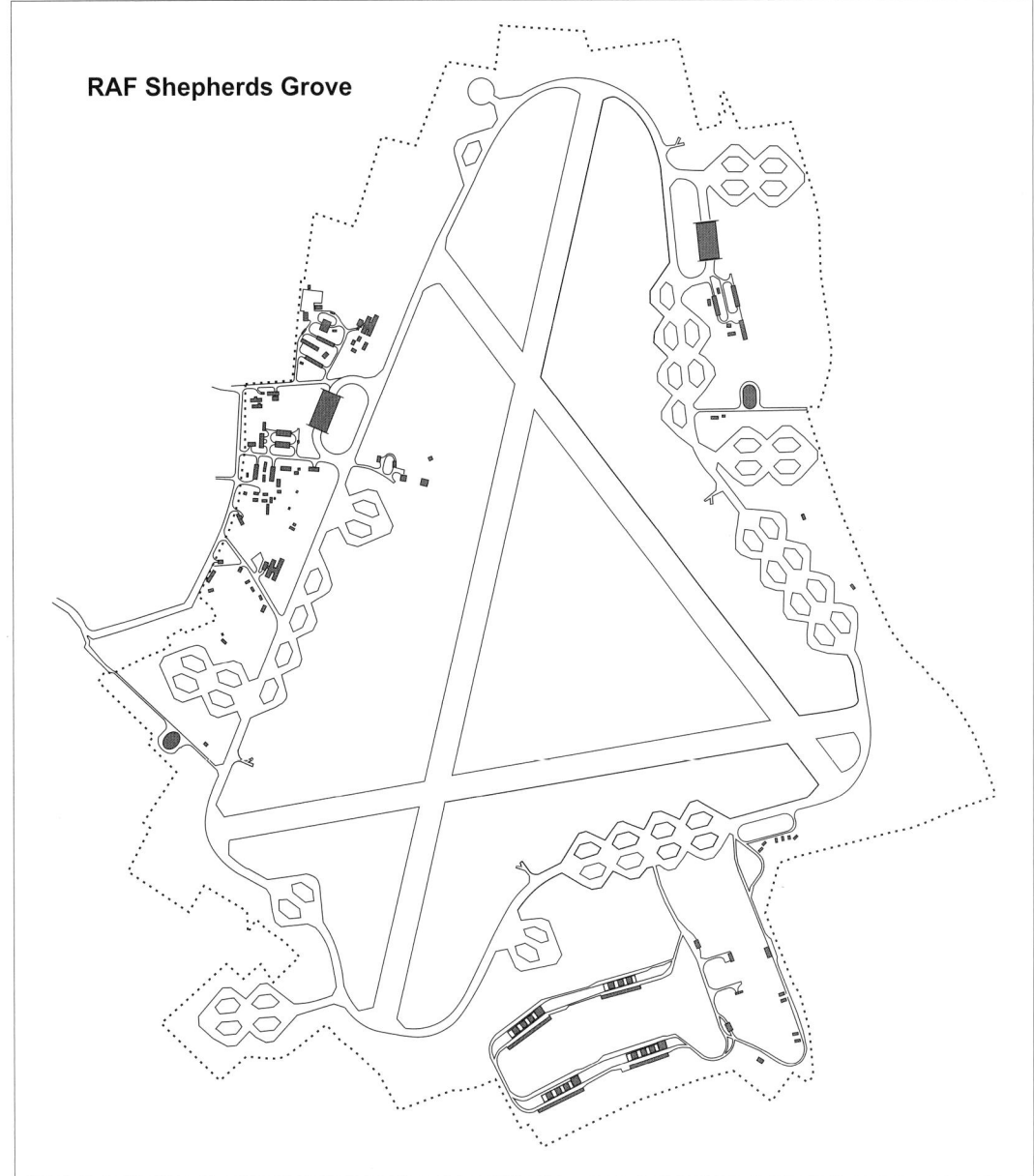

RAF Shepherds Grove

RAF Shepherds Grove, home of the CSE Flying Wing between June 1946 and December 1949. The three-runway layout better suited the CSE's intensive flying operations as it allowed operation under variable wind conditions and directions.

increase in establishment was short-lived as the original Anson (NK718) crashed on take-off on 1st May, happily without loss of life.

Noise Listening, June 46-June 47

While the first Lancaster (PD381) was being fitted-out, discussions took place between the RWE and the Air Ministry regarding future Elint operations by Y Wing. A representative from Deputy Director of Signals (B) visited the RWE towards the end of May 1946 and the Commandant of the RWE made a return visit to the Air Ministry a few days later. The finalised Elint programme, referred to as a 'Noise Listening' programme, was issued in June 1946.

The term 'noise' was signals intelligence jargon for non-voice radio emissions and covered radar, radio navigation and other radio frequency sources. The new programme was in effect a general assay of the Northern European non-communications signals environment, covering the UK and the Continent using ten numbered cross-country routes. These were: (1) UK East Coast, (2) UK South Coast, (3) North Germany, (4) Central Germany, (5) Central Germany, (6) South Germany, (7) Paris, (8) Eindhoven, (9) Berlin, and (10) Berlin. Although the primary targets of the programme were undoubtedly signals emanating from the Soviet Zone of Germany there is a suspicion that the Soviets were not the sole target. For example, route 8 covered Eindhoven, the location of the Dutch technology giant Philips'

LISTENING IN

R&D facilities, while route 7 covered Amiens, home of the ESIEE, a French university specialising in electronics.

Prosecution of the new programme required co-ordination between two elements of the RWE located on separate airfields. Planning, and subsequent analysis of results of the sorties, was carried out by RWE Y Wing, based at Watton, while the Y Lancaster was operated by the RWE Flying Wing, Development Squadron, based at Shepherds Grove.

Operations began in July 1946 when Lancaster PD381 flew a number of sorties along the South Coast, over France and Germany, and: 'proved the "Y" prototype installation to be entirely satisfactory'. The first phase of the Noise programme was completed in August with a further six sorties over Germany, four at high-level and two at low-level.

The Noise investigation flights were carried out using the same techniques developed by 192 Squadron during the Second World War. The SO would tune his receivers over the frequency range of interest (both the ARR-5 and APR-4 receivers had an automatic scan facility), listening in on his headphones. Since most radar pulse repetition frequencies (PRFs) fell in the audio frequency range an intercepted radar signal would be detected as an audio tone (intermittent if the radar was scanning), with the pitch indicating PRF. Once a signal had been picked up the operator could halt the automatic scan and then use manual tuning to determine the exact frequency and strength of a signal. Further information on the radar signal could be obtained by feeding it into the oscilloscope to determine the PRF (by mixing it with an audio oscillator signal), pulse-width and pulse shape. Signal polarisation could be determined by switching between the aircraft's horizontally and vertically-polarised aerials. The operator would manually record all intercepted signals in his log, noting the time of intercept, signal strength, frequency, duration, and pulse characteristics. Signals of particular interest would be recorded on a magnetic wire recorder and photographs taken of the oscilloscope to record pulse characteristics. Although the Lancaster's fit did not include a direction-finding (D/F) capability, and thus lacked the means to take a bearing on a received signal, the SO would be able to record the time at which a signal was intercepted and its strength, switching between port and starboard aerials to determine on which side of the aircraft the source of a signal lay. At the end of each flight the SO's log would be correlated with the Navigator's log to determine where each signal had been picked up, thus giving a rough location for the source of the signal.

One aim of the Noise investigation flights was to look for any signals in the 300-600MHz band previously used by Luftwaffe GCI and AI radars, presumably to determine if the Soviets were using captured wartime equipment. No unusual activity was picked up during the July sorties: 'the 300-600 mc/s [MHz] portion of the frequency spectrum ... is now completely devoid of signals', but those flown during August did reveal some unidentified transmissions requiring further investigation.

The first Y Wing Signals Interception Flight Report, recording the results of the Lancaster sorties during July and August was issued in September 1946 and copies were distributed to various parties, including the LSIB, LSIC and FORDE. The RWE were keen to utilise FORDE's expertise in identifying a number of unknown intercepts picked up by the Lancaster and, after some prompting, several useful observations and criticisms of the report were fed back via the Air Ministry.

The sorties in the Lancaster had also revealed a severe interference problem between the aircraft's H2S navigation radar and the Elint receivers, making it impossible to use both at the same time. This problem had previously been seen during wartime airborne Elint investigations and had been overcome by supressing the operation of receivers during the transmission of the H2S radar pulse. However, wartime specialist Elint aircraft had not normally carried H2S and so the only receivers modified were those intended for Main Force use. The interference during the flights in PD381 led to a new programme to suppress H2S interference in other receivers, with priority being given to the APR-4 and ARR-5.

September 1946 also saw a major administrative change when the RWE was merged with the Signals Flying Unit to form the new Central Signals Establishment (CSE) in 90 (Signals) Group. Based at Watton and Shepherds Grove, the new Establishment's responsibilities were very similar to that of its predecessor and included RCM, Calibration and Y tasks, the latter including: 'Maintaining a tactical Y section to ensure the development of the technique and equipment required to provide intelligence data for RCM'. The CSE was organised along much the same lines as the RWE with three wings (Technical Wing, Flying Wing and Y Wing), and two Schools (GCA School and Signals Training School). Primary responsibilities of the Y Wing were the continued development of Elint equip-

NOISE LISTENING

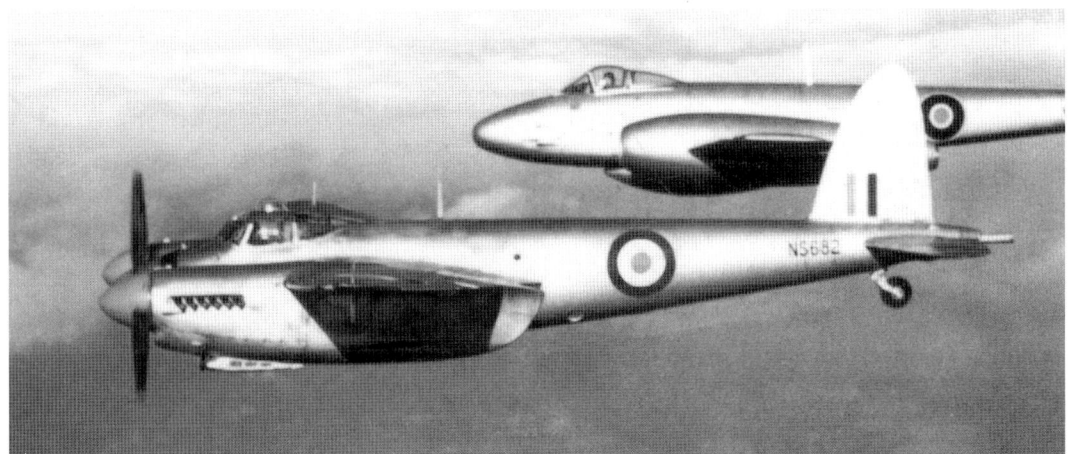

A de Havilland Mosquito PR.XVI. This version of the aircraft was originally built for photo-reconnaissance duties. In the Elint role the Perspex nose was painted and the camera ports (located in the lower fuselage behind the wing) faired over.
via Ron Henry

ment and of Y investigation techniques, and '… the planning and interpretation of such special air investigations as may be called for from time to time'. Other responsibilities comprised the training of Special Operators for ground and air Elint duties, the monitoring of RAF ground/air communications for security purposes, and ground 'noise' investigations. Although the CSE was administratively controlled via HQ, 90 (Signals) Group, an exception was made for Sigint tasking: its Organisation Memorandum noting that, in the interests of security or for urgent operational reasons, the Air Ministry might need to approach the CSE directly in connection with 'certain highly secret matters concerned with Signals Intelligence'.

The formation of the CSE was a largely administrative exercise and had little impact on the work of the Y Wing and the on-going Noise Listening programme. In the first phase of that programme Lancaster PD381 had ranged over Northern and Central Europe but in November 1946 the Air Ministry mandated that Y sorties should only be conducted over British-occupied territory. The CSE were not particularly pleased, noting that this ruling: '[limits] the work that can be done in the general investigation of the frequency spectrum that is being carried out'.

In November the second Y Lancaster (RE121) became available for operations and started flying on the programme alongside PD381. This aircraft was fitted with a centimetric aerial and carried an AN/APR-5 receiver on some sorties to look for centimetric radars. By the end of 1946 the two Lancasters had completed 33 sorties on the Noise investigation programme, covering routes 1-8. No sorties had been flown on the two Berlin routes (9 and 10) and, at the beginning of 1947, instructions were received from the Air Ministry to hold over these flights until further notice, presumably due to their risky nature.

Noise sorties completed by January 1947

Route	Locality	Band	Flights
1	East Coast	90-500MHz	4
2	South Coast	90-500MHz	4
3	North Germany	30-190MHz	2
4	Central Germany	30-90MHz	1
5	Central Germany	30-190MHz	2
6	South Germany	30-300MHz	3
7	Paris	30MHz 5GHz	9
8	Eindhoven	30-500MHz	9
9	Berlin	-	Nil
10	Berlin	-	Nil

The Air Ministry issued its second Noise Listening programme towards the end of December 1946; this included a number of follow-up investigations of 'interesting' signals identified in the CSE's first Noise Listening report. By then the CSE had modified Mosquito PR.XVI NS809 as a third Y aircraft, specifically for the follow-up investigation task. Although the Mosquito could only carry a single receiver, this was sufficient for investigations into previously-identified frequencies. The Mosquito was also more agile than the Lancasters and rather cheaper to operate. The first Mosquito sorties were conducted during January 1947, when NS809 flew ten sorties over routes 2, 7 and 8, all follow-up flights. The sorties were apparently looking, amongst other things, for previously-seen signals in the 300-500MHz band but none were picked up.

The Third Lancaster

Air Ministry plans had originally called for the conversion of three Lancasters for Y duties by the CSE but in mid-1946 it was decided, for reasons of economy, that future Y aircraft would be dual-role Radio Counter Measures (RCM)/Y aircraft with common fixed fittings (that is, aeri-

LISTENING IN

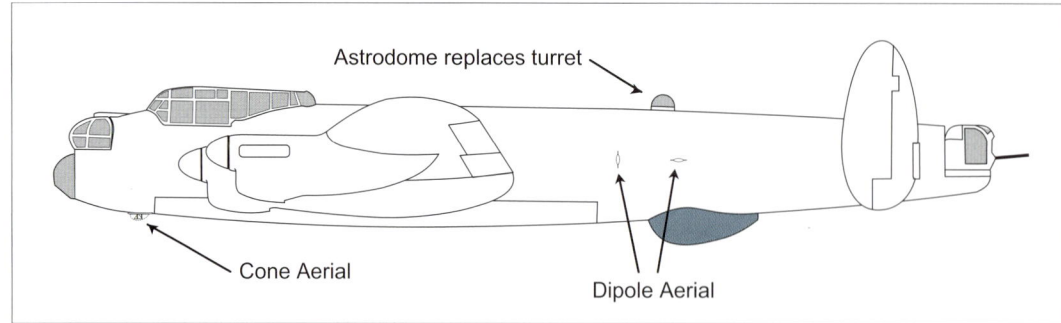

The third Y Lancaster with the dorsal turret removed and replaced with an astrodome. This installation applied the lessons learned from the earlier Lancasters.

als, racking and wiring and power supplies) to allow either RCM or Y equipment to be installed. The RWE/CSE was responsible for both Y and RCM trials and a dual-role aircraft could be switched between tasks as required. The fact that the Air Ministry did not consider it worthwhile to maintain dedicated Y aircraft gives some indication of the sparseness of the Elint task at that time, and also the lack of priority of airborne Elint operations.

As a result of the dual-role ruling, work got under way at Watton in July 1946 to modify the third Y Lancaster, B.I PA232 (4S-D), as a 'prototype' RCM/Y aircraft. By then the terminology was changing, possibly for reasons of security, and the new Lancaster was officially designated as a 'Monitor' aircraft, rather than a Y aircraft, although in fact, both designations continued to be used interchangeably for some while afterwards. The conversion was a variation on the earlier two Lancaster fits with additional racking, aerials and switching installed to allow the aircraft to take an alternative RCM jammer fit when required. As in previous conversions, a SO's station was provided in the mid-fuselage, aft of the cockpit. The mid-upper gun turret was removed and the resulting hole partially faired over and topped with an astrodome to provide an observation position. Arrangements were also made to mount a cine camera on a swing mount to record the SO's signal analysis oscilloscope. By August most of the wiring and aerial fit had been completed but further work was then delayed by other, higher-priority, projects at Watton. The conversion was structurally complete by November 1946 and, following tests, it was inspected and approved by the Ministry of Supply in January 1947.

Unfortunately, by the time PA232 had been completed there had been a further change in CSE policy, with a decision to re-equip the Establishment with the later B.I(FE) version of the Lancaster. The B.I(FE) was a long-range conversion of the Lancaster B.I, originally designed for service in the Far East in the war against Japan, and featured a long-range fuel tank in the bomb-bay and an improved radio and radar fit. The decision to re-equip with the B.I(FE) meant that Lancaster B.I PA232 was more or less redundant before it had even entered service. Despite the decision to replace PA232 with a B.I(FE), it was decided to get some use out of the aircraft by carrying out full trials of its RCM/Y installation to obtain data applicable to a similar B.I(FE) installation. As a first step, work started on plotting the polar diagrams (that is, cover and sensitivity) of the various aerials fitted to the aircraft.

Arrangements were then made to carry out a dual RCM/Y installation, similar to that applied to Lancaster B.I PA232, in a Lancaster B.I(FE). This would serve as the pattern for the conversion of a further two aircraft, providing a total of three dual-role RCM/Y Lancaster B.I(FE)s. Work on the 'prototype' RCM/Y fit in Lancaster B.I(FE) PA478 got under way at the CSE in February 1947 as Task 642. The installation was basically similar to that applied to the Lancaster B.Is, although some redesign was required to work around the long-range fuel tank in the bomb-bay. In addition to ARR-5 and APR-4 the Elint fit also include an R.1645 (*Blonde*) automatic receiver for full-scale trials of that system. Although the SO's station was complete by July 1947 the conversion programme was repeatedly delayed by higher-priority projects, and work on the aircraft was not completed until November 1947.

The First Lincoln

The decision to replace the Lancaster B.Is with B.I(FE) aircraft was an interim step: the long-term CSE plan was to follow the lead of Bomber Command and eventually replace all the Lancasters with Lincolns.

Interest in an Elint Lincoln had first emerged back in May 1946 when a Lincoln and a Lancaster B.VII were inspected by the RWE Technical Wing to assess their suitability as both Y and RCM platforms. The more modern Lincoln was clearly the better long-term choice, but the problem was where to accommodate the dual-

NOISE LISTENING

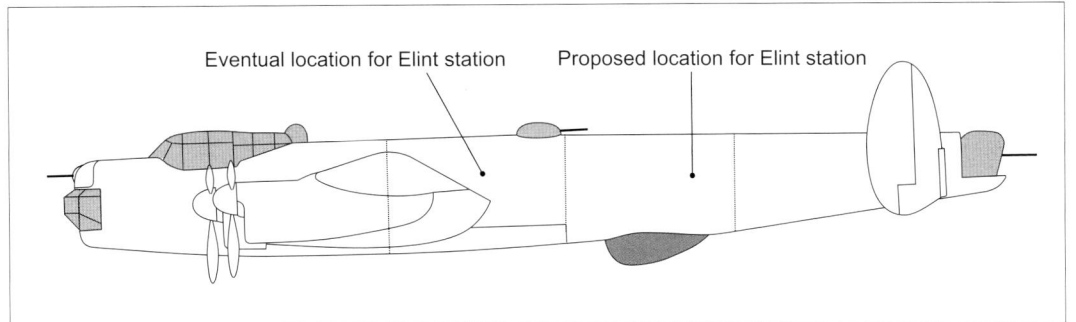

General arrangement of a Lincoln B.2/3G. The initial location proposed for the Elint station was aft of the dorsal turret but, following centre of gravity problems, it was moved forward.

role RCM/Y SO's station. In the Lancaster the SO had been installed in the mid-fuselage immediately behind the main spar but in the Lincoln that space was occupied by a variety of radio and radar equipment. The RWE decided that the most suitable space for the SO in the Lincoln was in the rear fuselage, aft of the mid-upper turret. However, there was some concern that positioning equipment that far to the rear would adversely affect the aircraft's centre-of-gravity and so, as a first step, it was decided to carry out a mock-up installation in a Lincoln to determine the weight distribution.

Lincoln B.2/3G RE395 (the 3G suffix indicating H2S Mk.3G radar) was delivered to the CSE for use as an RCM/Y prototype in August 1946, immediately going to the Technical Wing at Watton for mock-up installation work. The mock-up fit was completed by the end of September 1946 and in mid-October the aircraft was inspected by a representative of the Ministry of Supply who determined that the aircraft's centre of gravity had been unacceptably displaced aft. After further investigation it was concluded that a suitable installation could only be achieved by removing the mid-upper turret and installing the SO's station in the mid-fuselage section over the wing, although this would require re-arrangement of the equipment already installed in that space.

Authorisation for the modification of the Lincoln was granted by 90 Group in January 1947 and work started shortly afterwards under Task No.663, 'Prototype Installation of quantity four selected receivers with ancillary equipment into Lincoln RE395'. Unfortunately the modification of the Lincoln was much-interrupted, and work on the aircraft dragged on into 1948.

The loss of PD381

The winter of 1946-47 was particularly harsh, with heavy snow towards the end of January 1947 followed by freezing conditions through February. The bad weather impacted operations from Shepherds Grove and only one Noise investigation flight was flown in February (a five hour sortie by RE121 on Route No.4 over Düsseldorf, Frankfurt and Kassel). Further heavy snow fell at the beginning of March before temperatures rose mid-month. Unfortunately the thaw was accompanied by widespread flooding, heavy rain, and gales. On the night of 15/16th March 1947 a severe gale, gusting to 80-90 knots (148-166km/h), swept across the UK, damaging eleven CSE Lancasters standing in open dispersals at Shepherds Grove, including the two Y Lancasters, PD381 and RE121. Subsequent inspection showed PD381 had been so severely damaged it would have to be written off. While the remaining Lancaster (RE121) was being repaired, operations continued using Mosquito NS809, with Noise investigation sorties over the UK, France (Paris) and the Netherlands (Eindhoven) in April and May. Arrangements were also made to bring the RCM/Y Lancaster B.I (PA232), then undergoing aerial trials, into service as a replacement for PD381.

Lancaster RE121 resumed Noise Investigation sorties in May and was joined on operations by PA232 in June. By mid-1947 however, the CSE were questioning the usefulness of continuing the Noise monitoring programme, suggesting that: '[the] work of the CSE on RCM and Y must be more clearly related to the future and that our current effort on present tasks should be curtailed'. This appears to have been a push for a Y programme more relevant to future offensive operations by Bomber Command. A meeting at the Air Ministry on 9th July seems to have reached a similar conclusion and the Noise investigation programme was terminated with immediate effect.

Little or no Elint flying was carried out over the next three months. The Y fit in Lancaster PA232 was removed and the aircraft fitted with RCM equipment for proving trials in the RCM role. Lancaster RE121 was flown on trials of the Decca navigation system and of modifications to eliminate H2S interference in search receivers. The only Elint-related flying during the period took place in June, as part of a North Sea

17

LISTENING IN

maritime exercise involving units of the Home Fleet returning from Scandinavia. The CSE supplied both a mobile Y vehicle and a Y aircraft to support Blue Force (Coastal Command, Fighter Command and Bomber Command) by providing signals intelligence on the movement of the fleet. Lancaster PA232 flew a 6 hour 20 minute sortie against the Fleet on 26th June, monitoring signals in the 30MHz – 6GHz band.

During 1947 a number of administrative changes took place at the CSE and by the end of the year the Y Wing, responsible for the planning and analysis of Elint operations, had been renamed Y Squadron and subsumed into the new CSE Technical Wing. These changes had no effect on the flying side of Y operations, which remained the responsibility of the CSE Flying Wing Development Squadron.

The CSE continued to run Special Operator training courses during 1946-47. Operators were trained for both ground duties, at various intercept stations in the UK and overseas, and for airborne duties with the CSE Flying Wing. Standards were high and it was not unusual for trainees to fail the course. For example, only two out of six operators passed the course ending in November 1946. The initial course was classroom-based and those passing were dispatched to Shepherds Grove for further air training and operational flights. Training sorties were carried out in the Anson (NK540 V7-Q) and occasionally in the Mosquito and Lancasters when available. By November 1947 improved training facilities had been constructed at Watton, built by the SOs themselves. The training room featured three complete Y stations (equipped in a similar manner to the aircraft SO's position) with aerials mounted on the roof of the building.

A New Y Programme, September 1947

By mid-September 1947 the Air Ministry and the CSE had agreed a new, more relevant, airborne Y programme, comprising around four sorties a month over Northern Europe. Operations initially concentrated on the Amiens area of Northern France and the Lübeck area of Northern Germany, close to the Baltic coast and the border with the Soviet Zone. The first two sorties in the vicinity of Lübeck took place in late October/early November and were flown partly in the dark of the early morning, possibly reflecting the sensitivity of the area, or perhaps investigating the state of Soviet defences in the early hours. These were in fact the first post-war Y sorties flown in darkness. A third Lübeck sortie in mid-December was flown in daylight. Most of the September-December 1947 programme was flown using Lancaster RE121 since Lancaster PA232 was primarily employed on trials flying.

In November 1947 the RCM/Y installation in the Lancaster B.I(FE) PA478 was completed and, after servicing, the aircraft became available for operations in January 1948. Thus by the beginning of 1948 the Y resources of the CSE Development Squadron, comprised two Lancaster B.Is (RE121, PA232), a recently-converted Lancaster B.I(FE) (PA478) and a Mosquito PR.XVI (NS809). The Lincoln B.2/3G (RE395), delivered in August 1946, was still under conversion by the Technical Wing at Watton.

The new Y programme continued into January/February 1948 with eight sorties over northern, central and southern Germany. Most were flown in daylight, but a follow-up sortie in the Lübeck area investigating signals picked up on previous flights, was again flown in the dark of the early morning. The Y investigation programme begun in September 1947 came to an end in March 1948 with four sorties by RE121 over central and southern Germany.

H2S interference

One long-running problem with the Lancaster Elint fit was the interference produced by the H2S navigation radar. This was considered a serious problem by the Air Ministry, since it meant that H2S radar could not be used during Elint operations, adversely impacting navigation of the aircraft.

Finding a suitable method of suppressing the H2S interference took some time due to a shortage of manpower at the CSE and difficulties obtaining the necessary electronic components. The eventual solution was to generate 'blanking' pulses from the radar, suppressing receiver operation during the period of each H2S transmitter pulse. This scheme required the modification of each receiver to incorporate a blanking circuit and control, and the addition of a new plug to connect the receiver to the H2S Modulator. Flight trials of a modified receiver were flown in the first quarter of 1948 and showed the modifications were very successful, eliminating both interference on CRT displays and audio interference in headphones. By the end of the first quarter of 1948 the CSE Research Squadron had incorporated the suppression scheme in the most common Elint receivers (APR-4, APR-5, ARR-5) used by the CSE aircraft.

2 The Cold War Begins

'At present what is known is very meagre, but it would indeed be dangerous to assume that because little is known little exists. Means must be found to obtain such information, for without it RCM cannot be effective when it will be most needed i.e. at the very outset of hostilities'.
Air Ministry RCM Committee, February 1949

During 1947 relations with the Soviet Union had gradually deteriorated as the major powers argued over the future of Germany. Then in February 1948 a Soviet-sponsored coup overthrew the government of Czechoslovakia and established a communist regime in that country. The shock of the Czechoslovakian coup, and concerns regarding the likelihood of war with the Soviet Union, resulted in the UK's airborne Elint programme assuming a new importance. Unfortunately the low priority afforded Elint since the war had taken its toll. As the CSE Progress Report for March 1948 noted:

'... although all concerned with the research and development of RCM and Y equipment are imbued with the sense of urgency which the present situation demands, much still remains to be done, and the present rate of progress will not be greatly speeded up until our strength of scientific and technical personnel is greatly increased and the aircraft establishment filled with the types of aircraft needed'.

In response to the deteriorating international situation, plans were now laid for a more aggressive airborne 'Ferret' programme, concentrating on the borders and coastlines of the Soviet Union and its satellites. The use of the 'Ferret' designation (a wartime term for Y sorties) suggests this programme was intended to be rather more 'operational' in character than the previous Y investigation.

The first operations in the new Ferret programme, probably the first truly operational series of Elint sorties since the end of the war, took place in June 1948. The programme

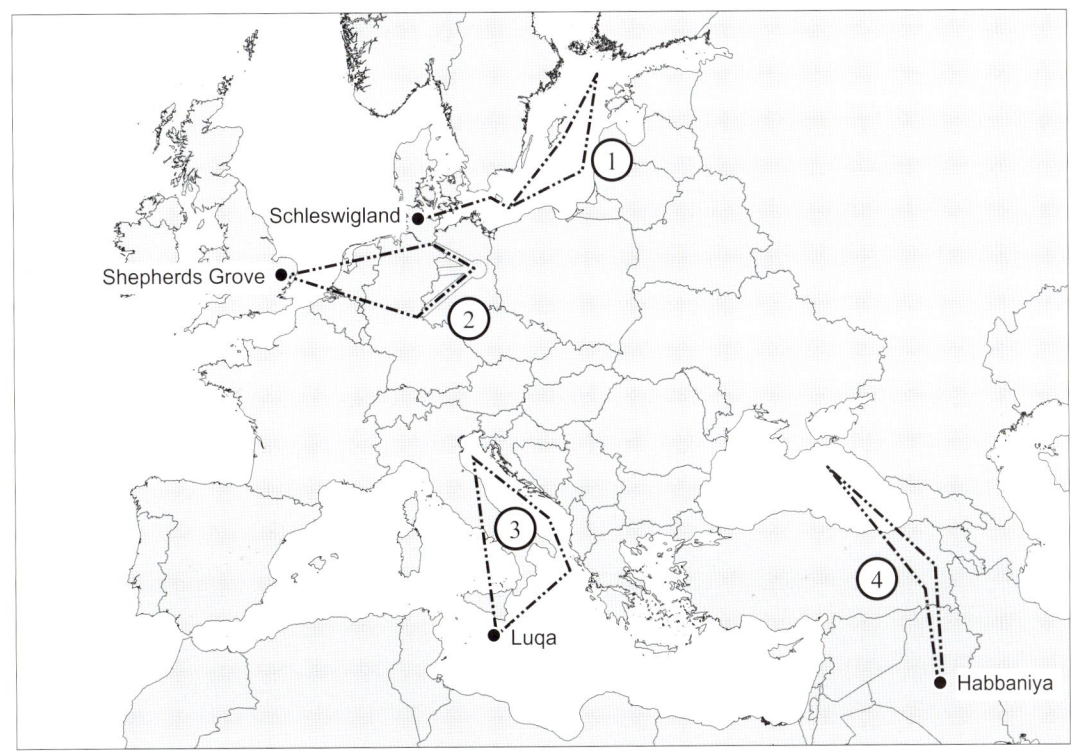

This map shows the areas investigated by the 1948 Ferret programme.
1. The Baltic
2. German borders and Berlin corridors
3. The Adriatic
4. The Trans-Caucasus and Eastern Black Sea.

19

LISTENING IN

planned by the CSE Monitoring Squadron (renamed from the Y Squadron in February for security reasons) included operations into the Baltic, over the Adriatic, and into the Black Sea. The exact circumstances under which the programme was authorised are unclear. Unlike previous flights, the aircraft would be operating over international waters rather than friendly airspace and there was clearly the possibility of an incident. It thus seems certain that it would have been approved by the Secretary of State for Air (Arthur Henderson) and possibly also, the Foreign Secretary (Ernest Bevin) and the Prime Minister (Clement Attlee).

Operations began on 7th June when Lancaster B.I(FE) PA478 was detached to the former Luftwaffe airfield of RAF Schleswigland in Schleswig-Holstein, Northern Germany, about 21nm (40km) from the German-Danish border. Schleswigland was then in use as a training airfield for RAF transport aircraft but its main attraction for the CSE was its close proximity to the Baltic coast. The first sortie in the new programme took place on 9th June and was a 6 hour 50 minute flight over the Baltic, flying in daylight along the German, Polish and Soviet coasts, locating and cataloguing Soviet radar stations. A second sortie, of similar length, was flown on 11th June, this time at night, during the dark new moon period. A follow-up detachment to Schleswigland using the same aircraft took place at the end of the month when one successful 6 hour 30 minute sortie, and two shorter (possibly abortive) sorties were carried out.

At the end of June 1948 the Soviet Union closed all rail and road links to Berlin, beginning an eleven month blockade of the city. This dramatic deterioration in east-west relations further emphasised the increased need for intelligence on Soviet defences and could well have been the justification for a marathon 10 hour night sortie by PA478 over Germany in July. The sortie, flown from Shepherds Grove, was carried out in the dark new moon period and may have been routed to Berlin and back via the air corridors. The Ferret programme continued in August when Lancaster PA478 was detached to Malta for two 8 hour 30 minute sorties (one day, one night) into the Adriatic, along the coasts of Albania and Yugoslavia. The night sortie was conducted during a full moon, presumably to aid navigation.

Despite the fairly intensive Y programme the Y Lancasters were still earmarked as dual-role RCM/Y aircraft and at the end of August all three Elint aircraft (PA232, PA478 and RE395) were refitted for the RCM role for Exercise *Dagger*, a national RCM exercise. All Elint equipment was removed, extra aerials and control panels fitted, and a variety of jammers installed, following which the aircraft were flown as airborne jammers in early September. Immediately following *Dagger*, PA478 was refitted in the Elint role for the next Ferret operation.

The third week in September saw Lancaster PA478 detached to Habbaniya, Iraq for two Y sorties along the Turco-Soviet border and into the Black Sea: the first an 8 hour 30 minute day

RAF Habbaniya in the late 1940s with visiting Avro Lincolns and Yorks. Located 55 miles (89km) west of Baghdad, Habbaniya was a major staging post for RAF aircraft transiting to the Far East. It was also ideal for Elint operations against the Soviet Union. *Norman Roberts via RAF Habbaniya Association*

operation, the second a 7 hour night flight. Habbaniya, located just west of Baghdad, was the home of 276 Signals Unit, an RAF Sigint station tasked with listening in to Soviet signals traffic, and the Lancaster sorties were co-ordinated with that unit. These were the first sorties flown along the southern borders of the Soviet Union and were vitally important in establishing whether the air defence system observed along the German border and Baltic Coast was representative of defences in other regions. The final operations in the 1948 Ferret programme were two daylight sorties over the Baltic by Mosquito NS809 from Schleswigland.

Co-operation With The US

The UK had co-operated closely with the US in the signals intelligence field during the Second World War and this co-operation continued in the post-war period under a series of so-called UK-USA Sigint agreements. However, the agreements initially covered only co-operation in the communications field and did not include Elint.

USAAF airborne Elint operations originally concentrated on the Soviet Far East as this was the direction from which a Soviet air attack on the United States was initially expected. However, in August 1946 a US Army Douglas C-47 transport was shot down over Yugoslavia after straying off course. The attack happened in poor weather and the US suspected the Yugoslavian fighters had been vectored onto the C-47 by GCI radar. In order to investigate further, a Boeing B-17 of the 10th Photo Reconnaissance Group at Fürstenfeldbruck was fitted with APR-4 search receivers and direction finders and flown on an Elint sortie off the Yugoslavian coast. The B-17 picked up a radar operating on the *Würzburg* frequency and the direction-finding plots showed the radar based at a wartime German radar school. Following the successful Yugoslavian sortie the 10th Photo Reconnaissance Group began operating in a dual photographic reconnaissance/Elint role over Germany and in November 1948 the unit was reformed as the 7499th Air Force Squadron. The Elint B-17s of the unit flew regular sorties along the length of the German and Austrian internal borders as well as missions out into the Baltic.

Although the UK-USA Sigint agreements initially only covered communications intelligence there is some evidence of co-operation in the Elint field as well. In February 1947 Dr Louis Tordella of the American Scientific Staff visited the CSE Y Wing on a technical liaison visit in company with Wing Commander Butler of the Foreign Office. Arrangements were made during the visit for Dr Tordella to keep the CSE up-to-date with US Elint development via the LSIC. Extending the UK-USA agreement to encompass Elint clearly made a lot of sense to the LSIC/GCHQ, since it would provide access to data gathered by the large US Elint programme. Consequently, in the first quarter of 1948, the UK formally proposed that the UK-USA Sigint agreements should be extended to include co-operation in the Elint field and that UK-US Ferret flights should be co-ordinated. It thus seems likely that the UK's 1948 Elint programme was planned to complement US operations, including sorties by the 7499th Squadron, as part of a joint intelligence effort. Reports on the 1948 sorties by the CSE were passed to the US in November 1948 and it appears that reports on US sorties were similarly passed to the UK authorities.

Soviet Air Defences

The 1948 Ferret programme (supplemented by US reports) revealed a developing, but still fairly basic Soviet Air Defence system. In fact, the various components of the Soviet air defence system had been organised into an independent branch of the Soviet armed forces, known as the *Protivo Vozdushnaya Oborona* (PVO) *Strany* (Anti-Air Defence of the Nation) in June of that year. United under its command were fighters, anti-aircraft guns, radar units, ground observers, sonic detection stations, searchlight units, barrage balloon units, and other specialised forces.

Soviet air defence radars comprised a trio of Soviet-designed truck-mounted early warning sets supplemented by a motley collection of wartime UK and US lend-lease radars. The three Soviet sets were the wartime RUS-2 and *Pegmatit* and the new P-3 *Dumbo*, all working in the 60-85MHz metric band. These radars were fairly basic by Western standards, with limited range and poor angular discrimination. The lend-lease radars still in use included the US SCR-527 (200MHz) and the AMES Type 6 Light Warning set (176MHz). Radar coverage extended along the Baltic coast, along the border of the Soviet Zone in Germany and Austria, down the Adriatic coast of Yugoslavia and into the Black Sea and Caspian areas.

Fighter defences comprised propeller-driven day fighters, principally the Yakovlev Yak-9 *Frank*, Lavochkin La-7 *Fin*, La-9 *Fritz* and La-11

LISTENING IN

A scientific intelligence model of the Soviet RUS-2 truck-mounted early-warning radar. Although basic by Western standards these radars were highly mobile and could be produced in large numbers. *Author's collection*

Fang, using obsolescent HF R/T sets to communicate with ground-controllers. The Soviets had no AI radar and thus had a very limited all-weather capability. The fighter defences were backed up by wartime 37mm and 85mm AA guns, directed via Soviet-built SON-2 *Fire Shield* fire-control radars, copies of the UK lend-lease GL Mk.II (54-85MHz) radar.

Although the Ferret flights suggested a fairly primitive air defence system the UK intelligence authorities were aware that other equipment might be under development deep inside the Soviet Union, where their signals could not be intercepted.

The RCM/Y Lincoln B.2/3G finally arrives

Work on the modification of Lincoln B.2/3G (RE395) as a dual-role RCM/Y aircraft had started at the CSE in January 1947 but progress on the conversion had been slow due to the lack of manpower and higher-priority tasks, and work appears to have come to a halt on a number of occasions. The mid-upper turret was removed in February 1947 but the main installation task did not really get going until October 1947.

Preliminary work involved stripping out all existing equipment in the mid-fuselage compartment, from the main spar back to the rear step of the bomb-bay. Following this the empty mid-upper turret position was blanked off and provided with an astrodome for observation purposes. Extra power for the fit was provided via four additional engine-driven alternators with the electrical switchgear for the alternators mounted in the rear fuselage aft of the bomb-bay.

A Special Operator's station, fitted with a table, chair and equipment racks was then installed on the port side of the mid-fuselage. The operator faced to port and his chair was mounted on rails, allowing it to be stowed under the table when not in use, leaving a free passage along the gangway on the starboard side. Y/RCM equipment was mounted on and below the operator's table, secured by quick-release fittings to allow the interchange of equipment when the aircraft switched roles. Three control panels for Y, RCM and power switching were mounted forward of the operator while the rotary converter panel was mounted to his rear on the starboard fuselage side. Two windows were cut in the starboard fuselage behind the Special Operator (SO) to provide illumination, with black-out curtains provided for night operations. The existing radar and navigation equipment, previously stripped from the compartment, was re-installed around the SO's station.

The aircraft was fitted with a comprehensive Elint aerial array: horizontally and vertically-polarised aerials were fitted to the outer surfaces of both tail fins, below the fuselage and in the nose of the aircraft. A rotary switch at the SO's position allowed any of these aerials to be connected to his Elint receivers. For RCM purposes additional aerials were fitted above and below, and to port and starboard of, the fuselage and on the top of the wings. The search receiver fit in the Lincoln was broadly similar to that applied to the Lancasters, that is, ARR-5, APR-4 and APR-5 although provision was made to carry four receivers. Like the Lancaster installations the fit was search-only, with no direction-finding facilities.

All installation work on the Lincoln was completed in February 1948 and, following major servicing and flight trials, the aircraft was finally made available to the CSE Flying Wing on 17th August 1948. After acceptance checks the Lincoln flew its first Y sortie in November when it carried out a short flight from Shepherds Grove, probably along the South Coast of the UK. In December the Lincoln flew its first major operational sortie, part of the 1948 Ferret programme, when the aircraft left RAF Lyneham for Malta, carrying out a 7 hour operation en-route, running down the Adriatic coastlines of

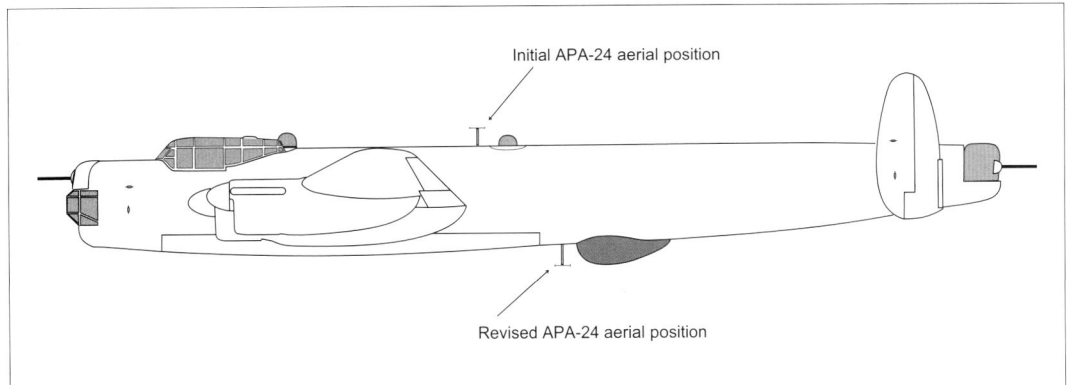

Lincoln B.2/3G Elint conversion. This programme included the conversion of RE395 to the dual-role RCM/Y, which required additional generators on the engines and alternators in the bomb bay.

Yugoslavia and Albania. From Malta the Lincoln was flown to Fayid, Egypt for a series of calibration sorties, returning to the UK via Malta in mid-month.

The Malta operation using RE395 marked the end of the 1948 Ferret programme. No further Elint operations were flown by CSE aircraft until August 1949.

Fleet Exercises During 1948

In addition to the Y investigation and subsequent Ferret programme, the Elint Lancasters of the CSE were also flown on a couple of naval exercises during 1948. In March the Admiralty approached the CSE to discuss the possibility of CSE aircraft participating in a series of naval jamming and monitor trials. The Admiralty were particularly anxious to determine the range at which ships and submarines could be detected from their radar transmissions. The CSE agreed to accept the task and the Air Council subsequently approved the use of the CSE in Fleet radio warfare trials.

The first Elint investigation at the request of the Admiralty took place in September 1948 during Exercise *One Step*. Lancaster PA232 was detached to RAF St Eval, in Cornwall, from where it flew a 7-hour sortie against the Home Fleet in the Western Approaches. At the end of October 1948, in a complementary exercise to *One Step*, Lancaster PA478 was detached to the Joint Anti-Submarine School, RAF Ballykelly for radar intercept trials against a submarine to establish the range at which the submarine's X-band (10GHz) sea guard and metric air guard radars could be detected. Since the CSE lacked an X-band intercept capability, the trials were confined to the metric band only. The Lancaster flew three sorties investigating the intercept ranges achievable against the submarine's 214MHz air warning radar with the submarine both running on the surface and snorkelling.

1949: The Need For Intelligence

Increasing tension between the Soviet Union and the Western Powers, and a revival in the fortunes of Radio Counter Measures (RCM) in the UK, led to a renewed focus on Elint during the early months of 1949.

In the immediate post-war years the Air Ministry had shown little interest in the application of RCM, either defensively or offensively. However, the continued efforts of the CSE to champion the cause, and provide practical demonstrations of the worth of RCM, had begun to pay off during 1948. By then the various RAF Commands were beginning to realise that RCM could not be ignored, and Bomber Command in particular seems to have got the message that RCM could make a significant contribution to the success of its operations. The upshot of the renewed interest in RCM was the formation in January 1949 of the Air Ministry RCM Committee 'to decide upon the policy of the employment of RCM by the RAF'.

The first meeting of the AM RCM Committee took place in the last week of February 1949. Item No.3 on the agenda was 'The Urgent Need for High Grade Scientific Intelligence on Radio'. It was clear to the committee that future RCM operational requirements would be completely dependent on access to scientific intelligence on Soviet air defences. Offensive RCM equipment was intended to support the bomber force by disrupting the enemy's air defence system, jamming early warning, fire-control, and AI radars, and interfering with the enemy's communications, particularly the GCI control of fighters. However, as wartime experience had shown only too well, considerable knowledge of the enemy's systems was required before effective radar and communications jamming equipment could be developed. Some intelligence data was available from the ADI (Science) ground listening stations in Germany and from the 'Ferret' flights

LISTENING IN

carried out by the CSE during 1948. In addition, an officer had been established on the Air Ministry intelligence staff to collate all the available technical data to produce an electronic order of battle of the Soviet Bloc. However, the existing flow of intelligence was clearly insufficient. Complaints regarding the relative paucity of intelligence on Soviet air defences were subsequently echoed by Bomber Command, the Air-Officer-Commanding (Air Marshal Sir Aubrey Ellwood) making the point that: 'virtually no information on our potential enemy's radar systems has been supplied to this Headquarters'. In response to these concerns the AM RCM Committee submitted a paper to the Air Staff pointing out the urgent requirement for improved intelligence on Soviet radio and radar, making the point that: 'At present what is known is very meagre, but it would indeed be dangerous to assume that because little is known little exists. Means must be found to obtain such information, for without it RCM cannot be effective when it will be most needed that is, at the very outset of hostilities'. The paper seems to have been well received within the Air Ministry, with general agreement regarding the need for an improvement in intelligence collection, and by March 1949 plans were being laid for an expansion of the Elint programme. On 9th March DCAS, Air Marshal Sir Hugh S P Walmsley, wrote to a number of overseas RAF Commands, informing them that there would be an increase in Elint operations and that: 'Because it is impossible to collect this type of intelligence by other means, it will therefore be necessary to dispatch Ferret aircraft to overseas Commands more frequently than hitherto in order that investigations may be carried out in areas beyond the range of aircraft based in the United Kingdom'.

Other Tasks

Although planning for a new series of Elint sorties took place in the first quarter of 1949, it appears to have taken some time to get the necessary authorisation. Thus the CSE Elint aircraft were employed on a number of tasks during the first half of 1949, including radio/radar intercept trials with the Fleet and the monitoring of RCM exercises. In the absence of Elint tasks the dual-role aircraft were also employed on RCM duties as jamming aircraft. Lancaster PA478 was also flown on practice Monitor flights (as Y flights were now known) during the first half of 1949 in order to keep the crews in practice for the upcoming operational sorties.

As part of the training programme two sorties were flown on Route 2 (between the Thames Estuary and the Wash) listening to non-communications transmissions in the 500MHz to 1GHz band.

The maritime sorties flown by the CSE during the first half of 1949 were essentially a continuation of trials started in 1948 to determine the ranges at which a fleet could be tracked by Elint aircraft. Under Exercise *Foxtrot* the CSE deployed a Lincoln (RE395), Lancaster (PA478) and Mosquito (NS809) during late January 1949 to monitor naval radio and radar transmissions as the Fleet left the South West Approaches at the start of its annual spring cruise. The aircraft monitored a variety of frequencies to determine which transmissions could be intercepted by aircraft flying at low and medium altitudes. A follow-on exercise, *Lancer*, took place towards the end of March when Lincoln RE395 flew further sorties monitoring the radar transmissions of the Fleet on their return from their spring cruise. These exercises showed that the Type 960 and Type 281 metric-band naval air warning radars in use by the RN could be intercepted at a range of 110nm (203km) at 20,000ft (6,096m), this figure reducing to 80nm (148km) at 10,000ft (3,048m). The S-band Type 277 surface warning radar was intercepted at 25nm (46km) range at 5,000ft (1,524m), this figure apparently (and surprisingly) increasing to 40 miles (74km) at the lower altitude of 1,000ft (304m). This latter result suggested that the interception of centimetric naval radars would need to be conducted at low altitudes.

Automatic Y

The first quarter of 1949 also saw the (unsuccessful) culmination of the CSE's first attempt at developing an automatic airborne Elint capability. One of RWE/CSE's initial tasks, laid down in 1946, had been the development of a so-called 'Automatic Y' capability. The aim was to produce an unattended system which could automatically intercept and record radar and other transmissions over a wide range of frequencies. The advantage of such a system was its ability, in theory at least, to capture a wider sample of signals than was possible by manual means. Such a system could also be used by operational bomber aircraft to record the enemy signals environment during bombing raids.

In March 1947 the CSE received instructions to progress Task 503 ('Design of a full range of Automatic Search Receivers for Y purposes').

Although two types of automatic receiver (*Bagful* and *Blonde*) existed, these only covered a limited range of metric frequencies. Task 503 instructed the CSE to investigate whether the operational requirements could be met by extending the frequency range of the two receivers, and to arrange for the design of new systems if this was not possible.

Blonde and *Bagful* were two complementary automatic scanning/recording superheterodyne receivers developed by the TRE during the Second World War. *Bagful* (R.1622) scanned a preset frequency band and recorded the frequency and time-of-intercept of received signals onto an electrically-sensitised paper tape. *Blonde* (R.1645) scanned a much narrower frequency band and used a photographic recording technique to capture the PRF and scan characteristics of intercepted signals. The *Bagful* system had been used operationally during the war, while the *Blonde* system was still under development at the end of the war.

As a first step towards meeting the Automatic Y requirement, the CSE took over responsibility for the completion of development of *Blonde*, and for the design of a new RF head (RF Unit 123, 1.8-3.7GHz) which was designed to extend both receivers' frequency range into the centimetric band. However, analysis suggested that the long-term goal of providing a complete range of automatic recording receivers could only be fully met by new equipment, and that might take five years to develop. In the first half of 1947 a further 'automatic Y' requirement was issued (Task 505), calling for the installation of a suite of four *Bagful* and four *Blonde* receivers in a Lancaster. Unfortunately a previous installation of *Blonde* and *Bagful* in a Y Search truck had revealed a mutual interference problem, produced by the receivers' local oscillators, when they were mounted in close proximity to each other. Further investigation revealed that a partial fix was possible at frequencies under 200MHz but fixing the problem above that frequency was extremely difficult, and would require a partial redesign of the receivers. This would be expensive and outside the resources of the CSE: in fact it was concluded that it would make more sense to get a contractor to redesign the system from scratch using modern techniques.

In November 1947 it was decided to proceed with the Task 505 Lancaster installation despite the interference problem, since it was felt that some operational use could be made of it as an interim measure. Work on an installation of four receivers in Lancaster B.I(FE) PA444

The R.1622 *Bagful* automatic receiver/recorder. The frequency of intercepted signals was recorded on a paper roll. *By kind permission of the Malvern Radar And Technology History Society*

was begun towards the end of March 1948 and, after some delay due to other commitments, air tests of the installation were carried out in late 1948. By the first quarter of 1949 a full range of automatic monitoring receiver aerials, including centimetric band, had been tested in PA444. It appears however that the trials were not successful and the Automatic Y concept using *Bagful* and *Blonde* was not considered a practical proposition, given the limitations of the receivers. Following the completion of the aerial trials PA444 was allotted as a third 'manual' Y Lancaster, supplementing PA232 and PA478.

Despite the failure of the Automatic Y investigation, the Air Ministry retained an interest in the concept and an Operational Requirement (OR.3522) was later issued for a suite of automatic Elint receivers for both ground and airborne use. There were however significant technical problems to be overcome in order to develop a practical system and, since it was not considered a priority, no further work was done on the project by the CSE for almost ten years.

Increased Resources

The increased priority afforded Elint during 1949 was reflected in a change of policy regarding the provision and fit of specialist Elint aircraft. Previous CSE policy had been to modify aircraft as dual-role RCM/Elint platforms with fixed fittings and aerials for both roles, allowing the aircraft to be used for either task, depending on the demands of the Establishment. Although the dual-role policy was theoretically very economical, maximising the use of individual aircraft, the problem was that, in practice, the continual change-over of the aircraft from one role to another was a time-consuming business. It also made less sense if the number of Elint operations was to increase. As a result of

LISTENING IN

the increased interest in Elint, it was decided during 1949 to permanently allot three CSE Lancasters aircraft to the Elint role. These were B.I(FE) PA478 and B.I PA232 (previously dual-role RCM/Y) and PA444 (previously the 'automatic Y' aircraft). Work commenced in April 1949 on replacing the *Blonde* and *Bagful* equipment in PA444 with manual receivers, and the conversion was completed at the end of August. The installation in PA444 is thought to have been very similar to that applied to PA478.

Allotment of the three Lancasters as dedicated Elint platforms was really an interim measure, to provide a reasonable operational Elint capability in the short-term. The long-term plan was to replace all the Elint Lancasters and the lone Lincoln B.2/3G with three Lincoln B.2/4As. However, this would take some time to achieve.

The single-role policy was also applied to the Elint Mosquito. Plans for the replacement of the CSE's elderly Mosquito PR.XVI NS809 had been made in late 1948 and PR.34 RG232 was delivered to the CSE at the end of that year. Following discussions regarding the aircraft fit, work on the conversion of the PR.34 to a dual-role RCM/Elint platform began at the end of April 1949. However the installation was amended in June when the aircraft was re-designated as a dedicated Elint platform. The change to a single role must have come as something as a relief to the Technical Wing as fitting even a dedicated Elint installation into a Mosquito was a challenge enough.

The aircraft installation required the addition of six external aerials, including a number of quarter-wave stubs and a cone aerial on the aircraft underside. A new sprayed-metal supressed tail fin antenna, (based on a TRE design for a suppressed Gee aerial for the Canberra) was fitted to the aircraft, covering the range 50-150MHz. A new, compact, pulse analysis unit was also developed for the aircraft to provide the same sort of pulse analysis facilities as those available in the Lincoln and Lancasters.

Providing Airborne D/F

1949 also saw a major breakthrough in the provision of a direction-finding (D/F) capability for the CSE Elint aircraft. The lack of an airborne D/F facility that is, the ability to obtain a bearing on a received signal, was a major operational shortcoming: direction-finding was vital to pinpoint the location of radar stations, and thus plot the distribution of enemy radar defences. Without D/F, airborne Elint could only provide rough data on the geographical distribution of Soviet Bloc radar stations. The lack of airborne D/F equipment was due to a number of factors. Developing an accurate airborne wide-band D/F set operating at metric radar frequencies was a difficult technical challenge, the main problem being that an aircraft airframe was electromagnetically resonant at metric radar wavelengths, and so reflection and re-radiation of a received signal produced large directional errors (of the order of 20° at frequencies in the range 30-500MHz). Despite the difficulties, sporadic attempts had been made by the TRE to develop airborne D/F equipment during the war, but resources were limited and the project had never received sufficient priority to make much progress. An urgent (and very high-priority) D/F requirement had arisen prior to D-Day to plot accurately the position of German early warning radars on the French coast in order that they could be destroyed prior to the invasion. However, signals from these radar stations could be intercepted on the UK coast and it made more sense to plot them using a ground-based D/F equipment known as *Ping-Pong*. The only real attempt at providing a British airborne D/F capability came at the end of the war, when a centimetric D/F system known as *Coalscuttle* was developed by the TRE and installed in one or two Halifax of 192 Squadron to investigate the German use of centimetric frequencies. This system apparently used the aircraft's H2S aerial for direction-finding, but does not seem to have been widely used and was abandoned at the end of the war.

Post-war, the development of an airborne D/F capability became a Priority 1 Air Ministry requirement. Unfortunately, the CSE was overworked and under-staffed and, although airborne D/F had a high priority, there were even higher priority RCM projects to progress. The provision of D/F was further complicated by the fact that it would need to cover a wide range of frequencies. Although Soviet air defence radars in the late 1940s were all in the metric band, the expectation was that centimetric equipment would emerge in due course and thus D/F equipment would be needed in both the metric and centimetric radar bands. The Establishment noted in 1948: '[a] very firm requirement of the Air Ministry is the provision of Airborne D/F. At present it is thought that all round D/F on all frequency bands is almost impossible owing to a number of limitations which are placed on the designer by the airframe and by aerodynamic considerations. It should however be possible to provide D/F over certain sectors …'. Unfortunately, due to staff shortages at the CSE, work on the D/F requirement proceeded at a slow pace and often came to a halt due to the demands of more urgent projects. Nevertheless, during 1949 significant progress was made in the field, resulting in the development and delivery of both metric and centimetric D/F equipment. It seems likely that the extra effort put into D/F development was a direct result of the increased priority afforded Elint during the year.

The CSE's centimetric D/F programme actually emerged from its airborne homing programme. During the late 1940s one of the top priority programmes at the CSE was the development of equipment to allow fighters to home-in on Soviet bomber radio and radar transmissions. Since no single equipment could cover the full range of frequencies, a number of homing systems, each covering a portion of the metric frequency band were produced. In mid-1947 intelligence emerged that the Soviet Union was investing heavily in radar research in the centimetric band and, as a result of this information, the Air Ministry directed the CSE to concentrate on a centimetric homer working in the S-band (2-4GHz). By the beginning of 1948, the CSE had developed a centimetric homing system using crystal detector-video (DV) technology. The system used a waveguide antenna in either wing, each feeding a rectifying crystal semiconductor detector. The output from each detector was fed through a separate delay-line and amplifier before being displayed as a trace on a small CRT. The relative amplitude of each trace on the CRT gave a rough indication of signal bearing, with 'dead-ahead' being indicated when both traces were the same size. In September 1948 the centimetric homer was nearly ready for installation in a Mosquito for trials. By then the Establishment had realised that the techniques used in the centimetric homer were directly applicable to D/F and had begun actively to investigate the development of a centimetric D/F system using DV technology. The proposed system was intended to give a high-probability of intercept and D/F over 60° azimuth in the 3-3.75GHz band.

By April 1949 a bench model of the airborne centimetric D/F receiver was giving good results under test on the roof of the laboratory. It had been found possible to fit the majority of the system (amplifiers, delay network, power pack and display unit) into a box measuring 9ins x 8ins x 18ins (23cm x 20cm x 56cm) with the only external components being the waveguide aerials and their detectors, and the connecting co-axial cables. Probably due to its compact nature the new system was nicknamed 'The Watton Box'. Following the success of the initial ground trials, an airborne laboratory model was constructed and bench-tested during mid-1949 before being installed in Lancaster PA232 for flight trials.

The system was flown in PA232 on trials against a centimetric radar at RAF Bard Hill, Norfolk in late October/early November in order to provide SOs with practice in using the set. Following these initial trials PA232 was detached to St Eval in the first week of November for Exercise *Porcupine*, during which the Watton Box was used to locate submarines by obtaining bearings on their centimetric air-guard radar transmissions. The report on the exercise noted that: 'The position lines obtained by these aircraft were remarkably accurate and would have enabled the enemy to have been tracked satisfactorily without other information'. The CSE conclusion was that the 'Watton Box' equipment had great possibilities as a search aid and was worthy of intensive future development and study. Following Exercise *Porcupine* the Watton Box was removed from PA232 and installed in Lancaster PA478 for use on an operational detachment to Iraq.

The CSE began serious work to tackle the 'formidable' problem of metric D/F during early 1949, in parallel with development of the Watton Box. A new technique was devised using a fixed aerial which it was hoped would provide reliable D/F, albeit it in a fairly narrow frequency band, over any part of the VHF band. Despite work on the system being delayed by a lack of staff, by September 1949 an experimental version of the system had been constructed for ground trials. By then, however, the CSE had acquired a US-built AN/APA-24 aerial. The APA-24 was actually a range of wartime directional aerials, designed to provide homing or D/F in the lower metric band. When used for D/F, the aerial was rotated until the minimum received signal point was detected, this indicating the direction of the emitter. In October 1949 a CSE-built copy of an APA-24 was installed in Lincoln RE395 feeding an APR-4 receiver with the aerial, covering the 75MHz band used by the majority of Soviet radars, mounted on a mast protruding through the top of the fuselage.

It seems curious that it had taken the CSE so long to adopt the APA-24 since the Establishment had been made aware of its existence by the TRE back in 1946. It is possible that this was due to an absence of supplies in the UK and dif-

LISTENING IN

ficulties in obtaining the system from the Americans. The fact that the CSE built copies of the device suggests it was indeed in short supply. In any event, since the APA-24 appeared to meet the immediate metric D/F need, work on the CSE-designed fixed aerial system was abandoned. The UK subsequently put in a bid for six APA-24 systems (plus spares) from the US in early 1950 under the Mutual Defence Aid Program (MDAP), a US government programme under which the US supplied military aid to friendly nations.

Ferret Flights 1949

Although plans for a new series of Ferret flights were initially made in the spring of 1949, it was not until August 1949 that the commencement of a new series of Ferret flights was announced. The delay may have been due to the need to obtain political approval for what were undoubtedly risky operations.

The new series of flights covered the Baltic, the Adriatic and the Caspian Sea areas and were intended to establish the number and types of radar in use by the Soviets, to check on their geographic distribution, and to look for new radars, particularly in the centimetric band. To ensure the maximum value was obtained from each sortie the programme was co-ordinated with that of the Americans.

Operations began in August with five day sorties from Schleswigland over the Baltic, the first on 8th August when Lancaster PA478 completed a 9 hour 35 minute daylight sortie. A further two sorties, each of 10 hours duration, were flown from the same base on 16th August, this time using Lancaster PA478 and Lincoln RE395. The final two sorties of the series took place on 26th August using the same two aircraft. The operations were apparently very productive with 'excellent results' obtained and in fact the sorties produced so much data on Soviet radar defences along the Baltic coasts of Germany, Poland and the Soviet Union that one officer at the CSE was fully engaged on correlation and analysis of the results of the flights for a number of weeks.

In October, in the second phase of the programme, Lincoln RE395 and Lancaster PA444 were detached to Malta, from where two dual sorties (each of around 9 hours duration) were flown into the Adriatic, looking for radar activity along the coast of Albania and Yugoslavia. The third and final series of Ferret sorties were flown in December, this time from Habbaniya, Iraq with Lincoln RE395 and Lancaster PA478 deploying via Malta at the end of November. The Lancaster had been fitted with the new Watton Box centimetric D/F prototype, while the Lincoln carried the APA-24 metric D/F system, making this the first post-war UK Ferret operation in which the aircraft were capable of plotting the position of Soviet radar stations with any degree of accuracy. A member of the Habbaniya ATC staff recalls that PA478 'bristled with more aerials than a porcupine'. Unfortunately the Lincoln suffered an engine problem and a replacement engine had to be flown out from the UK. While the Lincoln was awaiting repairs the Lancaster carried out five sorties, each of 8 hours 30 minutes duration, departing Habbaniya very early in the morning 'having requested Met forecasts for two different areas and without filing a flight plan', and returning at lunchtime. The sorties most likely went along the Azerbaijan border and into the Caspian Sea, attempting to locate Soviet radars in the area. The Lancaster returned to the UK in mid-month on the completion of its part of the programme. Following repairs, the Lincoln successfully carried out a single sortie on the 21st December before returning to the UK the following day.

The 1949 Ferret programme revealed a reorganisation of Soviet air defence radars: a lot of the wartime Lend-Lease radars had been withdrawn and a smaller number of Soviet systems (mainly RUS-2 and *Dumbo*) had been substituted in their place. There were a couple of competing theories for the change: one was that the technically-superior Lend-Lease systems had been withdrawn from the periphery to guard important targets within the Soviet Union, while another suggested that the ageing Lend-Lease systems were proving difficult to maintain and were being withdrawn for that reason. In any case, the Air Ministry remained suspicious of what the Soviets might be developing and how their air defences might evolve. As VCAS, Air Marshal Sir Arthur P M Sanders, noted in December 1949: 'Radio countermeasures are a kind of counter-attack and depend on detailed knowledge of the enemy's practice for instance, of the frequencies he will use. ... We know something of the present rather backward Russian radar organisation, but we know little of their research, and nothing of their staff requirements for the future. So we can only plan on the supposition ... that they will be equipped to our standards. But we cannot assume that they will be equipped like us; they will probably use quite different frequencies and methods'.

The CSE Flying Wing had originally moved to Shepherds Grove from Watton due to congestion of that airfield's single runway and weather-based flying restrictions. However, splitting the CSE across two airfields was an expensive arrangement which became increasingly hard to justify. In December 1949 the Establishment bowed to financial pressures and moved the Flying Wing back to Watton. The move produced another significant benefit by reuniting The Flying Wing with the various ground elements of the CSE. In particular it allowed much greater liaison between the aircrews and the scientific staff of the Research Squadron.

Shortly after the move, in January 1950 Lancaster PA478 was detached to St Eval to take part in another exercise with the Home Fleet. The aircraft shadowed the ships of the fleet on the outward trip of their spring cruise (en-route to the Mediterranean) via their S-band transmissions to further assess the Watton Box and to provide experience for SOs in its use. An assessment was also carried out on the accuracy of the APA-24 system in Lincoln RE395. Airborne calibration of the system in January/February 1950 showed the position of the D/F aerial was unsuitable, with unpredictable errors of large magnitude occurring on the port and starboard bow and directly astern. The errors appeared to be related to the dorsal aerial position and there were suggestions that another position, possibly underneath the aircraft, might be better. The aerial was subsequently fitted under the fuselage on a retractable mast, just forward of the H2S radome, where it was found that acceptable D/F accuracy could be obtained in a number of sectors.

Radio Proving Flights 1950

During the period 1946-49 the Elint sorties flown by CSE aircraft were variously referred to as 'Y' sorties, 'Special Signals' sorties and 'Ferret' flights. By 1950 these rather-too-descriptive terms for an Elint sortie had been abandoned in favour of the more ambiguous 'Radio Proving Flight' (RPF) with the equally opaque 'Air Ministry Operation' (AMO) also used on occasion.

In early 1950 a further Ferret/Radio Proving programme was approved, apparently covering the same areas as those visited during 1949. Following calibration of the repositioned APA-24 aerial, Lincoln RE395, along with Lancaster PA478 (fitted with the Watton Box) and Mosquito RG232, were readied for operations in the third week of February 1950. All three aircraft were detached to RAF Wünstorf, Germany on the 21st February and the aircraft subsequently flew co-ordinated long-range Elint daylight sorties (8 hours for the Lancaster, 5 hours for the Mosquito) into the Baltic. The Lancaster and Mosquito flew a second joint operation a couple of days later before all three aircraft returned to Watton on the 28th February. As it turned out, this was the last RPF flown by the CSE during 1950.

Special Operator's station in a Lincoln. Although this is an ECM Lincoln, the layout is similar to that employed in Elint Lincolns, and shows the cramped conditions. View forward shows equipment racking on port wall and main spar at forward end of compartment. View aft shows SO's seat, table and equipment rack. Immediately above table is faired-over dorsal turret position. Aft of the rear equipment rack is the step down into the rear fuselage. *via Ron Henry*

3 Upping the Stakes

'Our intelligence on Russia is most inadequate and ... high priority should be given to measures to improve it. The search for radio signals, both communications and non-communications, is of the highest importance' Report by the Working Party on Radio Warfare, April 1951.

The deterioration in relations between the Soviet Union and the western powers during the late 1940s gave rise to occasional incidents of aggression. However, despite the increasing tension, the Soviet Union had allowed Elint sorties by UK and US aircraft over the Baltic to proceed unmolested. It therefore came as something as a shock when, on 8th April 1950, a US Navy Consolidated PB4Y-2 Privateer (59645, *'The Turbulent Turtle'*) was attacked by Soviet fighters and shot down off the coast of Latvia. The Privateer, belonging to VP-26, the US Navy's Sigint squadron, had been detached to West Germany from its normal base in Morocco for an Elint operation over the Baltic. Taking off from Wiesbaden, near Frankfurt, it had flown north to the coast at Bremerhaven, then turning east, following the coasts of Germany, Poland, Lithuania and Latvia. There is some suggestion that the aircraft may have temporarily infringed Soviet territorial waters during the sortie, but in any case the Privateer was intercepted at 1739hrs by four Soviet La-11 *Fang* fighters in international waters off Liepaja, Latvia and shot down, killing all ten crew. Although the Soviets later claimed the aircraft had fired on the intercepting fighters it was, in fact, unarmed.

The available evidence, which included GCI voice intercepts by a Swedish Elint station, suggested that the shooting down was a deliberate act by the Soviet Union and may have been intended to send a warning to the US that it would no longer tolerate intrusive Elint sorties. The loss of the Privateer did in fact lead to a change in US policy, with Elint flights in sensitive areas restricted to the hours of darkness and prohibited from approaching closer than 20nm (37km) to the Soviet coast. However, aircraft were also armed and authorised to return fire if attacked.

The effect of the shoot-down on UK airborne Elint operations was immediate and drastic: all sorties were suspended and a review was

The *Turbulent Turtle* was an Elint conversion of the Consolidated PB4Y-2 Privateer. Developed as a dedicated long-endurance patrol bomber for the Pacific War, the Privateer was a stretched B-24 Liberator with a single fin. It made an ideal Elint platform. *US Navy*

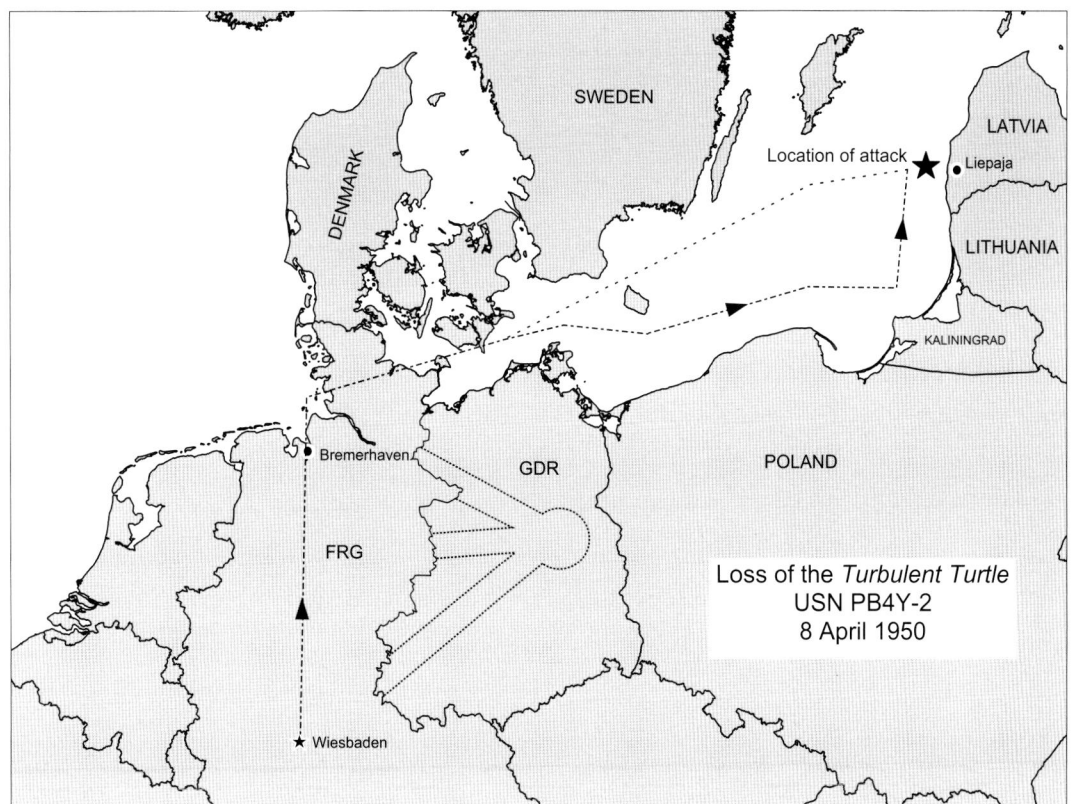

The loss of the US Navy's VP-26 PB4Y-2 Privateer *Turbulent Turtle* was a turning point in the conduct of Elint sorties into the Baltic.

launched into how these operations were planned and conducted. The conclusions of the review seem to have been similar to those reached by the Americans that is, that it would be safer to fly future operations in sensitive areas such as the Baltic during the hours of darkness, to protect the aircraft from interception. It also appears that the existing CSE Lancaster and Lincoln B.2/3G Elint aircraft were ruled operationally unsuitable and barred from future operations. The capabilities of the replacement Lincoln B.2/4A aircraft, then under conversion, also came under scrutiny, especially their navigation fit. While the Air Ministry, CSE, LSIB and government discussed the conduct of future operations Elint operations were put on hold and no sorties were flown for the remaining nine months of 1950.

During 1950 Bomber Command remained dissatisfied with both the lack of intelligence on Soviet air defences and progress in the provision of RCM for bombers. In July the AOC, Bomber Command (ACM Sir Hugh Lloyd) visited the CSE to examine the resources allotted to both Elint and RCM and was not impressed. In a letter to VCAS (ACM Sir Ralph Cochrane) he complained about the small research effort allotted to both fields, noting: 'Clearly CSE must continue its Y service tasks in "ferreting", which is now limited to four aircraft or less but is this on a sufficient scale? ... I am interested not only that it is done adequately but, what is equally important, is that something is done as a result of such information as may be acquired'. Presumably Sir Hugh would have been even less impressed if he had discovered that no RPFs were then being flown. In September 1950 Sir Hugh went on to suggest that operational control of 'radio reconnaissance' operations be transferred to his command, allowing him to treat photographic and radio reconnaissance in a similar manner. The suggestion was rebuffed by VCAS who explained that the existing system had 'certain overwhelming advantages'. Bomber Command were not the only ones dissatisfied with the lack of intelligence, and in November ACAS(Ops) (Air Vice-Marshal Charles Guest) wrote to VCAS drawing attention to 'the vital need to watch closely Russian C and R [Control and Reporting] activities along the Iron Curtain'. In fact Elint operations had by then been reviewed by the LSIB and Joint Intelligence Committee and arrangements put in to improve matters. By the winter of 1950 the CSE and Air Ministry had made plans replace the Lincolns with three Boeing B-29 Superfortress aircraft (see Chapter Four) and arrangements were also put in hand to expand the ground Elint organisation in Germany. The latter involved the formation of 365 Signals Unit at RAF Uetersen, supplementing the existing ADI(Science) Overseas Party at

LISTENING IN

Obernkirchen. Discussions also took place regarding the resumption of RPF operations. However, before looking at the resulting RPF programme, the conversion of the Lincoln B.2/4As needs to be described.

The First Two Lincoln B.2/4As

From 1946 onwards Bomber Command had begun to replace their obsolete Lancasters with the Lincoln B.2. The CSE made similar plans to replace their Lancasters with Lincolns and a B.2/3G (RE395) was acquired for conversion to the RCM/Y role in August 1946. CSE's plan was to eventually convert three Lincoln B.2s for use in the RCM/Y role, with RE395 serving as the 'prototype' conversion. However, modification of RE395 was a drawn-out affair and by the time work was completed (in August 1948) Bomber Command were taking delivery of the Lincoln B.2/4A, carrying the more advanced H2S Mk.4A radar. For practical and operational reasons it made sense for the CSE to operate the later 4A variant of the Lincoln, which meant that the recently-completed RE395 would have to be replaced with a Lincoln B.2/4A. Unfortunately, the CSE 'prototype' RCM/Y Lincoln modification applied to RE395 was specific to the B.2/3G and could not be applied to the B.2/4A, which had a significantly different internal radio and radar equipment layout.

By the first half of 1949 the CSE had adjusted their Lincoln re-equipment programme to accommodate the switch from the B.2/3G to the B.2/4A. The plan was to replace both the Monitor Lancasters and Lincoln RE395 with three Lincoln B.2/4As. The first Lincoln B.2/4A (SX980) was received by the CSE on 13th May 1949 and work to convert the aircraft as a Monitor started shortly afterwards. The conversion required considerable design work: not only did the RCM/Y modification applied to RE395 have to be redesigned to fit into a B.2/4A, but the CSE now wanted to fit the B.2/4A with two Special Operator (SO) positions, rather than the single position fitted in the B.2/3G.

In the Lincoln B.2/4A the majority of the H2S installation was located in the mid-fuselage above the wing, exactly where the Special Operator positions had to be accommodated, creating a major headache for the CSE installation design team. At that time there were conflicting opinions regarding the usefulness of H2S in an operational Elint aircraft, with one view being that H2S provided a negligible improvement in navigational facilities at the expense of a much more complex Monitor installation. When the conversion of SX980 started the question had not been settled, and little guidance was forthcoming from the Air Ministry and Bomber Command. As a result the CSE, who were not convinced that H2S was really useful in the Elint role (presumably based on operational experience), decided to dispense with H2S, thus drastically simplifying the installation. This allowed them to install two fairly roomy and (by the standards of the day) comfortable SO stations in the mid fuselage.

The two positions were designed to provide an intercept, analysis and direction finding (D/F) capability in the medium frequency (MF), metric and centimetric bands using the standard ARR-5, APR-4 and APR-5/APR-5A receivers and oscilloscope plus oscillator signal analysers. D/F

Lincoln B.2/4A SX980, showing astrodome replacing dorsal turret and aerial mast under fuselage. Note the retention of the H2S radome that housed antennae for the Elint receivers, the H2S radar having been dispensed with. *Author's collection*

facilities were provided by a CSE-built copy of APA-24 and a Watton Box centimetric system, providing D/F in the important 75MHz and 200MHz bands (Soviet EW radars) as well as the 3GHz (S-band) and 10GHz (X-band) centimetric bands (GCI/Fire-control radars). As on RE395, the APA-24 aerial was mounted on a mast under the fuselage, while the Watton Box aerials were mounted on a mechanically-rotatable platform in the otherwise empty H2S radome. The provision of both metric and centimetric D/F facilities was described by the CSE as an 'outstanding feature'. The installation also provided the ability to intercept and record voice communications in the HF and VHF bands, recording onto a magnetic wire recorder. With this installation the aircraft would theoretically be able to intercept and plot the positions of Soviet metric radars, to monitor and record conversations between Soviet radar stations and fighters in the HF bands (and VHF bands if required) and to detect and plot any centimetric radars in the S and X-bands. At that time the presence of Soviet centimetric radars was suspected but not confirmed.

The conversion of SX980 was a relatively complicated affair and the initial conversion work was not completed until March 1950. However, it appears that the CSE had still not quite shaken off the 'dual use' RCM/Elint concept and the conversion programme was then extended by a further month or so to include an RCM capability, allowing the aircraft to be used on RCM exercises when not employed on Monitor duties. The installation work in SX980 was finally completed towards the end of May 1950 and the aircraft air tested on 2nd June. The initial reports from the Special Operators were extremely positive, due in part no doubt to the improved layout of the Special Operator positions. Unfortunately subsequent calibration checks of the APA-24 metric D/F system revealed serious and inconsistent errors in the system with discrepancies of up to 15° in bearing. Unlike the installation in RE395 there were no 'good' sectors where results were consistent.

In mid-June the Lincoln was detached to RAF Aldergrove to participate in the Flag Officer Submarines Summer War. This was considered part of the work-up of SX980, checking the intercept and D/F of both metric and centimetric radars. In July 1950 arrangements were made to further test the X-band cover of the Watton Box, using an RN submarine's surface search radar, which radiated at the required frequency. Lincoln SX980 was detached to Aldergrove from where it made a number of X-band D/F sorties against the submarines HMS *Anchorite* and *Ambush*.

The second Lincoln B.2/4A for Monitor conversion (SS715) was delivered to the CSE in the first half of 1950 and went straight into the Installation Flight hangar for modification to the same standard as SX980. The main modification task was completed on 28th July and the metric and centimetric D/F aerials (specially-constructed at Watton) were fitted shortly afterwards. The aircraft was little used during the following months due to the embargo on RPFs and did not make its Monitor debut until December when it was flown on an anti-submarine search exercise.

The introduction into service of the two Lincolns B.2/4As rendered the last remaining Elint Lancasters (PA232 and PA478) obsolete and both aircraft were withdrawn from service at the end of 1950.

1951: Border RPF programme

As described above, when the conversion of the first Lincoln B.2/4A got under way in mid-1949 the Air Ministry did not have a clear policy on H2S navigation radar in Monitor aircraft. The CSE, doubtful of its benefits, had decided to remove the radar to expedite the installation. Unfortunately, this proved to be a controversial decision. The shooting-down of the USN Privateer over the Baltic in the spring of 1950 had led to a cessation of RPFs and an investigation by the Air Ministry and other authorities into how these flights were conducted. By mid-1950 the Air Ministry had decided that H2S was in fact an operational requirement in Monitor aircraft, due to the additional navigational accuracy it provided. Unfortunately the decision came too late to change the fit of the first two Lincoln B.2/4As (SX980 and SS715). As a result these aircraft lacked H2S and thus did not fully meet the operational requirements laid down by the Air Ministry.

The Air Ministry's insistence on the use of H2S in Elint aircraft looked like having expensive consequences. It was not really practical to refit the radar into SX980 and SS715 and by the second half of 1950 there was some doubt as to whether the Air Ministry would allow the two Monitor Lincoln B.2/4As to be used operationally. In August 1950 the CSE surmised that: '... it seems inevitable that the two monitors already completed will be declared operationally unsatisfactory – or rather, that they will not be satisfactory in all respects for the full

LISTENING IN

operational commitment'. The CSE assessment proved correct: after some discussion, the Air Ministry decided in September 1950 that the two aircraft could only be operated under visual navigation conditions in daylight, where the chances of a navigational error were small, and only over Germany, where the risk of a hostile reaction was relatively small.

Although the Lincolns would be confined to operating over Germany, they would still be useful. The Air Ministry already operated a number of Elint ground stations in Germany to intercept and D/F Soviet radar and other non-communications signals, but these could only intercept relatively long-range signals. An aircraft flying at 20,000ft (6,096m) would be able to 'see' further across the German border and intercept short-range (that is, low power and high-frequency) signals missed by the ground-based receivers. Plans were drawn up for a monthly programme of RPF sorties over Germany, flying over the British Zone of West Germany along the East German border. Even though the sorties would take place over friendly territory they were restricted to a single RPF each week, in order to avoid alerting the Soviets to the programme and to avoid provoking a hostile reaction.

Despite the limited scope of the new RPF programme, it took some time to finalise arrangements and to obtain the necessary political approval, and thus operational flying did not begin until January 1951. It appears that, due to their relatively low risk, the weekly RPFs were not subject to individual approval by the Prime Minister or the Foreign Office, but were instead covered by a blanket authorisation. The first RPF since February 1950 was flown on 16th January 1951 when Lincoln SX980 carried out a 5½ hour border sortie over northern Germany. The sortie was flown in daylight at 20,000ft (6,096m) with the route taking the aircraft from Watton to just south of Hanover, northwards to Lübeck, and then back to Watton. A further sortie over the same route, but this time at 10,000ft (3,048m), was carried out nine days later.

Although the plan was for a weekly daylight Border sortie, the operations were subject to weather, aircraft unserviceability and other factors, and so were not always flown every week. The operations were normally carried out using a sole Lincoln B.2/4A, but the Mosquito PR.34 (RG232) was also used on a few occasions during the year, flying dual sorties with a Lincoln in March and April.

Coincidentally, in the same month that RPF sorties were resumed, a number of CSE aircrew connected with Ferret operations were gazetted for awards for past operations. Sqn Ldr R T Bainbridge (OC Development Squadron) was awarded the Air Force Cross, Engineer I T Kennedy received the Air Force Medal, and Flt Lt F Reid, Signaller I W E Lowther and Signaller IIB F C Slee received the King's Commendation for Valuable Service in the Air. The citations for the awards provide an interesting insight into the nature of some of the sorties carried out during 1948-1950. Sqn Ldr Bainbridge's citation notes: 'He has successfully undertaken eight "Ferret" flights, many of which have been of a very hazardous nature. Squadron Leader Bainbridge has at all times been a source of inspiration to the members of his squadron. His determination coupled with his ability as a pilot has enabled him to complete sorties which, but for his qualities of leadership, would never have been accomplished'. Flt Lt Reid served as both Navigator and Special Operator in Lincolns and Mosquitos, and his citation recorded: '[The Mosquito Ferret] task,

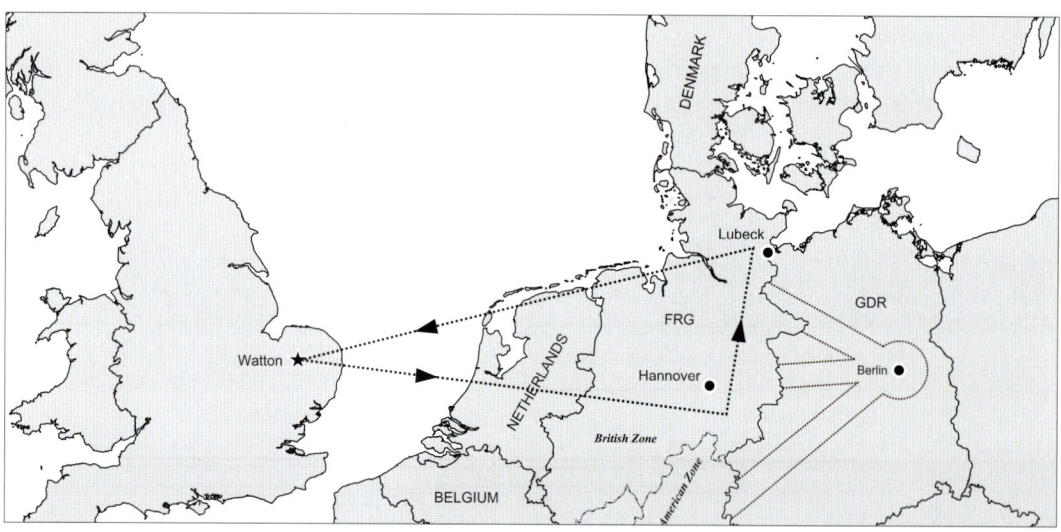

Route of the weekly Border sorties flown by the CSE during 1951. The sorties were confined to the British Zone of Germany.

A number of Douglas C-54 Skymaster transport aircraft were converted as Elint platforms for use by the USAF 7499th Sqn operating out of Wiesbaden. *USAF*

especially, called or a high degree of determination and skill for the satisfactory performance of the dual role of Navigator/Special Operator in areas and in aircraft where navigation aids were negligible, and where considerable hazards were encountered'.

The new programme of RPFs over Germany was co-ordinated with a similar series of operations by the USAF. In January 1950 the USAF's European theatre Elint unit, the 7499th Squadron, had absorbed the 7498th Squadron (a photographic reconnaissance unit) and its Douglas RB-26Cs to become the 7499th Composite Squadron, moving to Wiesbaden in August 1950. In 1951 the squadron began to re-equip with modified Douglas C-54 Skymaster transport aircraft, replacing the B-17s in the Elint role and the RB-26Cs in the photographic role. The re-equipment process was a drawn out affair and the squadron operated B-17s and C-54s side-by-side for a number of years. In November 1951 the 7499th Squadron made a liaison visit to Watton with one of their new C-54 Elint aircraft, accompanied by Majors Hill and Anderson, responsible for US Elint flights in Europe. The C-54 was inspected by a number of interested parties, including officers from the Air Ministry, HQ 90 Group, the CSE and by the Development Squadron Special Operators. Following the aircraft inspection a 'closed' meeting was held to discuss the operational side of Elint sorties, covering a number of topics of common interest.

Despite the resumption of limited RPFs during 1951 there was still a lack of intelligence on Soviet air defences. This was a cause of concern to the Air Ministry and Ministry of Defence, particularly the Defence Research Policy Committee (DRPC) Working Party on Radio Warfare, tasked with formulating radio warfare policy. The lack of intelligence meant that it was very difficult to draw up requirements for RCM equipment, especially jammers, for the new jet bombers. During discussions in April 1951, one of the Air Ministry representatives on the Working Party (Air Commodore H D Spreckley, Director of Operational Requirements(B)) suggested the solution was to employ one of the new Comet jet airliners, a prototype of which was then undergoing development flight trials, on a 'ferret flight deep into Russia'. The Chairman (Dr R Cockburn) agreed that the absence of adequate intelligence justified the most extreme measures, but also observed that 'this was essentially a political matter'. The final conclusions of the Working Party were more measured, concluding that: 'our intelligence on Russia is most inadequate and … high priority should be given to measures to improve it. The search for radio signals, both communications and non-communications, is of the highest importance'. Although there is no evidence the suggested overflight took place, the Comet was investigated as a possible Elint aircraft (see Chapter Four). Another result of the uncertainty regarding Soviet air defence systems was the recommendation to cancel a proposed specialist RCM variant of the Valiant bomber since 'in the absence of adequate intelligence on which to base the design, the aircraft would be of doubtful value when produced'.

Border RPF operations by the Lincolns and similar operations by the USAF's C-54s during 1951 seem to have picked up little of interest. A paper written in September of that year reported that the backbone of the Soviet EW and Control system continued to comprise Lend-Lease and former German radar equipment, supplemented by Soviet radars of comparatively simple design. The Soviet AF continued to use vulnerable HF R/T for its communications, with no evidence of a switch to VHF. However, doubts remained in some quarters as to whether this was the complete picture. A follow-on paper in November 1951 discussing Specialist RCM aircraft made the point that 'the evidence at present available on radio equipment of the Soviet Air Force and their Fighter Defence system is scanty'. It went on to warn against assuming that the Soviets would continue to depend on the fairly simple equipment so far observed in the East German, Poland and the Baltic areas, making the point that 'The Russians know that such areas have come under the observation of British and American radio search aircraft'. An obvious possibility was that the large-scale deployment of existing radar systems in these areas was an attempt to divert attention away from less accessible areas where the development of new equipment was taking place. An indication that this might be the case was that: 'although we have intercepted no signals from Russian microwave equipments as a result of our search in the Baltic areas, we have discovered that microwave valves are in production which could be employed in such equipment'.

Formation of 192 Squadron

By 1951 it was becoming clear that the CSE Development Squadron, originally established for development work, was increasingly being used both as an operational Elint force and as an RCM training force in exercises and demonstrations. These commitments were rapidly outstripping the Squadron's resources and interfering with its development flying tasks. In order to remedy the situation proposals were made to increase the Squadron's aircraft establishment and split it into three units, comprising an operational squadron for 'radio proving' Elint operations, a specialist radio warfare squadron providing RCM training and demonstrations and a development squadron, dedicated to research and development work.

The requested increase in establishment was approved in mid-1951 and plans for splitting the Squadron into three units were announced in the second half of 1951. The change took place on 1st November 1951 when 192 Squadron (radio proving) and 199 Squadron (RCM) were split off from the Development Squadron. The choice of 192 Squadron as the number for the new Elint squadron was not accidental: the previous incarnation of 192 Squadron had, of course, been as a 100 Group Elint unit during the Second World War. The formation of 192 Squadron regularised the flying side of Elint operations: responsibility for the planning of operations and the analysis of results, and for aircraft installations, continued to reside in the Radio Squadron of the CSE Technical Wing (renamed from the Radio Warfare Squadron the same month).

There were however changes in the control of Elint operations at a higher level. As previously recounted, back in 1946 GCHQ had not been particularly interested in 'noise' listening, preferring to concentrate on the interception of long-range 'strategic' HF communications traffic. As a result of this policy the direction of Elint operations had been effectively delegated to the Air Ministry. During 1951 however GCHQ began to take charge of all signals intelligence work, including Elint operations. By the end of the year the efforts of CSE and 192 Squadron were effectively directed by GCHQ in conjunction with the Directorate of Scientific Intelligence (DSI) via the LSIB, with tasking via the Air Ministry.

The Operational Monitor Lincoln

The Air Ministry's insistence on the inclusion of H2S in the third Lincoln B.2/4A meant that when the CSE began work on that aircraft, they had to disregard their original Lincoln B.2/4A Monitor modification (as applied to SX980 and SS715) and start again from scratch.

As already described, the problem was that in the Lincoln the main H2S units (apart from the scanner) were installed in the mid-fuselage and would have to be relocated to provide space for the two Special Operator positions. Additionally, the retention of H2S meant that the under-fuselage radome could not be used for the centimetric D/F aerial and an alternative site would have to be found. Following a detailed investigation of the problem in the first half of 1950, the CSE Radio Warfare Squadron concluded that producing a Lincoln Monitor installation which retained H2S was possible by relocating most of the radar equipment (everything apart from the scanner and transmitter) to the bomb-bay, and by fitting the Watton Box aerial under the aircraft's nose. In July the CSE then learned that the Air Ministry required the Monitor aircraft to be 'fully armed and fully navigable' and thus required the retention of both H2S and the dorsal gun turret. The armament requirement, like the requirement for H2S, stemmed from the shooting down of the USN Privateer in April 1950. After further study the CSE Installation Flight declared that, by relocating the H2S units forward, it would be just possible to incorporate a two-position Monitor fit, comprehensively equipped for analysis and recording, including both metric and centimetric D/F, in a Lincoln fitted with a dorsal turret and H2S. However, the installation would be cramped, accessibility would suffer, and servicing would consequently be more complex.

The CSE proposals for the Lincoln conversion were accepted by the Air Ministry at the end of November 1950. The CSE remarked: 'The acceptance of these proposals marks the end of a long period of discussion and work has now commenced on what promises to be a true monitor prototype devoid of any operational limitations'. Work started on the conversion of the third Lincoln B.2/4A (WD130) as an 'operational' Monitor aircraft in early December 1950 under Signals Order 651. At the start of the conversion the mid-fuselage position, from the bulkhead at the rear of the signaller's station to the aft end of the bomb bay, was stripped of all radio and radar equipment, racking and the crew rest seat. By then there appears to have been a rethink by the Air Ministry regarding the retention of the gun turret and this requirement was dropped, no doubt to the relief of the CSE. The dorsal gun turret was thus removed and the resulting hole covered by a transparent Perspex dome, providing a lookout position. Some thought was also given to crew comfort and a new bulkhead was built at the rear of the compartment, separating it from the rear fuselage, apertures were sealed to eliminate draughts, and heating was provided via the aircraft's heated air system.

Once the major structural work had been completed, the H2S units were reinstalled in the forward part of the mid-fuselage compartment on new racking above the oxygen bottle storage. The two SO stations were then installed with the layout based on that used in the previous Lincoln conversions but benefitting from operational experience in those aircraft. As a result of that experience, and by using improved construction methods, considerable reductions were made in the weight and size of the installation with a consequent improvement in crew comfort. Two equipment racks, mounting the search receivers, were fitted on the port side of the compartment, aft of the new H2S racking, with the SOs seated between them, back-to-back. A retractable table was built into each rack, along with an Anglepoise light, instrument lights and a general floodlight. Control panels were mounted on the port fuselage wall adjacent to the operating positions and other equipment installed aft of the racks in the remaining space on the port side. A gangway on the starboard side of the racks allowed crew members to pass the Special Operator's station. Each Special Operator's rack could accommodate two receivers, comprising an AN/ARR-8 communications receiver (500KHz to 32MHz) and either an ARR-5 (20-140MHz), APR-4 (38MHz to 1GHz) or APR-5A (1-6.23GHz). The forward SO station was additionally fitted with a Watton Box centimetric D/F receiver while the rear station was fitted with facilities to control a CSE-built APA-24 metric D/F system. Both stations were also fitted with an oscilloscope and oscillator for signal analysis, and a wire recorder to record received signals. An intercom system allowed the two Special Operators to communicate with each other, listen to each other's receiver audio output and talk to rest of the crew.

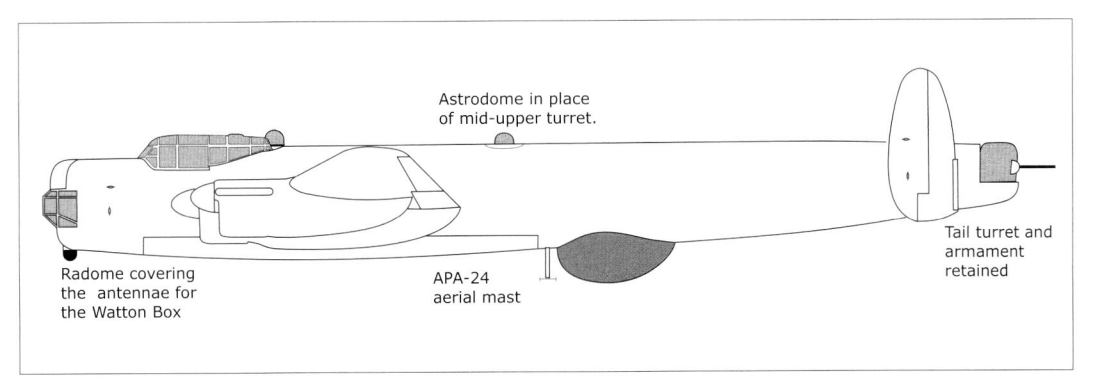

The 'operational' Elint conversion applied to Lincoln WD130. Note chin radome housing the Watton Box aerials, replacement of the dorsal turret by an astrodome, but retention of the tail turret.

LISTENING IN

The Lincoln Elint fit included the installation of some 20 aerials, many accommodated in specially-designed and manufactured 'blister' radomes. There were five main aerial suites: the Watton Box centimetric D/F aerial array, the metric D/F aerial, each operator's dedicated search aerials, common search aerials and sense aerials. In the previous Lincoln conversions the Watton Box aerial had been accommodated in the H2S radome but this was not an option in WD130 as the radome was still occupied by the H2S scanner. The solution was to mount the aerials (a pair of S-band horns and a pair of X-band horns) on a rotatable shaft protruding under the nose of the aircraft, covered by an inverted Perspex astrodome. The shaft was rotated electrically using a modified GEC Radio Compass drive unit and controls enabled the operator to continuously rotate the aerials in either direction at a rate of 1rpm or to 'inch' the aerial at about a third of that speed. The output from the centimetric D/F aerials was fed to a plugboard at the forward operator's position, allowing him to select the S-band or X-band output as desired. The operator was also provided with an accurate indication of aerial bearing and aircraft heading. The metric D/F system used an APA-24 aerial (75MHz or 200MHz) mounted beneath the aircraft, mechanically rotated under the control of the rear operator. The output from the aerial could be input into any suitable receiver at the rear operator's position via a plugboard. Like the centimetric installation, facilities were provided to display aerial bearing and aircraft heading to the operator.

Each operator was provided with three dedicated cone aerials for metric and centimetric search: one 200MHz to 2GHz vertically-polarised, one 200MHz to 2GHz 45°-polarised and one 1-6GHz. A suite of common metric search aerials (a 70-600MHz sword antenna, mounted at 45°, a horizontal whip aerial, a vertical whip aerial, and a 200MHz stub) were terminated at a plugboard and could be used by either operator. Finally, a number of vertical and horizontal whip aerials were mounted on the outer faces of the aircraft fins, terminated in a switch, and were intended for use comparatively for determining sense.

The Lincoln's H2S navigation radar was supplemented by a Decca navigator and flight log. The advantage of Decca over H2S was that it was a passive (non-transmitting) navigation aid; the disadvantage was that it was limited to use in areas covered by a Decca chain. Nonetheless, Decca meant that the aircraft could be navigated in Northern Europe with the H2S set turned off, eliminating any possibility of interference with the search receivers and minimising the likelihood of the aircraft being identified and tracked by Soviet ground plotting stations.

The conversion of Lincoln WD130 was completed and the installation air-tested by the end of August 1951. Like all Elint aircraft, the Lincoln required considerable calibration before it could be used operationally and since this was the most complex Elint fit yet attempted, the calibration effort for WD130 was expected to take some time. Although calibration of the S-band centimetric D/F was reasonably straightforward, significant problems were encountered at X-band which took some time to track down and fix. Calibration of the metric D/F installation began in September 1951. It had previously been thought that producing an error correction curve for metric D/F in the Lincoln was almost impossible since the errors varied with signal frequency, range to emitter, aircraft altitude and

Poor quality shot of Lincoln B.2/4A WD130. Note astrodome replacing dorsal turret, chin radome and APA-24 aerial mast forward of H2S radome.
Author's collection

aircraft aspect to the emitter. To overcome the problem the CSE Research Division developed a twin-channel CRT D/F system for WD130, initially working on 85MHz and by September 1951 an experimental model was ready for installation in the aircraft. However, by then a systematic investigation of D/F errors at 200MHz using the existing APA-24 rotating dipole aerial had revealed that bearing errors were not as large as had originally been thought, and that they appeared to be consistent for a particular frequency (±2MHz) and signal polarisation. This meant that it should be possible to produce error-correction charts for particular spot frequencies. As a result of these findings the development of the experimental CSE twin-channel system was abandoned and WD130 retained the APA-24 D/F fit.

Although calibration of the metric D/F system in WD130 was feasible, it was a time-consuming business and by the end of 1951 the task was dragging on with no end in sight. Since the aircraft was urgently required for operations it was decided to cut short the calibration effort and get the aircraft into service as soon as possible. The problem was, the aircraft would be unable to carry out accurate D/F if it picked up signals outside the calibrated frequency bands. To overcome this shortcoming it was decided to adopt a calibration-on-demand approach. When a Special Operator received a signal in an uncalibrated band he would first determine its exact frequency and then measure its polarisation by switching between the horizontal and vertical aerials on the tail. The operator would then take as many bearings as possible on the signal source over as wide an arc as possible. These would of course incorporate an unknown error. Back at base, the bearings would be corrected by setting up a transmitter on the same frequency and radiating at the same polarisation and determining the necessary correction curves via ground and air tests.

By the end of the first quarter of 1952 flight trials in WD130 of D/F in the important 60-90MHz band (used by Soviet EW radar) had shown that '[the] installation if used correctly can give valuable results'.

Improved Centimetric D/F

In early 1952 the Lincoln was also upgraded with an improved centimetric D/F system. The CSE had introduced the the Watton Box centimetric D/F system in 1949 but this was really a laboratory prototype device, albeit built for airborne use. Initially, only one model of the Watton Box was built, and this was transferred from aircraft to aircraft as required. Two additional units were subsequently produced by the CSE around 1950, but these were identical 'Chinese' copies of the original with no improvements.

In 1949 the CSE had also begun work on the design of a new Elint system to replace the US receivers then in use. The basic concept was to use four units to intercept, D/F and analyse signals. The first unit was a wideband DV receiver (working on the same principles as the Watton Box), providing a basic signal intercept and D/F capability of any signal in a 2GHz range. The second unit would comprise a number of narrow-band DV receivers, allowing the received signal frequency to be determined to within 200MHz. Signals from the first two stages would be fed into a superheterodyne receiver with a 200MHz spectrum analyzer. The superheterodyne receiver would be tuned to the 200MHz band indicated by the narrow-band DV receivers and would provide an exact frequency for a signal. The final unit was a signal analysis unit which could determine PRF and pulsewidth. The new Elint system was intended to cover a wide frequency range via a number of interchangeable units although initial work concentrated on the 2-4GHz S-band since this was considered one of the more important bands of interest and was technically less challenging than the higher-frequency X-band.

By the autumn of 1950 concerns regarding the lack of intelligence on Soviet radar and other systems had led to the new Elint system being prioritised as the top priority RCM development programme in the UK. The proposed CSE Elint system was a fairly ambitious programme and development was expected to span a number of years. (See Chapter Seven). However, by 1951 it had become clear that there was an urgent short-term requirement for a properly-engineered replacement for the Watton Box, providing basic intercept and D/F capabilities in the S and X-band. As the DRPC RW Working Party concluded in mid-year: 'The immediate needs for equipment for all users can be met by developing and producing in small quantities simple receivers, and we recommend that this development and production should be given the highest priority'. In response the CSE adapted the design of the proposed Elint system wideband DV receiver to produce a simple standalone system providing basic intercept and D/F capabilities. This system, codenamed *Flange*, used a pair of aerials to give D/F over an arc of 110° on either S-band or X-band and was designed to be suitable for use in the air or on the ground.

LISTENING IN

ARI 18021 *Flange* intercept and D/F system. *Flange* was designed for both ground and airborne use. The frequency band was selected via the tuned waveguide aerials. For D/F a pair of aerials was mounted on a turntable. *Author's collection*

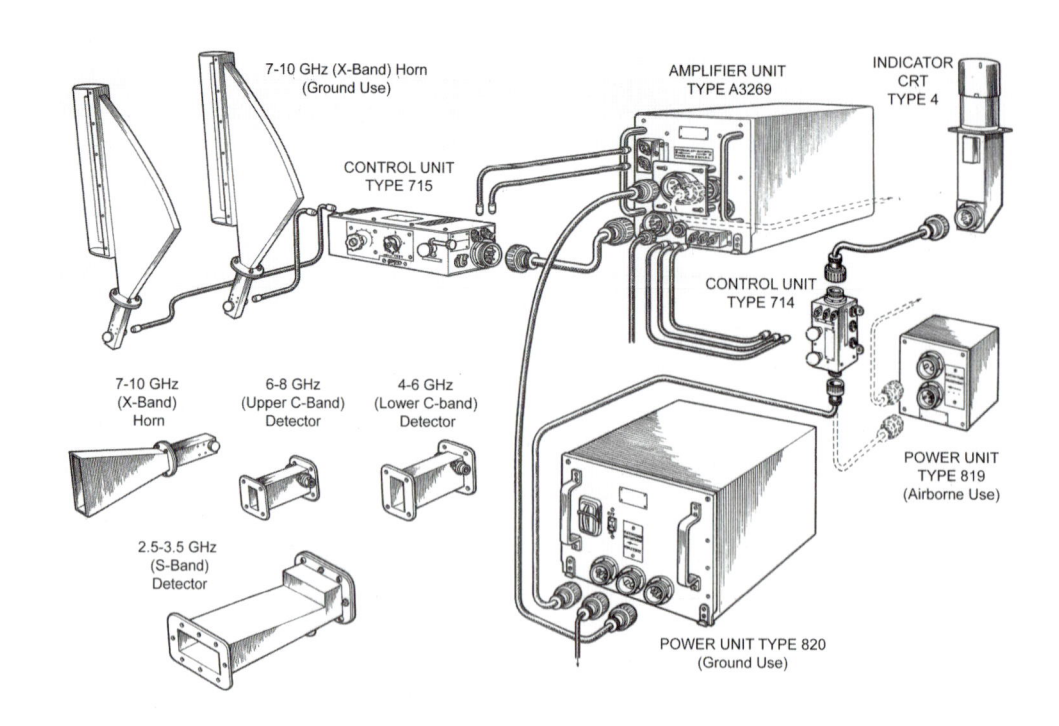

A small number of prototype *Flange* sets were produced by the CSE at the end of 1951 and large-scale production was begun at the Radio Engineering Unit (REU) Henlow in 1952. The receiver later entered service as the ARI 18021 (airborne) and MGRI 18023 (ground use). Lincoln WD130 was fitted with one of the prototype *Flange* sets in early 1952 and the other Elint Lincolns were similarly re-equipped.

During early 1952, 192 Squadron continued to fly its two 'semi-operational' Lincolns (SX980 and SS715) on the weekly programme of daylight Border sorties over Germany while the operational Lincoln (WD130) was working up. After intensive calibration and H2S/Decca trials Lincoln WD130 (coded '61') made its operational debut on 7th May 1952 when it carried out a 7 hour 40 minute day sortie over Germany as part of the existing Border programme. WD130 was subsequently flown on further Border sorties during June and July, taking turns with the other two Lincolns.

Bomber Command and 192 Squadron

Bomber Command had first complained about the lack of intelligence on the Soviet air defence system in 1949 and during 1950 had suggested they take control of Elint operations. By the beginning of 1952 the situation was still unsatisfactory and Bomber Command were beginning to worry about the provision of RCM equipment for the new Valiant bomber. In April the AOC, Bomber Command (ACM Sir Hugh Lloyd) wrote again to VCAS (ACM Sir Ralph Cochrane) complaining that: 'I am as dis-satisfied with the number of flights made for radio reconnaissance as with the results'. Sir Hugh considered the problem lay with the agencies controlling the Elint aircraft that is, GCHQ and the Directorate of Scientific Intelligence (DSI): '… in my view it is not possible for an outside agency to be as active and as energetic in obtaining this information as one of the users'. His proposed remedy was, once again, to transfer control of the task to Bomber Command. This time the proposal was to fit Bomber Command aircraft (Canberras and the new Valiant PR aircraft) with automatic Elint equipment, and have them gather the required intelligence during regular Bomber Command training sorties over Germany. This would provide Bomber Command with the required intelligence to develop RCM equipment, while also having the additional benefit of rendering specialist Elint aircraft redundant, thus making it unnecessary to re-equip 192 Squadron with Washingtons.

VCAS had some sympathy with Sir Hugh, noting that he was similarly dissatisfied with the number of RPF flights being carried out, and with the results obtained. Both the DSI and GCHQ wanted regular flights, the latter in particular having a requirement for weekly RPFs into 'various areas' of the Baltic and Black Sea. Unfortunately, their hands were tied: the Air Ministry had asked for permission to extend the limited Border RPF programme to other areas, but had not received the necessary political authorisation. The severe operational imitations to which these operations were subject was: '…the only reason for the small number of flights carried out by 192 Squadron, especially in the last six months. It was definitely not through any lack of enthusiasm on the part of the Squadron or 90 Group'. Despite his sympathy regarding the paucity of RPF operations, Sir Ralph had had significant objections to Sir Hugh's proposal. On a practical level, the problem was that the required automatic Elint equipment did not exist. An OR had already been issued for suitable equipment (OR.3522) but meeting it was a technical challenge, priority was low, and progress was unlikely for many years. VCAS also pointed out that, even if the necessary equipment existed, the intelligence take from Bomber Command Valiants and Canberras operating over Germany would be low. Long-range radio and radar signals were already being intercepted by ground Elint stations in Germany and Bomber Command aircraft on regular sorties, flying at some distance from the border, would be unlikely to receive much more. The 192 Squadron Lincoln Border sorties were intercepting short-range signals and thus had to fly relatively close to the German border to intercept them. Even if Bomber Command aircraft did intercept signals of interest, their automatic equipment would be unable to carry out the detailed measurement and analysis required to extract the maximum value from them. In any case, GCHQ was doubtful that much could be obtained from sorties along the German border since: 'GCHQ now knew the defensive organisation in that area and could get little more without closer penetration of the Russian frontiers'.

Sir Hugh was not entirely convinced by VCAS' reply and, temporarily abandoning the automatic Elint proposal, he suggested instead that control of 192 Squadron be transferred to Bomber Command. His argument was that Elint aircraft would be required to accompany his bombers in time of war to update intelligence on enemy defences and to assess enemy reactions to RCM measures. This proposal was discussed at a meeting attended by representatives from 90 Group, GCHQ, DSI, various branches of the Air Ministry, and Bomber Command in July 1952. The conclusions were that transferring control of 192 Squadron to Bomber Command was neither practical nor desirable. GCHQ noted that specialist Elint sorties would have to be flown independently (often in advance) of bomber raids. The major objection to the proposal was however that Bomber Command was only one user of the intelligence provided by RPFs, and RPFs were only one part of the overall intelligence picture. Efficient control of these flights, and the subsequent analysis of results (collated with information from other sources), could only be carried out by the DSI and GCHQ. The CSE was an impartial agent in the intelligence-gathering process, carrying out flights as directed by Deputy Director of Signals (B) on behalf of LSIB (in turn directed by GCHQ and the DSI). There was a concern that if Bomber Command assumed control, then the flights would focus on Bomber Command intelligence targets (primarily the identification of Soviet radar and communications for RCM purposes) to the detriment of other intelligence targets. Another major objection was the impact that transfer of control might have on the close relationship between 192 Squadron and the CSE. The squadron was highly dependent on the CSE for the manufacture, installation, modification and maintenance of Elint equipment carried in their aircraft and there were real concerns that moving control might have an adverse impact. Sir Hugh's proposal to assume control of 192 Squadron was thus rejected.

1952 Ferret Programme

The entry into service of Lincoln WD130 gave the CSE, for the first time since early 1951, a fully-operational Elint platform. The Air Ministry were anxious to utilise this new capability and in September 1952 proposed a new RPF programme covering the Baltic, Black Sea and Caspian Sea areas. The programme was first discussed by the Secretary of State for Air (Lord De L'Isle and Dudley) and the Foreign Secretary (Anthony Eden). The proposed programme included an RPF into the Baltic in mid-September 1952, planned to coincide with Exercise *Mainbrace*, a large-scale NATO naval exercise off the coast of Norway and in the entrance to the Baltic. The hope was that more Soviet radars would be active during *Mainbrace* and thus flying an RPF during the exercise produce a good intelligence 'take'. The sortie was scheduled for a dark night during the new moon to minimise the risk of interception and was co-ordinated with a similar operation planned by the USAF.

Flying the sortie on a 'no moon' night was a necessary precaution. There had been three attacks by Soviet fighters on Elint-related aircraft operating in the Baltic during 1952 and

LISTENING IN

the Air Ministry were anxious not to add to the tally. The incidents comprised an unsuccessful attack on a USN Martin P4M Mercator in January, the shooting down of a Swedish Elint Douglas C-47 (Tp79 79001) by Mikoyan-Gurevich MiG-15 *Fagots* in June and an attack on a Swedish Consolidated Catalina SAR aircraft looking for survivors from the C-47 a few days later. Flying on a dark night considerably reduced the chances of an interception: Soviet radar would of course be able to track the Lincoln, and might even be able to vector fighters onto it, but the Soviet Union did not then have a radar-equipped night fighter, which meant that the fighter pilot would have to acquire and attack the Lincoln visually, a difficult task on a pitch black night.

The Foreign Secretary was alarmed at the proposal for a Baltic flight during *Mainbrace*, noting that operations in the Baltic were the most provocative part of the exercise and thus Soviet fighters were more likely to be active. Concerned that 'Any incident would be most unfortunate' he therefore ruled that the flight be deferred until the following month. The Air Ministry protested the decision, citing the support of the Minister of Defence and describing the sortie as being of the highest operational importance, also noting: 'Such an opportunity is unlikely to recur for a long time, and certainly not this winter'. Despite their protestations the Foreign Secretary stood firm, replying that: 'It is my considered view that no operation should be sanctioned which would add to the already serious risks of operation *Mainbrace*' and the Baltic flight was thus deferred until October.

Following the amendment of the programme, the Foreign Secretary's agreement to the programme, and the subsequent agreement of the Minister of Defence (Earl Alexander of Tunis), a formal proposal was made to the Prime Minister (Winston Churchill) on 1st October. In his proposal Lord De L'Isle and Dudley explained that intelligence was needed on Soviet radar and radio communications, about which knowledge was still 'very limited'. The information was required in order to plan the air defence of the UK, plan bomber operations against the Soviet Union and evaluate the Soviet Union's preparedness for war. The sorties would also make a UK contribution to the joint Anglo-American intelligence pool, 'from which we expect valuable returns in kind'. He went on to explain that the only practical means to gather the intelligence was by flying aircraft along the Soviet border. He noted that the UK had in the past carried out a small number of such flights in co-ordination with the USAF and asked for permission for a new programme of similar flights. A total of four flights were proposed in the period October-December 1952, covering the Soviet borders of the Baltic Sea, the Black Sea and Caspian Sea.

All the sorties would be flown during the darkest period of the month and none would approach closer than 30nm (55km) to the

Operation *Gamash* on 19th October 1952 was the first RAF Elint sortie into the Baltic since the shooting down of the PB4Y-2 Privateer in 1950. This map shows the approximate track of Avro Lincoln WD130.

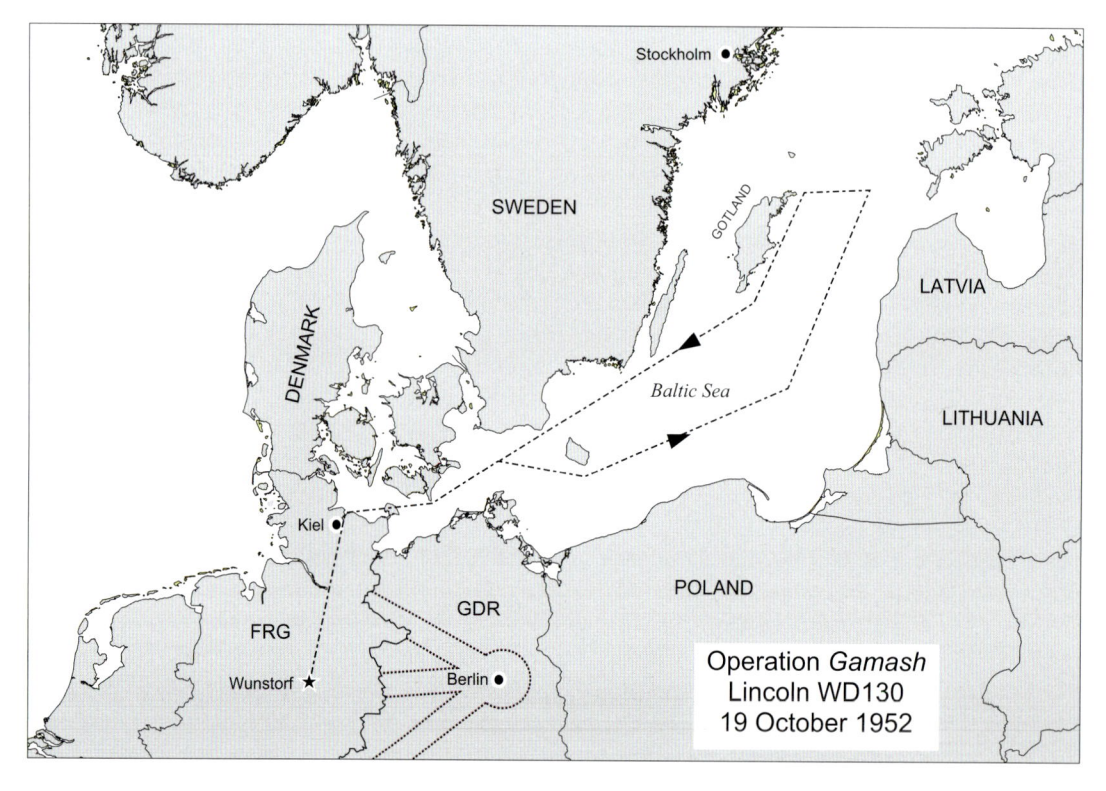

Soviet, or Soviet satellite, coastline. The aircraft would be detected by radar but, in the absence of Soviet fighters equipped with AI radar, the risk of interception was very small. The Secretary of State for Air went on to say that: 'The Foreign Secretary and I are satisfied that the risk of mishap is very small, and that such risk as there is must be taken because of the extreme value of these flights'. The programme appears to have been approved the following day.

The first flight of the new RPF programme took place on 19th October when Lincoln WD130 flew Operation *Gamash* into the Baltic from Wünstorf. The aircraft was fitted with APR-4 (38-95MHz), APR-5 and a wire-recorder at one position; *Flange*, APR-5A, an oscilloscope analyser and a wire recorder at the other, thus allowing the aircraft to monitor both metric and centimetric radars. The Lincoln (Flt Lt D E R Lang) took off from Wünstorf at 1820hrs and flew north to Kiel, then east over the Baltic, flying off the coast of East Germany and Poland, then north up the coast of Lithuania and Latvia. The Lincoln then turned west towards the island of Gotland before returning to Kiel. The nearest approach to Soviet territory (probably off the Latvian coast) was 40nm (74km) and total flight time was 5 hours 50 minutes.

In November Lincolns SX980 and WD130 were detached to Habbaniya, Iraq, via Luqa, Malta and Fayid, Egypt for the second phase of the programme. The detachment appears to have actually covered two tasks. The primary task was Operation *Mashie*, a series of two sorties into the eastern Black Sea and southern Caspian Sea as part of the 1952 RPF programme, while the secondary task, Operation *Hobble*, was apparently an investigation into the accuracy of HF/DF bearings obtainable in the area. Three stations (Nicosia, Habbaniya and Shaibah, near Basra) were organised into a fixer network for the detachment and on 13th November WD130 (Flt Lt Lang) flew an 8 hour night RPF (Operation *Mashie*) across Iraq and Turkey into the Eastern Black Sea, the sortie having been previously cleared with the Turkish authorities. On the same night SX980 (Flt Sgt McEachern) flew a 6 hour 30 minute night sortie (Operation *Hobble*) on the HF/DF task. A second operation was flown four days later. Early in the morning of 17th November SX980 (Flt Lt Lang) flew a 2 hour day sortie on *Hobble*. Later that day WD130 (Flt Sgt McEachern) flew a 5 hour night sortie on *Mashie* across Iran and into the Caspian Sea. This sortie was not cleared with the Iranian government, since Iran had no radar cover and thus the UK government considered that: '[they] will not know anyway that we are flying over their territory'. Both aircraft returned to the UK via Nicosia and Luqa on 18th November.

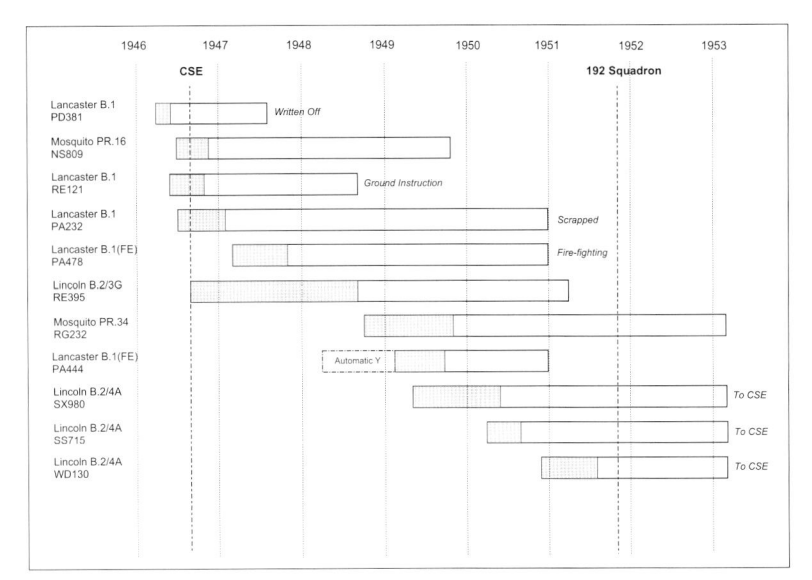

The last operation in the 1952 RPF programme was an RPF (Operation *Possum*) over the Black Sea in December 1952. By then the first of the B-29 Washingtons procured as replacements for the Lincolns had become available and the Air Ministry, apparently eager to try out the new Elint platform, used that aircraft for the operation. Following *Possum* all further RPF operations were put on hold pending the working-up of the new Washingtons. Operation *Mashie* in November 1952 thus marked the end of the Lincoln as an operational Elint platform. The original plan had been for 192 Squadron to retain two of the Lincoln B.2/4As for Special Operator training purposes, but this was cancelled at the end of 1952. Thus by the beginning of 1953 the three Lincoln B.2/4As of 192 Squadron had been effectively declared redundant and were retired from operations. In the end the 'operational' Elint Lincoln WD130 only had a useful life of some seven months, flying a handful of Border sorties and three RPFs. On the face of it that appears to be a poor return on the 18 months of effort put into the conversion and calibration of the aircraft by the CSE. However, the Lincolns, although short-lived, had allowed the CSE to develop a number of Elint techniques and equipment.

In February 1953 the Lincoln Flight of 192 Squadron (B Flight) was disestablished and the three aircraft (SS715, SX980 and WD130) were transferred to the CSE Development Squadron. By then the new Washingtons had taken over as the operational Elint aircraft.

Aircraft used by the CSE and 192 Squadron in the period 1946-1952. The shaded portion of the bars indicates the time taken for the fitting of Elint equipment and conversion to the role.

4 Washingtons

'...much useful information was obtained.'
Report on Operation *Possum*, December 1952

The CSE and Air Ministry had begun to investigate a replacement for the Elint Lincoln sometime in the second half of 1950. The problem with the Lincoln, obsolescence apart, was that it was simply too cramped to function as an effective Elint platform and could only accommodate two Special Operators (SOs). The long-term goal was to develop a jet Elint aircraft, ideally based on one of the new jet bombers then under development, but in 1950 a jet was some years away. There was thus a requirement for an interim Elint aircraft to replace the Lincoln in the short-term.

One candidate briefly considered for the role was the Handley Page Hastings transport aircraft, then equipping a number of squadrons in Transport Command. The Hastings had a reasonable range, slightly better speed and altitude performance than the Lincoln, and a spacious passenger cabin able to accommodate a large number of SOs and their equipment. On the downside, the aircraft was not pressurised, lacked navigational aids and could not be armed. A superior candidate then emerged in the form of the Boeing B-29 Superfortress. The UK had made arrangements with the US in late 1949 to obtain 87 of the B-29A variant of the Superfortress for use by Bomber Command, as an interim type pending the availability of the new jet bombers. The first RAF squadron formed in March 1950 and the B-29A eventually equipped eight squadrons of Bomber Command as the Washington B.1. The performance of the B-29 was better than that of the Lincoln but the main advantage, from the Elint point of view, was a large pressurised rear cabin, eminently suitable for use as an Elint compartment accommodating up to six SOs and their equipment. The USAF had already come to the same conclusion; a prototype Elint conversion on an RB-29A aircraft had been carried out in 1947 and by 1948 the USAF were operating six Elint RB-29As with the 324th Reconnaissance

The Handley Page Hastings was considered as an Elint platform before being rejected in favour of the RB-29A. Unlike the RB-29, the Hastings was unpressurised, lacked suitable navigation aids and carried no defensive armament.
via Ron Henry

44

Squadron, based at McGuire AFB, New Jersey.

In the second half of 1950 an attempt was made to obtain three of the B-29A Washingtons allotted to Bomber Command as replacements for the CSE Elint Lincolns. When that proved unsuccessful, an approach was then made directly to the US for the provision of three fully-equipped RB-29A aircraft for 'radio monitoring' purposes via a supplementary addition to the FY 1951 MDAP bid. In late 1951 the CSE learnt it had been allotted three RB-29A aircraft, previously converted for Elint duties, and immediately began to make arrangements for their arrival. Initial planning was hampered by a lack of information regarding the equipment state of the three aircraft as it was unclear whether the aircraft would be delivered fully-equipped with Elint equipment, delivered unfitted but accompanied by Elint equipment, or simply delivered as bare aircraft. In October the Air Ministry informed the CSE that the aircraft would be delivered in November 1951 'without any special radio equipment' and based on that, the CSE estimated it would take around 200 man-hours per aircraft to bring them up to an operational standard. The initial Air Ministry estimate of the delivery date proved optimistic and in the event the aircraft were not delivered until April 1952.

Planning the Washington Fit

The first two RB-29As (44-62283/WZ966, 44-62282/WZ967) arrived at Prestwick on their way from the USA during April 1952. Before delivery to Watton three painters were dispatched to Prestwick to erase the US markings and apply temporary RAF markings and a few days later both aircraft were delivered to Watton. The third aircraft (44-62286/WZ968) arrived on 21st May but was retained at Prestwick due to fuel leaks and was not delivered to Watton until 19th June. Following their delivery to the CSE, all three aircraft went to a second-line servicing unit at RAF Marham to receive the latest modifications before the start of their Elint conversions. The three RB-29As received the same 'Washington' designation applied to the B-29A bomber variants and, presumably for security reasons, were not allotted a different mark number.

Planning the Elint fit in the Washingtons had started in early 1952 with the CSE initially suggesting that the rear compartment of the RB-29As be fitted with four SO positions, covering VHF, Metric, S-band and X-band, and with two rest positions. The centimetric direction finding (D/F) requirement would be met by two ARI 18021 *Flange* sets (S-band and X-band) with their aerials mounted on rotating platforms projecting beneath the aircraft, covered by radomes. Since *Flange* was a wide-band receiver, providing little or no frequency information, it would need to be associated with a narrow-band superheterodyne receiver to determine the frequency of received signals. Previous installations in the Lincolns had used the APR-5A receiver for centimetric search, but that did not cover X-band. By the late 1940s however a new, and advanced, centimetric search receiver had been developed in the US. Designated the AN/APR-9, this was a scanning superheterodyne receiver with four interchangeable RF tuning units covering the range 1-10GHz and arrangements were made under MDAP for the UK to obtain a small number for use in the RB-29As. The metric search requirement would be met, as in the Lincolns, using the tried and trusted APR-4 receiver, but metric D/F would be provided by the AN/APA-17 system, rather than the APA-24 used in the Lincolns.

The APA-17, developed at the end of the Second World War, used a rapidly-rotating highly-directional 'spinner' aerial, and came with a number of interchangeable aerial heads, covering a range of frequencies. Signals received by the aerial were fed through an APR-4 search receiver, providing frequency determination, then into the APA-17 CRT display unit. The scan of the latter was electro-mechanically synchronised with the spinning aerial, allowing the bearing of a received signal to be determined and displayed as 'spoke' radiating from the centre of the display. The plan was to mount the APA-17 aerial underneath the fuselage, covered by a radome. It is something of a mystery why the CSE had not switched to the APA-17 earlier, although a subsequent investigation suggested that stocks of the set had only recently been discovered in the UK. There was also a suggestion that the set had been tried in the past but had proved to be unpopular due to the operating complexity. In any case, the APA-17 did not entirely solve the metric D/F problem: although the CSE thought that APA-17 could offer improvements over the APA-24 rotatable dipole aerial, particularly with regard to sense, it was not entirely optimistic, noting: 'it is felt that anyone who expects a substantial reduction of D/F errors induced by the airframe is likely to be disappointed'.

The VHF search requirement would be initially met using ARR-5, to be replaced by newer

LISTENING IN

Using the wartime era AN/APA-11 analyser an SO could determine pulse width and pulse repletion frequency (PRF) of a signal.
Author's collection

US-built VHF search receivers when available. There was also a requirement for signal analysis (determining PRF and pulse-width) and for some means of recording the audio output of the search receivers. The original plan was to use the CSE-modified Oscilloscope Type 10 as a pulse analyser but the US-built AN/APA-11, which had the advantage of being in large-scale production in the United States, was later substituted. For similar reasons, the US-built ANQ-1A magnetic wire-recorder was adopted to record signals.

It would take some time to acquire the necessary Elint equipment so the CSE proposed a staged installation, in order to get the Washington in service as quickly as possible. This involved initially fitting the aircraft for centimetric search and D/F only using *Flange*, then adding the APR-9 and finally adding the metric D/F system, with other equipment installed when available. The CSE submitted their proposals for the Washington Elint fit to the Air Ministry in mid-May 1952. Although the Air Ministry were in general agreement, they specified that six SO positions should be provided, rather than the four suggested by the CSE. A pre-installation mock-up conference was subsequently held at the CSE on 26th June 1952 and the proposed fit was accepted.

While the first Washington (WZ966) was in the installation hangar for its Elint fit, 192 Squadron started to train air and ground crews on the aircraft. By June a single Washington crew had been converted and a second crew arrived in July. During this period the second aircraft, WZ967, was used for crew continuation training and assessment trials. Unfortunately these came to an end in mid-June when a planned long-range fuel consumption sortie was cut short with engine problems and a subsequent engine change.

Converting the Washingtons

Modification of the first Washington, WZ966 by the CSE Installation Flight began in May 1952 and took about five months. The aircraft had previously been converted for the Elint role in the US and had already had its gun turrets removed and the fuselage side observation blisters faired over. The major work at the CSE comprised the installation of six SO stations in the rear compartment and the fit of a comprehensive aerial suite to feed the receivers. Provision was made for mounting three D/F aerials (two *Flange*, one APA-17) under the fuselage, each covered by a 'thimble' radome, and large mast aerials were fitted above and below the fuselage. However, not all the required Elint equipment was immediately available and WZ966 initially only had two positions fully fitted-out, both with *Flange*.

During August 1952, with the Elint fit in the first Washington approaching completion, the Air Ministry began to voice doubts regarding the wisdom of the project. A proposal had just been made to withdraw the B-29A Washington from Bomber Command in 1954 due to a shortage of spares and there were worries that the three RB-29As might therefore have a short operational life. 90 Group, however, stood by the Elint RB-29A programme, with the AOC (Air Vice-Marshal William Theak) reassuring the Air Ministry that the three Elint aircraft could be maintained for a considerable number of years beyond 1954, and in fact the withdrawal of the Washington from Bomber Command might make the task easier by throwing up redundant aircraft as sources of spare parts. Air Vice-Marshal Theak also pointed out that the only alternative to the Elint RB-29A was the operationally less-suitable Hastings, and if the latter was adopted, the CSE would have to start the design and installation process again from scratch, delaying the resumption of RPFs by six to nine months. This argument seems to have done the trick and the Air Ministry agreed to continue with the Elint Washington project.

The modification of WZ966 was completed at the end of September 1952 and a start was then made on intensive flying trials, with the goal of getting the aircraft operational by the end of November. Initial flights in the Washington showed promise but also revealed electrical interference problems, requiring a redesign of the rotating *Flange* aerial, thus delaying completion of the trials until early December.

In mid-December 1952, with the aircraft not yet fully equipped, Washington WZ966 was

WASHINGTONS

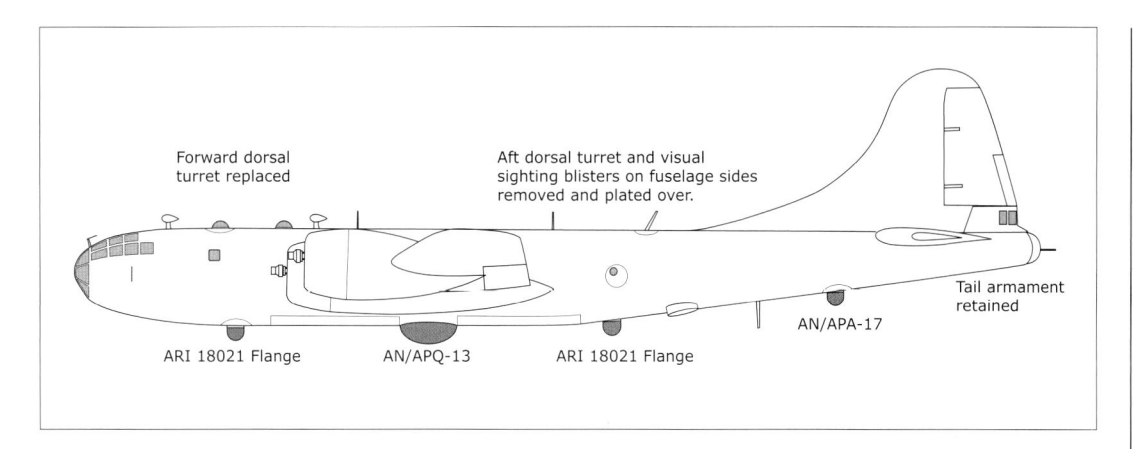

Diagram labels: Forward dorsal turret replaced; Aft dorsal turret and visual sighting blisters on fuselage sides removed and plated over; Tail armament retained; AN/APA-17; ARI 18021 Flange; AN/APQ-13; ARI 18021 Flange

The Elint Washington as operated by 192 Sqn from 1952. The aircraft were converted from USAF Elint Boeing RB-29As. Note three thimble radomes on underside of fuselage covering ARI 18021 and APA-17 aerials. The larger radome is for the APQ-13 navigation radar.

Washington WZ966 of 192 Sqn in a silver finish and coded '55' cleans up as it climbs out from Watton. The D/F radomes are visible beneath the aircraft's fuselage.
B Jordan via Paul Stancliffe

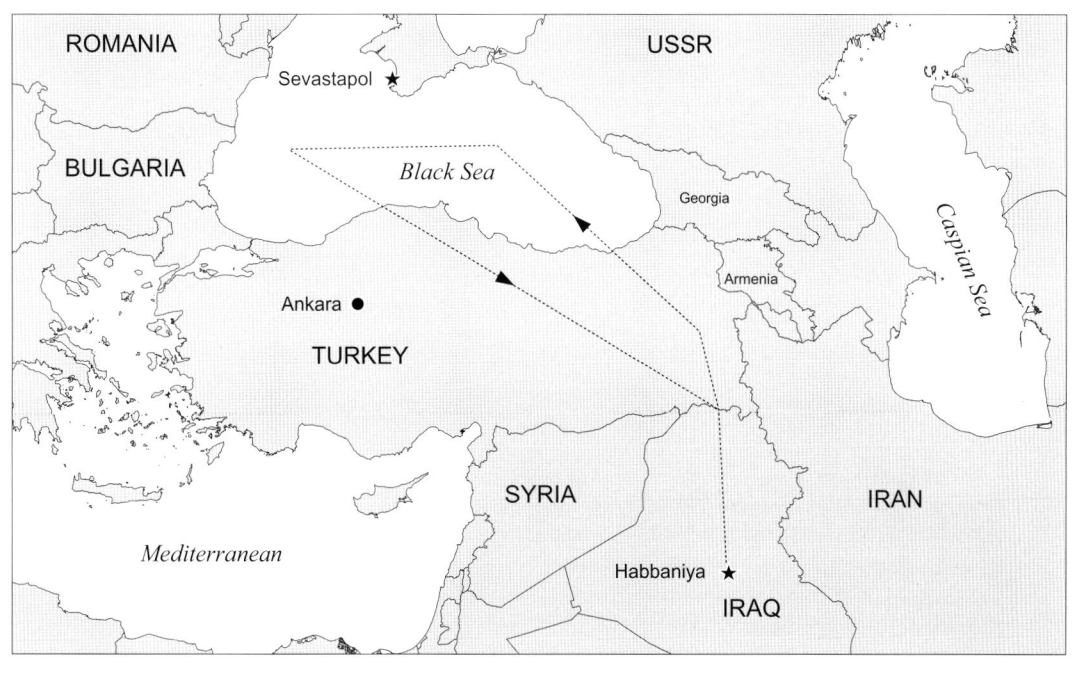

The first Washington radio proving flight (RPF) was Operation *Possum* from Habbaniya. The aim of the sortie was to obtain intelligence on the deployment of Soviet radar systems in the Black Sea area.

47

detached to Habbaniya, Iraq for Operation *Possum*. Ostensibly an investigation of HF/DF cover in the Middle East, this was in fact the last operation in the autumn 1952 RPF programme and was intended to obtain intelligence on the deployment of Soviet radar systems and the efficiency of the defences in the Black Sea area. The first attempt at a sortie on 16th December was abortive due to an engine vibration problem but a second attempt two nights later was more successful and 'much useful information was obtained' from the 6 hour 30 minute flight over Iraq and Turkey into the Western Black Sea. Operation *Possum* was the final flight in the 1952-53 Elint programme and no further Elint operations were planned until all three Washingtons had been fully fitted and readied for operations.

The CSE Installation Flight had started work on the conversion of the remaining two Washingtons in September 1952. Both these conversions benefitted from the earlier experience with WZ966 and were completed in mid and late January 1953 respectively. Thus by February 1953 all three Monitor Washingtons had been converted to the Elint role and delivered to 192 Squadron. There was still, however, a lot of work required to bring the Washingtons up to a fully-operational standard.

Exercise *Jungle King*

In March 1953 the Air Ministry took advantage of a major Bomber Command exercise (*Jungle King*) to evaluate the reliability and efficiency of the Washingtons and their crews in the Elint role. By then the necessary Elint equipment had been acquired to fit all six SO positions in the aircraft. The fit for *Jungle King* comprised: (1) VHF communications (ARR-5, ANQ-1A); (2) Metric radar (APR-4, APA-11, ANQ-1A); (3) S-band radar (*Flange*, APR-9, APA-11, ANQ-1A); (4) X-band radar (*Flange*, APR-9, APA-11, ANQ-1A); (5) L-band radar (APR-5, ANQ-1A) and (6) HF communications (BC.348). The above illustrates what an advance the Washington was over the Lincoln: while the former could only cover two frequency bands, the Washington could cover HF through to X-band. The combination of *Flange*, APR-9 and APA-11 at positions 3 and 4 gave the operators the ability to intercept and obtain a bearing on a centimetric radar signal using *Flange*, determine its frequency using the APR-9 and determine its characteristics (PRF, pulse-width) using the APA-11 analyser.

Since the exercise was also used to test the reliability of the Washingtons, 192 Squadron were required to have all three aircraft available for the duration of the exercise. A single Washington operated from Watton on each night of the exercise, typically flying a 4-5 hour sortie while a second aircraft was maintained at take-off readiness as a backup during each a sortie. The aircraft were intercepted by fighters from 2 TAF on two sorties, but no claims were made. Operations were apparently successful and valuable training was obtained by the aircraft crews. *Jungle King* also exercised the ground elements of the CSE as officers of the Flying Wing and Operations Section gained experience in setting up and running a centralised Operations Room to control CSE operations during the exercise, and in briefing and debrief-

'A' Flight 192 Sqn in front of one of the Elint Washingtons. The aerials either side of the cockpit are for the SCR.729 blind approach beacon system (BABS).
Brian O'Riordan via Paul Stancliffe

Washington B.1 WW346 of 192 Sqn, a standard B-29A, was acquired for flight crew training purposes, also serving as a detachment support aircraft. Note the retention of turrets and guns. *Author's collection*

ing the crews. The ground crews also benefited from the experience of servicing aircraft under operational conditions, at times working on a 'round-the-clock' basis to keep aircraft serviceable.

The Training Washington

Initial planning for the re-equipment of 192 Squadron with Washingtons had assumed that two of the redundant Elint Lincolns would be retained for SO training duties. However, in November 1952 the CSE submitted an ambitious proposal to replace both these aircraft with a single standard B-29A Washington with the intention that the aircraft would initially be used for crew continuation training but that it would, over time, be modified by the CSE for the Elint role. This would allow it to be eventually used for flight crew training, SO training and as a reserve operational aircraft. Disposing of the Lincolns would also improve the maintenance situation, by reducing the number of types on 192 Squadron charge.

The CSE submission was approved and the establishment of 192 Squadron was officially changed on 13th February 1953, deleting the two Lincolns and substituting a fourth Washington. A standard B-29A Washington (WW346) was subsequently acquired in April 1953, initially for pilot continuation training and the two Lincolns were transferred to the CSE Development Squadron for other duties. Unfortunately the plan to convert the training Washington to an Elint aircraft proved unrealistic. In May the CSE managed to obtain the drawings from the US detailing the conversion of a B-29A to an Elint RB-29A. These came as something of a shock and showed the project was well beyond the capabilities of the establishment. It also appears the job was either too big or too expensive to be sub-contracted to industry and in the end the aircraft was retained in its standard bomber configuration. Despite this setback, WW346 proved extremely useful for pilot continuation training, relieving the three operational aircraft of that task, and was also used as a transport to support overseas detachments.

Working up the Washingtons

During the first half of 1953 the squadron devoted its efforts to the training of Washington crews and the installation and trials of equipment in the aircraft. The training Washington was used for pilot continuation training, allowing the Elint aircraft to be used for SO and navigator training. Training SOs was a particular problem since the Washington carried three times the number previously carried by the squadron's Lincolns. Efforts were made to combine Navigator and SO training sorties whenever possible to get the most out of the sorties and training flights were carried out to the Middle East. Unfortunately these gave limited SO training, due to an absence of information on radars in the region.

The initial SO training regime was based on an optimal sortie time of 4-6 hours but the poor serviceability of the Washington affected the number of sorties which could be flown. Investigation then showed that a Washington that managed to take off would normally complete a sortie of any length. As a result the training sortie length was extended and two SO crews were carried, each crew taking turns to man the equipment. This training regime proved a success and by the third quarter of 1953 all SOs had been trained.

5 Jet Elint and the Canberra

'There seems very little hope that [the Canberra] could undertake any "ferreting" …' Air Ministry Operational Requirements Branch paper, May 1951

In early 1950 the CSE began to consider the requirements for a future Elint aircraft to support Bomber Command's new jet bombers, which were scheduled to enter service in the mid-1950s. During wartime, Elint aircraft would be expected to accompany the bombers on raids and would thus need a comparable range, altitude and speed performance. Elint aircraft might also have to operate alone over enemy territory, where they would need to fly high and fast in order to survive. The conclusion was that the future Elint aircraft would also need to be a jet.

One obvious solution was to develop Elint conversions of the new jet bombers, in the same way that Elint conversions of the Lancaster and Lincoln had been developed. Accordingly in August 1950 the CSE looked at what would be required to produce Monitor conversions of the new B9/48 (Valiant) and B35/46 (Victor and Vulcan) jet bombers, which were then still under development. The resulting paper suggested that although, with bomb-aiming equipment removed, the aircraft could each accommodate four search receivers and two Special Operators (SOs), the major challenge would be fitting a wide-range of aerials without introducing aerodynamic problems. The CSE suggested this needed to be tackled during the aircraft design stage, possibly by the use of suppressed aerials and interchangeable nose and fuselage panels, allowing different aerials to be installed according to the sortie requirements. By late 1950 the Air Ministry had approved plans for a specialist RCM/Monitor conversion of the B9/48.

A formal requirement for a long-range high-speed, high-altitude jet Monitor aircraft subsequently emerged in 1951: 90 Group issued an initial requirement in April 1951, and this was followed by draft Operational Requirement (OR) from the Air Ministry in the second half of the year. By then however, the view was that an Elint aircraft needed at least six (SO) positions in order to monitor efficiently the widest range possible of an enemy's radio emissions, effectively ruling out the use of the B9/48 and B35/46 in the role and raising a requirement for an aircraft with more internal accommodation. A meeting to discuss the OR, and the choice of aircraft, was convened at the end of January 1952 at the Ministry of Supply and seems to have been fairly positive, with 90 Group reporting that: 'It was apparent that high priority had been allotted to the requirement …'. The conclusion was that the new Comet jet airliner, then just about to enter service with BOAC, was a suitable Elint platform, having the required performance and sufficient internal space. It was suggested that one aircraft could be made available in early 1953, with a projected in-service date of December 1953. Unfortunately this proved overly-optimistic as it was later discovered that all the Comet production was spoken for, making it extremely unlikely that an aircraft could be obtained for conversion to the Elint role in the required timescales. By then however, the Air Ministry had turned their attention to what looked like an even better candidate for a jet Monitor aircraft: the new Vickers V.1000 transport. In May 1951 the Air Staff had decided to replace the Hastings transport aircraft with a jet, the requirement being for an aircraft capable of carrying over 12,000lb (5,443kg) of payload over a range of 3,000 miles (5,556km) at a cruising speed of 450kts (833km/h). To expedite development it was suggested that the aircraft should be based on an existing design. By November 1951 the choice had been narrowed down to a development of the Comet (de Havilland) or a design loosely based on the Valiant bomber (Vickers). The Air Staff chose the Vickers design, apparently swayed by BOAC who regarded the aircraft as the better choice for civil operations, and a draft operational requirement (OR.315) was issued for the new transport in May 1952

The Vickers V.1000 transport aircraft, conceived as a replacement for the Hastings. The entire programme suffered delays and changes to the requirement, OR.315, but its greatest enemy was weight gain. Sir George Cox

The proposed Elint version of the V.1000, intended to replace 192 Sqn's Washingtons. The terms 'Position 1' to 'Position 6' refer the Special Operators' positions within the cabin.

JET ELINT AND THE CANBERRA

with a target in-service date of late 1956. The Vickers' proposal, designated the V.1000, was a large aircraft with plenty of internal space for an Elint fit plus good altitude, speed and range performance. By mid-1952 plans had been drawn up for an Elint version of the V.1000, accommodating up to 14 Special Operators. The only problem was that the Elint V.1000 was unlikely to enter service until sometime in 1957.

Discussions regarding the future Elint aircraft took place against a background of increased demands for Elint operations and by July 1952 there were growing calls from the intelligence community for an increased number of 'Ferret flights'. The Deputy Director of Scientific Intelligence noted that the existing facilities were being used to the full, suggesting that more could be done if additional resources, including more aircraft, could be made available.

In 1952 the CSE/192 Squadron were operating two aircraft types in the Elint role: a heavy aircraft (Lincoln) for the main collection task, and a smaller, more agile, aircraft (Mosquito PR.34A) for the 'follow-up' task. The RB.29A Washingtons had been procured as Lincoln replacements but little thought appears to have been given to a Mosquito replacement. The call for an increase in Elint operations seems to have led the Air Ministry to examine the case for a Mosquito replacement, to supplement the Washingtons until the V.1000 entered service. In many ways the obvious candidate to replace the Mosquito was the English Electric Canberra jet bomber: both were small, fast and manoeuvrable aircraft. Using a Canberra in the Elint role would also provide some experience in jet Elint operations in advance of the V.1000. Unfortunately the Canberra, like the Mosquito, lacked sufficient internal space for a really effective Elint installation although, unlike the Mosquito, it did provide accommodation for a dedicated SO. A prototype Canberra had in fact been examined by the CSE in the summer of 1950 to assess its usefulness as an Elint platform and the initial conclusion was moderately optimistic, noting that: 'A monitoring installation slightly better than that at present fitted in Mosquito

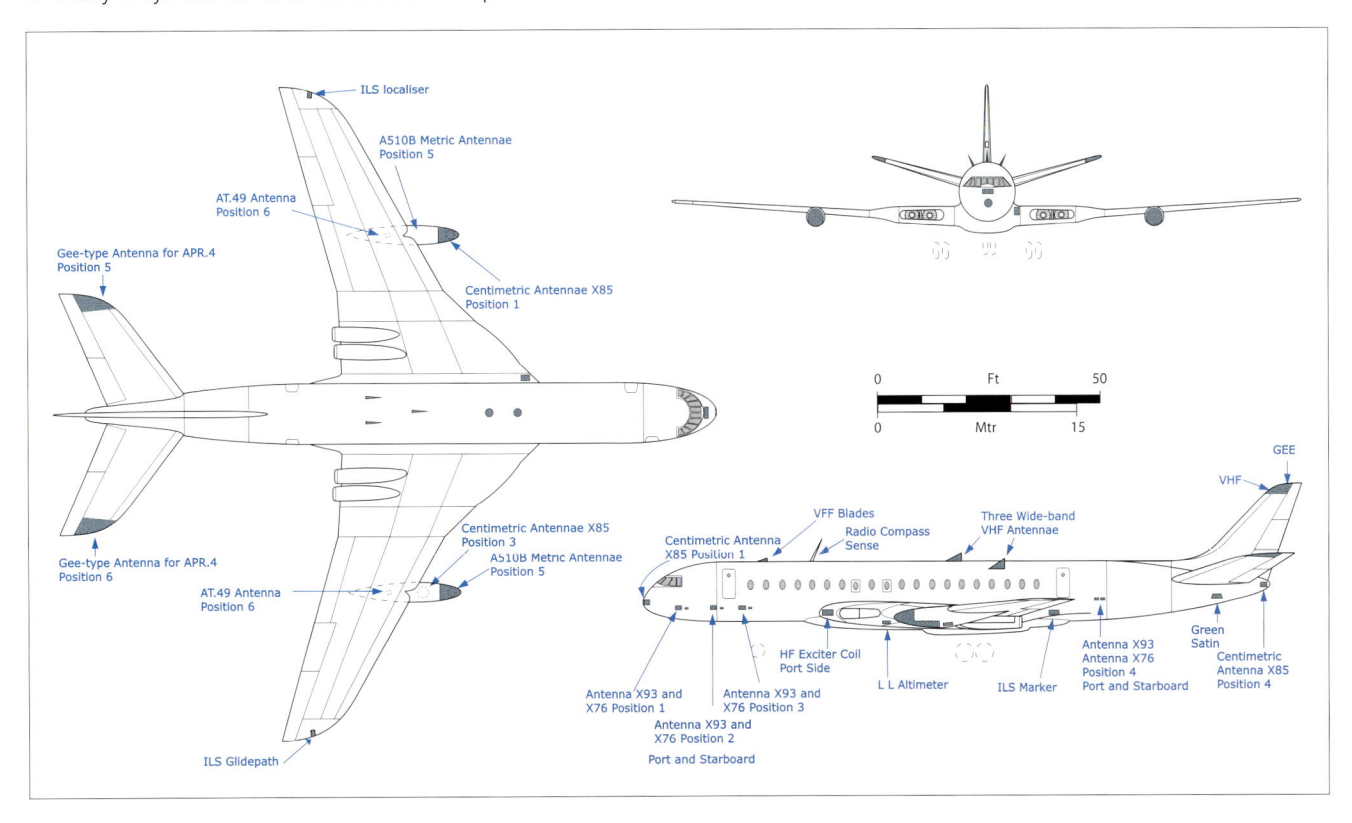

51

LISTENING IN

A possible mock-up of the Specialist Elint Canberra. Note the offset canopy layout and extended nose section with fixed seats for Special Operators.
via Ron Henry

The Canberra B.2 Elint conversion was fitted with a nose radome covering an ARI 18021 aerial array, APR-9 aerials either side of nose, a Blue Shadow side-looking airborne radar (SLAR) aerial on starboard fuselage side and an ARI 18021 aerial in the tail cone radome.

aircraft, and including centimetric or metric DF, should be possible'. By January 1951 the tone was more pessimistic, the Air Ministry RCM Committee reporting that: 'The use of the Canberra in the Specialist [RCM and Monitoring] role was impracticable because of the lack of adequate power supplies and space'. A later assessment by the Operational Requirements branch of the Air Ministry in May 1951, concurred, noting: 'There seems very little hope that [the Canberra] could undertake any "ferreting" …'. Although the view that the Canberra was too small to function as an effective Elint platform seems to have been generally accepted by the Air Ministry and the scientific intelligence authorities, some work on drawing up the specification for an Elint Canberra appears to have taken place. By early 1952 a draft standard of preparation had been drawn up for a specialised Elint Canberra conversion accommodating four SOs. The cabin layout was based on the 'New Look' OR.302 Canberra (as later used for the Canberra B(I).8) with the pilot seated at the rear to port, the navigator to his right and below. The four SOs were to be accommodated in the nose of the aircraft, seated facing aft on conventional aircraft seats. Each special operator position would be equipped with a single search receiver (APR-4, APR-9 or APR-5 depending on sortie requirements), a panoramic adaptor, a pulse analyser, wire recorder and camera. Metric direction finding (D/F) would be provided via two APA-17 and centimetric D/F via *Flange*. The conversion would have required an extended nose section to accommodate the four SOs and their equipment but even then, the layout would have been cramped and escape arrangements would have been rudimentary, with only the pilot provided with an ejection seat. In the event the Special Elint Canberra never got past the specification stage with one of the major stumbling blocks probably being the projected date of entry into service. In 1952 the English Electric Company (EECo) was heavily loaded with design work and the proposed Elint modification, in effect a new mark of Canberra, would have taken significant resources to develop and it thus seems unlikely that it could have been completed for some years.

Approval for an Elint Canberra

Despite the lack of interest in the proposed 'New Look' Elint Canberra, the fact remained that the only possible Elint Mosquito replacement was a Canberra of some description. The only practical way to meet the requirement within a reasonable timeframe would be to carry out a fairly simple Elint installation in an existing mark of Canberra, accepting a single SO's position. The Air Ministry seem to have come to the same conclusion and, in the second half of 1952, gave the CSE approval for the conversion of two Canberra B.2s as interim Monitor aircraft.

There were however a number of practical problems to be overcome to produce a suitable Canberra Elint conversion with one of the most obvious being that of space. Existing Elint equipment, mostly of wartime vintage, was not designed for high-altitude operation and would have to be installed in the Canberra's already-cramped pressure cabin. Another problem was fitting aerials, as the majority of those fitted to the Lincolns and Washingtons were unsuitable for use at high speed and new, aerodynamically-faired, aerials would be required. An Elint platform also needed a good navigation suite for fairly obvious reasons: to facilitate navigation to and from the operational area in various inhospitable parts of the world, to prevent the aircraft inadvertently straying into foreign airspace during an operation, and to provide an accurate reference point when plotting the location of radar stations and facilities. Unfortunately the Canberra was lacking in this respect. It carried no self-contained navigation aids; relied on the Gee/Gee-H aids, which were only available in Northern Europe, and was a difficult aircraft to navigate visually. Finally there was the question of providing sufficient power to run the Elint equipment. The Canberra's engine-driven Type P3 DC generators were not particularly efficient and did not provide a large excess of power. AC power was also in relatively short supply.

At the end of October 1952 the CSE submitted their preliminary proposals for the Canberra installation to 90 Group. Informal discussions had been held with EECo with a view to the company carrying out the majority of the conversion work, with the CSE only installing the Elint equipment. The plan was to produce a Monitor Canberra with a single Special Operator's position occupying the right hand side of the rear cabin, replacing the bomb aimer. The Elint fit would comprise a *Flange* receiver, providing intercept and D/F on S and X-bands, an S-band APR-5 search receiver, and a pulse analysis unit. Provision would be made to replace the *Flange* and APR-5 with VHF or HF search receivers when required, thus giving the aircraft a dual-role Elint/Comint capability. A rotatable *Flange* aerial would be installed in the nose of the Canberra, with a fibreglass radome replacing the aircraft's Perspex nose cone, and a fixed aerial installed in the tail cone. The navigation fit would be enhanced by the addition of a *Blue Shadow* sideways-looking airborne radar (SLAR) and Gee Mk.3, and a 300-gallon (1,135-litre) bomb-bay fuel tank would be fitted to extend the aircraft's range. The CSE estimated, somewhat optimistically, that if EECo carried out most of the modifications then it should only take about six weeks after delivery to Watton to install the Elint equipment.

Verbal approval for the CSE proposal was received from the Air Ministry in the third week of November 1952 and Air Staff approval quickly followed. Because of the problems likely to be encountered in converting the Canberra for Elint operations it was decided to carry out a prototype Monitor installation on one aircraft. Lessons learnt from this installation would then be used to develop a production installation which would be applied to the second Canberra, and then retrofitted to the first. The Air Staff requirements were thus to have two Canberra B.2s converted for the Monitor role with the first conversion, to an interim standard, to be completed by the end of February 1953, and the second, carrying a more comprehensive Monitor fit, by July 1953.

There was some unease in the Air Ministry regarding the timescales for the project and the ability of EECo to carry out the work, so to address these matters a meeting was convened at Warton in December 1952, attended by representatives from the company, the Air Ministry, the Ministry of Supply and the CSE. One of the major problems was how to accommodate both the new navigation suite and the Elint fit in the cramped confines of the Canberra cabin. Unfortunately, due to security concerns, the company could not be given details of the Elint equipment, which made it almost impossible for them to come up with proposals for the navigation system layout. English Electric were also pessimistic about dates, declaring the target for the first conversion unrealistic, and suggesting the Air Ministry should consider having the work done by the CSE if it had to be met. The company would however be able to fit the first aircraft with the necessary fixed-fittings to take an overload fuel tank in the bomb-bay.

During November/December 1952 the CSE made arrangements to select two Canberras from the EECo production line for conversion. There was some urgency since aircraft from the 163rd onwards would have a new pitot tube arrangement with the head protruding through the Perspex nose cone and would thus be incompatible with the nose-mounted *Flange* aerial system. Two suitable aircraft (WH670 and WH698) were selected in the last week of November and were allotted to 192 Squadron in December 1952, with the first arriving in January 1953 and the second a month later. By then it had been decided to carry out both Canberra conversions 'in-house' at the CSE and TRE, rather than contract the work to EECo.

Fitting out WH670

The CSE started work on the conversion of WH670 in early January 1953 with a target completion date, set by the Air Ministry 'for operational reasons' of 1st April. As previously

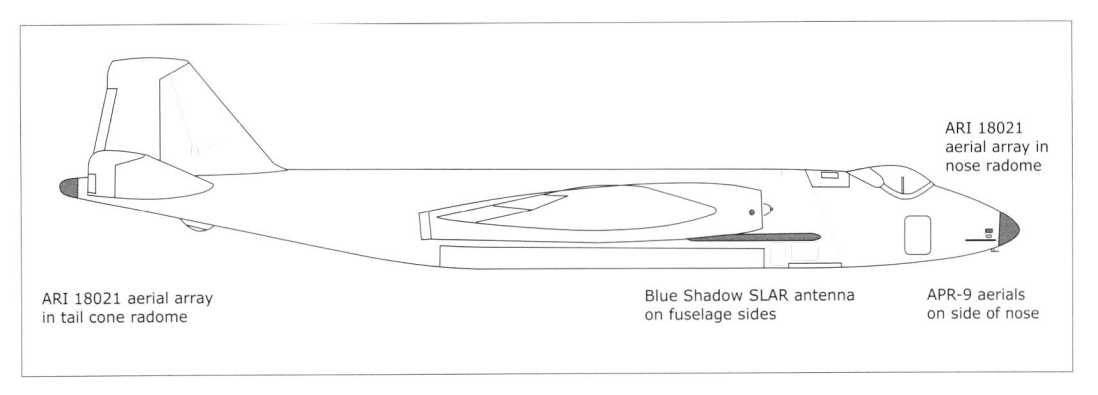

ARI 18021 aerial array in nose radome

ARI 18021 aerial array in tail cone radome

Blue Shadow SLAR antenna on fuselage sides

APR-9 aerials on side of nose

LISTENING IN

An Elint Canberra B.2 parked alongside a Washington on the ramp at Watton showing its black radome nose. The radome was fibreglass covered in a weatherproof neoprene finish. *via Paul Stancliffe*

described, there was considerable pressure to increase the number of RPFs and the Air Ministry wanted the Canberra operational as quickly as possible. By the time work started the preparation standard for the first Monitor Canberra had been refined and issued as Service Radio Installation Modification (SRIM) 1280. This specified an Elint suite comprising *Flange*, APR-9 and APA-11, and a navigation suite comprising Gee Mk.3, *Blue Shadow* and ARI 5428 Radio Compass. The navigation suite, although crude by today's standards, was the best then available. Gee Mk.3 (a miniaturised Gee receiver produced for use in fighters) would provide accurate navigation over Europe, while the combination of a *Blue Shadow* SLAR, which produced a radar picture of the terrain to one side of the aircraft, and Radio Compass would allow the Canberra to operate in areas outside of Gee cover. The Elint suite provided search, D/F and analysis facilities at centimetric frequencies. Centimetric search facilities were provided by the APR-9 receiver, with flush aerials fitted either side of the aircraft's nose. The APR-9 could take either an S-band or X-band tuning unit as required and its output could be fed into an APA-11A analyser, allowing PRF and pulse-width to be measured. The D/F portion of the Elint suite was provided by an ARI 18021 *Flange*, connected to aerials in the nose and tail of the aircraft. The nose aerials were mounted on an electrically-driven turntable and could be rotated to obtain a bearing on an intercepted radar signal while the fixed tail aerial would provide both an additional intercept capability and a passive tail-warning system, able to detect the approach of an AI-equipped fighter. The APR-9 and *Flange* systems were interconnected, allowing the SO to distinguish a signal picked up by the APR-9 on the *Flange* display. To give the *Flange* aerials a decent field-of-view the Canberra's nose cone would be removed and replaced with an electrically-transparent fibreglass radome designed by the TRE.

Before any new equipment was fitted, WH670 was gutted of nearly all of the existing radio, navigation and bombing equipment at Watton. The aircraft then went to TRE Defford in early February where it was fitted with the *Blue Shadow* radar and a Radio Compass. Following the completion of the navigation fit the Canberra returned to Watton for its Elint installation. The right-hand position in the rear cabin was re-worked as the SO's station, with most of the Elint equipment arranged within easy reach while other equipment was installed in the nose and tail of the aircraft. Work proceeded at a frenetic pace and the installation was completed in the first week of March, the only missing items being the nose radome and Radio Compass aerial. Despite the absence of these two items, the aircraft installation was cleared by the Ministry of Supply in mid-March.

WH670 was then air-tested and despatched to TRE Defford (still with its original Perspex nose) in early April for *Blue Shadow* trials, returning to Watton at the end of the month for the nose radome fit. The aircraft was now complete with the exception of the Radio Compass aerial. Since the Canberra was now nearly a month late the Air Ministry agreed to accept WH670 temporarily without a working Radio Compass in order to get it into service as soon as possible. The goal was now to get the aircraft ready for operations by 1st June but unfortunately further technical problems then emerged which took the CSE almost a month to fix, delaying delivery to 192 Squadron, and making it impossible to meet the new goal. The snags were finally fixed at the end of May 1953, following which WH670 was transferred to 192 Squadron where it began an intensive trials and training programme to prepare two fully-trained crews for operations in as short a time as possible.

Working-up WH670

Operational training of the Canberra crews began at the end of May 1953 with a series of joint Elint/navigation training flights to allow both navigator and SO to develop the most effective way of using their equipment. The Canberra was a fast, high-flying aircraft and the workload of both the navigator and the SO was expected to be high. Signal analysis in particular could be a slow business and a new APA-11 operating technique had to be evolved to cope with the fast rate of arrival of signals.

The plan was to complete the operational phase of the work-up in an intensive programme of eight cross-country flights but the first flights in WH670 using *Flange* and APR-9 showed the receivers were picking up considerable electrical noise, thought to emanate from the carbon brushes on the *Blue Shadow* blower motor. The trials programme was temporarily suspended while the interference was tracked down and eliminated. Flying on the operational trials resumed in mid-June, but the first flight revealed that the *Blue Shadow* radar was not producing the expected range, giving only 35nm (65km) against the expected 45-50nm (83-92km). Accurate navigation was vitally important and so the Canberra was immediately returned to TRE Defford for an investigation. Unfortunately the TRE were unable to identify the cause of the problem and so, in an effort to meet the target date, WH670 resumed operational trials in July while the TRE continued to work on the *Blue Shadow* range issue using another aircraft.

By the end of July the CSE had tested and amended the Elint operating technique for the Canberra and, in August 1953, a series of three SO training flights were carried out over the Continent to prove the new procedures. Unfortunately only one was deemed successful due to further electrical problems. There were also 'discrepancies' with the bearings produced by the Radio Compass and concerns still existed regarding the limited range obtained from *Blue Shadow*. To add to the problems, a second *Blue Shadow* navigator training sortie had to be aborted when the surface of the aircraft's nose radome started to degrade. Despite this catalogue of troubles, and after heroic efforts by the CSE, the aircraft was finally declared fit for its first operational sortie early in September 1953.

Planning a new RPF programme

A new programme of RPFs to investigate the Soviet air defence organisation in the Baltic, Caspian and Black Sea areas had first been proposed by the Air Ministry in March 1953. At that time the Washingtons were still working-up and the first Canberra was still under modification.

The details of the proposed flights were worked out by the Air Staff in conjunction with the Foreign Office, thus ensuring the programme was diplomatically acceptable. Protocol required the plans be cleared by the Foreign Secretary and Secretary of State for Air (Lord De L'Isle and Dudley) before being formally submitted to the Prime Minister (Winston Churchill) for final approval. Since the flights were intended to obtain military intelligence, the formal submission to the PM would normally have come from the Secretary of State for Air but Winston Churchill was at that time also acting as Foreign Secretary. Thus Selwyn Lloyd, the Minister of State for Foreign Affairs, submitted the proposals to the Prime Minister in May 1953 in his capacity as acting Foreign Secretary. Unfortunately, due to a 'difficulty over a technical detail', the Air Staff submission to the Secretary of State for Air was delayed and occurred after the papers had been passed to the PM. This drew a strong complaint from Lord De L'Isle and Dudley to Selwyn Lloyd, pointing out that he had been left out of the loop, and had not had a chance to comment on the plans. To prevent this happening again, instructions were subsequently issued to the Air Ministry that all proposals for Radio Proving Flights should first be presented to the Secretary of State for Air before being discussed with the Foreign Office.

The Prime Minister decided that the proposal for the new programme of RPFs was important enough to be discussed by the Cabinet and a paper on the subject was prepared by the Air Ministry. This ruffled a few feathers,

A close-up of the *Blue Shadow* SLAR aerial on the starboard side of a Canberra fuselage. One of the limitations of the system was that it only produced a radar picture of the ground to one side of the aircraft. *via Ron Henry*

LISTENING IN

The operations of 192 Sqn Washingtons were limited to the dark phases of the Moon. There were great concerns about the Washington's exhaust flames making the aircraft easier to see in the final phases of a GCI-controlled interception by Soviet fighters.
Adrian Mann

with CAS (Marshal of the RAF Sir William Dickson), in particular, expressing his concern regarding the security implications of so sensitive a subject receiving such a wide circulation. In the event the paper was actually discussed by a small group of Ministers, comprising the Chancellor, the Acting Foreign Secretary, the Minister of Defence, the Secretary of State for Air, the Colonial Secretary and the Secretary to the Cabinet on 30th July 1953.

The proposal was approved, conditional on the agreement of the Prime Minister and subject to a number of conditions. These were that: flights should be made in darkness, they should not approach closer than 30nm (55.5km) to the coastline of the Soviet Union or its satellites, Swedish airspace should not be violated during the Baltic flights, the flights over Turkey to the Black Sea be subject to the consent of the Turkish government, the Foreign Office be notified as far in advance as possible of each flight, and finally, that the Foreign Secretary be able to withdraw permission for, or change the date of, any flight as he saw fit. There was apparently no attempt to obtain the consent of the Iranian government for flights over Iran into the Caspian Sea. At that time UK-Iranian relations were poor, diplomatic relations having been broken off the previous year. However, the Iranian Air Force had no radar defences and was equipped with obsolete Republic P-47 Thunderbolt and Hawker Hurricane fighters with an almost non-existent night capability. In the view of the Air Ministry, the Iranians would be unaware that overflights were taking place.

On 8th August the Lord De L'Isle and Dudley asked the PM to approve the programme, asserting: 'On the successful prosecution of these essential series of flights, much of the safety of the Country depends'. The new RPF programme was approved by the Prime Minister on 10th August 1953.

Vulnerability of the Washington

In early 1953, even before the new RPF programme had been authorised, the Air Ministry began to have doubts regarding the suitability of the Washington for use on Radio Proving Flights. The Washington was a large and fairly slow aircraft, restricted to medium altitudes. Its only protection against interception was to operate in darkness and the intention was to fly RPFs on dark 'no moon' nights for this very reason. However, there were concerns that, due to poor shrouding of the engine exhaust stubs, the aircraft would be visible for some miles at night. Although there was no evidence that the Soviet Union had night fighters equipped with AI radar the aircraft might still be vulnerable to a GCI-guided visual intercept. If the Washington was intercepted by a hostile fighter, it had little chance of defending itself as its only armament was the tail gun position, and trials showed the tail guns would be of limited effectiveness in fending off a fighter due to their restricted traverse angle.

The Air Ministry were so concerned about the survivability of the Washington that they ruled, in early 1953, that the aircraft could not be flown on RPFs in regions where there was a significant danger of interception. This included the Baltic Sea, Black Sea and Caspian Sea, effectively restricting the Washington to Border operations over Germany. In mid-1953, with both aircraft yet to fly on operations, the CSE made efforts to obtain data on the visibility of both the Canberra and the Washington at night to better assess the vulnerability of both aircraft. Remarkably, no figures were initially available for the Canberra and only estimates were available for the Washington which, worryingly, suggested that, thanks to the exhaust glow, the Washington was actually more visible on a moonless night than on a moonlit night. The Air Ministry ruling on the Washingtons meant that the aircraft could not be used in the main RPF operational areas. It also meant that the Canberra, originally conceived as being supplementary to the Washington, now became the primary Elint platform.

Operation *Reason*

The first operational Elint sortie carried out during 1953 was actually a maritime operation, additional to the RPF programme proposed in March. In 1949 the Soviet Union had begun work on an improved class of light cruiser, designated the Type 68*bis*. The first example of the type, the 17,000 ton *Sverdlov*, was launched in mid-1950 and completed her fitting out in May 1952 and a number of these cruisers were eventually completed. The *Sverdlov*-class cruisers were the Soviet Navy's most modern naval units and thus of considerable interest to British Naval Intelligence. In January 1952 the Admiralty obtained Foreign Office approval in principle for photographic reconnaissance operations against the vessels (Operation *Reason*) when and if they emerged from Soviet waters into the North Sea on exercise. An Avro Shackleton MR.1 of Coastal Command was maintained at readiness for the operation but in the event no opportunities arose during 1952. The Foreign Office renewed their political approval for the operation for a further six months in December 1952.

During 1953, plans were made to supplement photographic reconnaissance with an Elint operation by 192 Squadron to investigate the ship's radar fit and a Washington and crew were placed on 24-hour readiness (that is, ready to fly within 24-hours of notification) for Operation *Reason* from February/March 1953 onwards. By May 1953 it looked like the long wait might be over. At the beginning of 1953 the Soviet Navy had been invited to attend the Spithead naval review planned as part of the 1953 Coronation celebrations. This was the first official post-war visit of a Soviet naval vessel to the UK and the Soviet Union took it seriously, announcing in the spring of 1953 that they were planning to send the *Sverdlov*, probably its most prestigious naval unit.

The visit of the *Sverdlov* to the UK for the Coronation Review looked like a golden opportunity to carry out Operations *Reason*. The Admiralty suspected the *Sverdlov* would leave its home port of Baltiysk, in the Soviet enclave of Kaliningrad in the Baltic, en-route for the UK in company with sister ships of the same class bound for northern waters. The plan was to photograph the latter off the Norwegian Coast, while an Elint operation would be conducted against the *Sverdlov* herself as she cruised south across the North Sea towards the UK.

Since the political approval for Operation *Reason* had expired at the beginning of June 1953, the Admiralty applied for a new approval early in the month. The Secretary of State for Air refused to sanction the proposed photographic operation as the photographic aircraft would have to make a close approach to the Soviet vessels and he considered the operation would be contrary to the UK policy of seeking better relations with the Soviet Union. The Elint operation was considered less controversial, since the aircraft would stand off at a reasonable distance from the *Sverdlov*. The Foreign Minister (Selwyn Lloyd) saw no objections to either operation and thus, despite the reservations of the Secretary of State for Air, a new authorisation

The *Sverdlov*-class cruisers were amongst the most modern units in the Soviet Navy in the early 1950s and constituted threat to NATO convoys. This close-up view shows an array of search, air warning and gunnery control radar aerials. *Author's collection*

request, covering both operations, was submitted to the Prime Minister (Winston Churchill) in early June 1953. Churchill however refused the request, responding that: 'The operations are unsuitable at this juncture and also it would not be in accordance with British ideas to take advantage of a ship visiting this country to salute the Queen'. The visit of the *Sverdlov* to the UK passed off without incident and, after a week moored at Spithead, the cruiser returned to the Baltic on 17th June.

A further opportunity to mount Operation *Reason* occurred at the beginning of August when the *Sverdlov* was spotted leaving the Baltic en-route to the Barents Sea. This time there were no political objections. 192 Squadron was alerted on 2nd August, when the ship passed the northernmost tip of Denmark, and Prime Ministerial approval for an Elint sortie was obtained the following day. The conditions laid down for the conduct of the operation were strict. The Washington would be allowed one approach to a distance of 8nm (15km) from the *Sverdlov* for visual identification, after which the aircraft had to stand off at least 15nm (28km) from the vessel whilst attempting to intercept radar transmissions.

Washington WZ966 (Flt Lt Lang and crew) took off from Watton on Monday 3rd August at 1043hrs and flew across the North Sea and up the coast of Norway to the predicted position of the *Sverdlov*, based on previous aircraft and ship reports. The six Elint positions were fitted for VHF intercept (two positions), metric radar intercept (two positions), S-band radar intercept and X-band radar intercept. Following a short search the cruiser was located about 35nm (64km) off the Norwegian coast at 1455hrs. After an initial approach to eight miles in excellent visibility to identify the target the Washington stood off 40-60nm (74-111km) and remained in signals contact for about 3½ hours. Signals in a variety of bands were intercepted, including the first, and very brief, confirmation of the use of X-band fire-control radar. The cruiser kept the Washington under radar surveillance for the duration of the operation. The aircraft landed back at Marham at 2130hrs, having been in the air for 10 hours 45 minutes.

The operation was apparently very successful with the Secretary of State for Air sending his thanks for the squadron's: '… skill and devotion in successfully completing the operation …'. In fact, the results of Operation *Reason* were considered so valuable that the Air Ministry sought permission for further maritime operations, subject to the same safeguards, whenever the opportunity presented itself. The Prime Minister agreed in principle but asked that he be notified before each flight. By the end of August 1953 the maritime Elint commitment had been renamed Operation *Claret* and operational control of the task transferred from Coastal Command to 90 Group with arrangements were made to keep a 192 Squadron Washington at 24-hour notice for these operations.

Operation *Reason* was the first maritime RPF carried out against a unit of the Soviet Navy. The *Sverdlovsk* was repositioning from Baltysk on the Baltic coast to the Northern Fleet at Severomorsk near Murmansk.

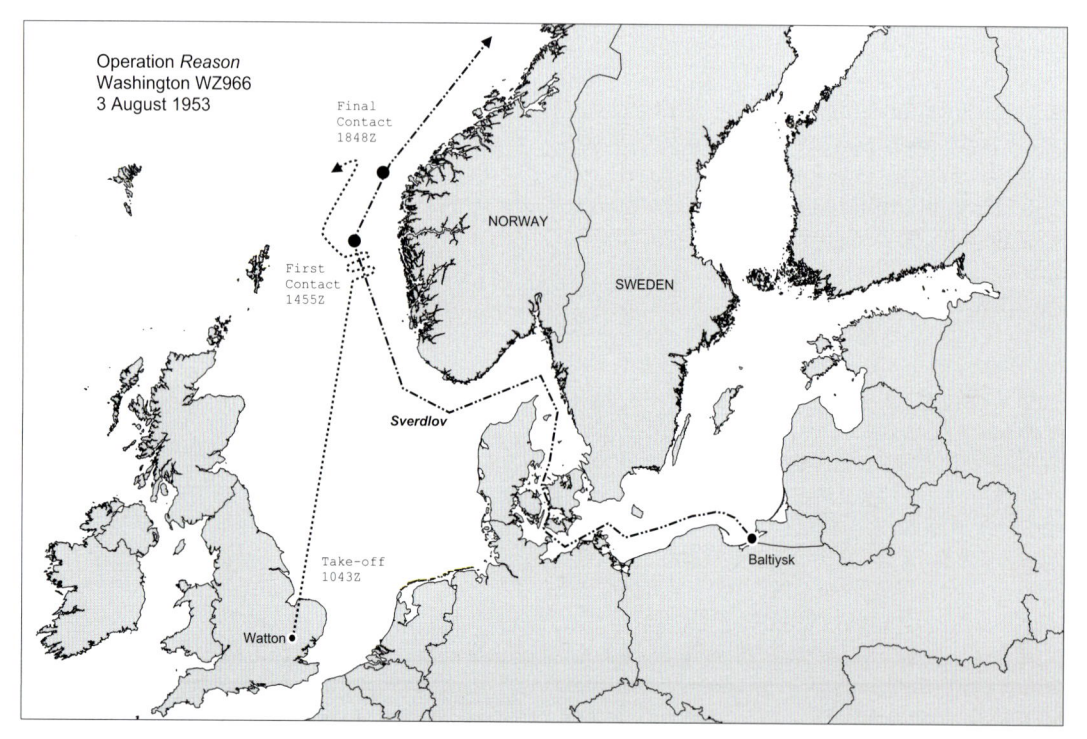

6 Probing the PVO Strany

'These flights have taken place and have obtained valuable information for us. The Russian Control and Reporting System has reacted on each occasion, but the Russian Air Force has not, so far as we can discover, attempted interceptions'. VCAS, Air Chief Marshal Sir John W Baker, 1953.

By the autumn of 1953 the CSE and 192 Squadron were ready to fly the first sortie in the 1953 RPF programme. The aim was to investigate and map the Soviet Air Defence system, identifying ground radars, intercepting ground-to-air communications, and gathering general information on its operation, strengths and weaknesses.

Some information on Soviet air defences already existed from the previous Lancaster and Lincoln RPF programmes, and from ground-based Elint surveillance operations carried out by 646 Signals Unit in Germany and 276 Signals Unit in Iraq. In addition, the US shared the results of its Elint sorties with the UK. However, an airborne RPF programme would facilitate the interception of signals outside the range of ground stations and would also represent an important UK contribution to the common US/UK intelligence pool.

By 1953 the Soviet Air Defence system was slowly evolving and a chain of Early Warning (EW), Ground-Controlled Intercept (GCI) and Anti-Aircraft Fire Control (FC) radars now stretched along the Soviet bloc borders, from the Baltic to the Caspian. The majority of the EW radars were Soviet-designed mobile truck-mounted sets working in the 70MHz metric band, supplemented by a few wartime German and US/UK lend-lease radars. The primary Soviet radars were the wartime-vintage RUS-2, the later P-3 *Dumbo* and the P-8 *Knife Rest A*, first seen in 1951. The RUS-2 and P-3 only had a range of around 74 miles (120km), poor azimuth discrimination, and limited height-finding capabilities but the later P-8 radar had a much improved range and better display facilities.

The EW stations were backed up by a chain of GCI radars, used to direct fighters onto intruding bombers. Both the P-3 and P-8 radars were also used in this role, supplemented by increasing numbers of the new S-band P-20 *Token* radar (first seen in 1952). The P-20 was effectively a copy of the US CPS-6 radar and was the most capable set then in Soviet service, with a range of about 155 miles (250km) and effective up to 60,000ft (18,280m) with good height-finding capabilities. Other radars were associated with anti-aircraft gun defences, the primary AA fire-control radars being wartime UK and US lend-lease equipment (Gun Laying Mk.2, 50-80MHz, and SCR-584, 3GHz) supplemented by the SON-4, a Soviet-built copy of the SCR-584.

The final elements in the Soviet air defence system were the fighters: by 1953 these mainly comprised MiG-15 *Fagot* jets which had good altitude performance and high speed but lacked AI radar and thus had no all-weather capability. In poor weather or at night the fighter would need to rely on GCI to bring it within visual range of its target. The GCI/fighter communications link was initially MF and HF R/T but by 1953 the Soviet Union was moving to the more secure VHF band.

RPF Operations Start

The first RPFs in the 1953 programme were two sorties over the Baltic in September. Since the Washingtons had been barred from flying in risky operational areas these were conducted by Canberra WH670 and, because of the relatively short range of the aircraft, were flown from a forward base in Germany. Thus WH670, accompanied by Washington WW346 carrying ground crew and spares, deployed to Wünstorf, near Hanover, in West Germany on 7th September. The Canberra was fitted for centimetric search and direction finding (D/F), suggesting its task was the interception and plotting of *Token* radars along the Baltic coast of the Soviet Bloc.

It appears that the sorties had been co-ordinated with the UK Elint organisation in Ger-

LISTENING IN

Locations of RAF ground Elint stations in West Germany. Most were controlled by 646 Signals Unit.

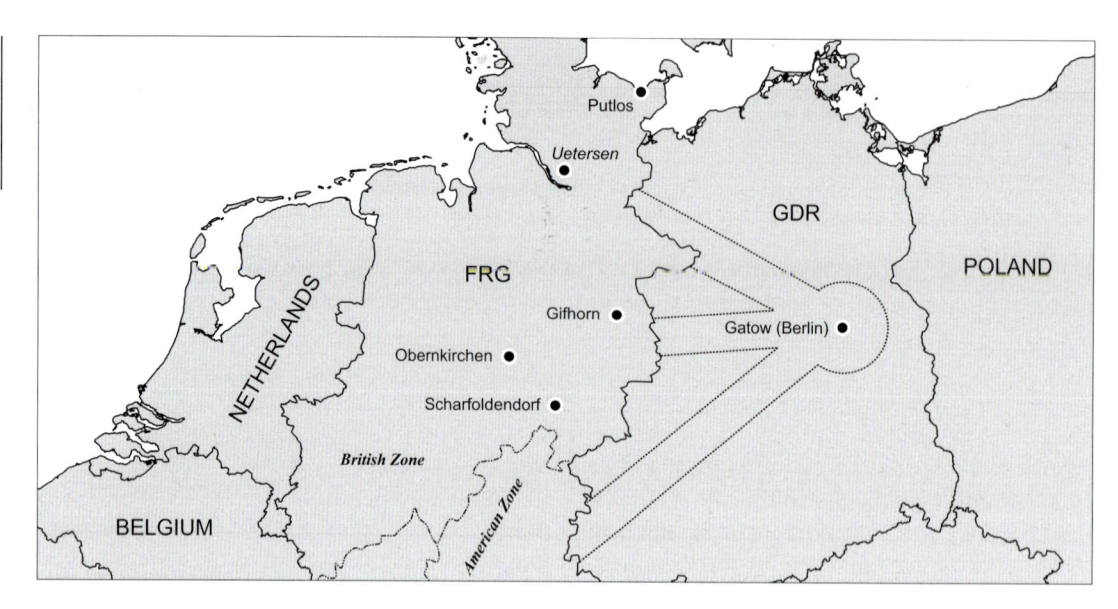

many since representatives of the CSE and 90 Group had visited the headquarters of 646 Signals Unit (SU) at Obernkirchen, near Hanover to discuss the task in July. 646 SU, formed from the ADI(Science) Elint unit during 1952, maintained a number of Elint ground-stations in West Germany, including one at Putlos on the Baltic coast. The Canberra sortie would almost certainly provoke a response from the Soviet Bloc air defences, providing the ground stations with signals interception opportunities as Soviet radars tracked the Canberra and reported its presence to the air defence organisation.

Both sorties were scheduled for the new moon period, guaranteeing absolute darkness. At 0100hrs on the 10th September Canberra WH670 (Sqn Ldr Emmett) took-off from Wünstorf on the first RPF of the new series (Operation *Bonaparte*), flying north towards the German coast, then east out over the Baltic. The Canberra had only just been declared operational after a prolonged working-up period beset by various electrical problems and unfortunately the electrical problems resurfaced shortly after take-off, forcing the crew to abort the sortie and return to Wünstorf.

After repairs and tests, the Canberra was readied for a further attempt at the sortie on the night of 14th September. This time things proceeded more smoothly and WH670 and its crew carried out a 3 hour RPF over the Baltic, flying along the coasts of Germany, Poland and

Blue Shadow Sideways Looking Airborne Radar (SLAR). The radar mapped a narrow swathe of ground to one side of the aircraft using a fixed linear aerial, the forward motion of the aircraft producing a continuous radar picture. A common complaint was that the radar showed 'where you had been, rather than where you were going'.

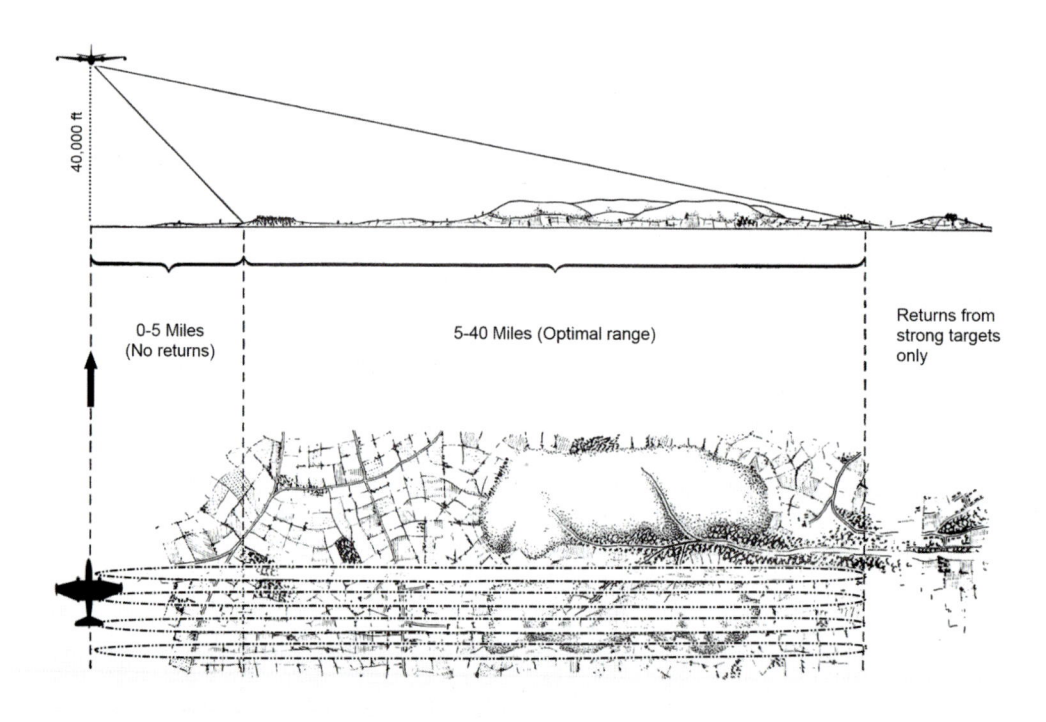

60

the Soviet Union, before safely returning to Wünstorf. Operation *Bonaparte* was followed by a similar second RPF (Operation *Jetsam*) the following night, although navigational difficulties were experienced during this second operation, exposing the limitations of the Canberra's navigation fit. The main navigation aids were the Radio Compass and *Blue Shadow* radar. In theory *Blue Shadow* had a range of around 40-50nm (74-92km) at 48,000ft (14,630m), allowing the Canberra to stand off the required distance from the Baltic coastline and still obtain a fix. In practice the Radio Compass was not particularly accurate and the *Blue Shadow* range was below specification. The Air Ministry closely monitored the conduct of both sorties and the navigation difficulties seems to have generated some concern. Intermittent electrical problems were also encountered with the Elint equipment, preventing the correlation of *Flange* D/F indications with APR-9 intercepts, thus degrading the quality of intelligence obtained. Despite these problems, Operations *Bonaparte* and *Jetsam* were a major step forward for 192 Squadron and were almost certainly the first operational jet Elint sorties conducted by any country.

The next RPFs in the programme were two sorties over Iraq and Turkey into the Black Sea in October 1953. By the first week in October the restrictions on Washington operations had been relaxed, allowing the aircraft to fly over the Black Sea, although operations in the Baltic and Caspian Sea were still prohibited. Despite the concern at the Air Ministry, the CSE and 90 Group had remained fairly sanguine regarding the Washington interception risk. At a meeting in August, during which the problem was discussed, the CSE Operations staff suggested that the Air Ministry were paying too much attention to the risk and that: 'Exhaust flames will only assist the attacking aircraft in the final stages of the kill, after a preliminary interception by AI. This small operational risk must be accepted'. By the beginning of October the CSE had prepared a draft letter to the Air Ministry: '… urging that the exhaust flames visibility risk with the Washington should be accepted since there appeared to be a tendency to exaggerate the implications'. The letter was not sent, probably because the Air Ministry had already come to the same conclusion, at least regarding the risk of Washington operations over the Black Sea. The Black Sea may have been considered less risky than other areas because the Washingtons could keep a long way from the Soviet coast and had friendly territory and airfields (in Turkey) within easy reach.

As part of the preparations for the detachment, the CSE Wing Commander (Operations) and officers of the Flying Wing visited Nicosia to make arrangements for the local control of the aircraft and for 'flight briefing of the crews in the theatre prior to each exercise'. Two Washingtons, accompanied by WW346 as a support aircraft, were detached to Nicosia at the beginning of October. Two operational Elint sorties were flown over the Black Sea during the detachment (7th October, Operation *Probate*/Air Ministry Operation No.63; 10th October, Operation *Camelia*/Air Ministry Operation No.64), both using Washington WZ968, fitted for HF, VHF, metric radar and centimetric radar search and D/F. Unfortunately the second sortie did not go well and navigation errors were made. Although the aircraft may not have intruded into Soviet airspace, the deviation from course was sufficiently serious for the Air Ministry to ask the CSE to: '… give assurance that suitable action is being taken to prevent the re-occurrence of avoidable errors that were made during the Operation'.

While the Washingtons were detached at Nicosia, efforts were being made to prepare the Canberra for the next operation in the series: two sorties over the Caspian Sea. In the first week of October 1953, Canberra WH670 was detached to Habbaniya in Iraq on a training detachment. The RAF airfield at Habbaniya, some 60 miles north-west of Baghdad, was ideally placed for sorties over Iran into the Caspian Sea and along the southern borders of the Soviet Union. The RAF also maintained a ground Elint station there (276 Signals Unit) to intercept Soviet communications and radar signals. Only two navigational training flights were carried out by WH670 from Habbaniya (plus a return sortie south to Shaibah, near Basra) before the detachment was cut short due to unserviceability.

The routes for a series of Canberra RPFs from Habbaniya during November were initially approved at a meeting at the Air Ministry on 9th October but as a result of 'navigational difficulties with the Canberra' (possibly highlighted by the training detachment) the task was reviewed at a further meeting on 14th October. The number of sorties was reduced to two and the RPF routes were amended, presumably to simplify navigation and to guard against any inadvertent incursion into Soviet airspace.

At the end of October the Canberra was fitted with an ANQ-1A wire recorder to allow signals picked up by *Flange* or the APR-9 to be recorded for later analysis. A week later, fitted with S-band *Flange* aerials and S-band APR-9 tuning units, the Canberra was again detached to Habbaniya for its first RPF in the Middle East. The detachment was commanded by the CO of the squadron and accompanied by the CSE Operations Officer with ground crew and spares ferried out in Washington WW346. This time, having learnt from previous experience, the detachment included a servicing team to cope with any equipment failures. Two night RPFs were flown, almost certainly over the Caspian Sea, looking for *Token* radars in the area. The first (Air Ministry Operation No.66) was a 3 hour 20 minute night sortie. An attempt at a second RPF (Air Ministry Operation No.67) on 11th November was abandoned due to VHF radio failure but the operation was successfully completed the following night. The Soviet reaction to both sorties was monitored by 276 Signals Unit (SU) at Habbaniya.

November also saw the resumption of regular 'Border' Elint flights over Germany, these having been suspended at the end of 1952 when the Lincolns had been withdrawn from operations. The first Border flights using the Washington were combined with two specific investigations: 'vertical polar diagrams' and 'beamed communications'. The vertical polar diagram task was most likely an attempt to obtain technical intelligence on the cover of individual radars along the German border at varying heights whereas the beam communication task was apparently an attempt to intercept signals from the point-to-point radio communications links widely used by the Soviet Bloc forces. A combined flight plan was prepared for the regular monthly task, including optional 'excursions' to cover both the special tasks. The operational order for the first operation (Operation *Adjunct*) noted that: 'Aircraft are to be fuelled for maximum endurance. Continuous navigation problems are likely to be encountered during the flight and operational fatigue must be considered by the flight commander'. An initial series of four sorties was planned, with one in November and three in December and the first sortie, a 7½ hour night flight, probably along the German internal border, possibly including the Berlin air corridors, was successfully flown on 26th November. A subsequent report noted that all the special equipment worked very well and both excursions had been flown.

Routes for a further series of RPFs over the Baltic during December were approved by the Air Ministry at the end of November. There was initially some doubt as to whether the Air Ministry could obtain clearance for the operations in time, due to 'pending repercussions' from the navigation problems experienced during the Canberra's first Baltic foray in September. However, the flights were approved and the Canberra was detached to Wünstorf at the beginning of December for a series of three RPFs over the Baltic, a Washington ferrying out a supporting ground team. As before, the Canberra was fitted for S-band search. The first sortie from Wünstorf on 8th December was successful but an APR-9 failure was experienced on the second the following night. After a delay waiting for suitable weather, a third sortie was flown on 15th December but had to be abandoned due to intermittent operation of the *Blue Shadow* radar. No further sorties were possible as the dark new moon period had now ended. Despite only two sorties being completed the detachment had gone reasonably well: this time navigation was satisfactory and the Air Ministry 'expressed appreciation of the conduct of the operations'.

December 1953 saw attempts to fly the remaining three extended Border sorties planned in November. The first sortie (Operation *Brimstone*) was flown on 1st December but returned early due to Decca problems. A second sortie on 8th December was abandoned even earlier for the same reason. The use of Decca for navigation along the German border and in the Berlin air corridors was vital since use of the Washington's APQ-13 radar would immediately identify the aircraft to Soviet radar controllers. The final operation in the series was cancelled due to poor weather.

On 1st January 1954 the Canberra returned to Habbaniya for a series of two RPFs (Operations *Barrow* and *Mourne*) with Washington WW346 once again carrying ground-crew and spares. The navigation equipment problems experienced the previous month had clearly not been fully resolved as the first of these sorties, on 4th January, was terminated early due to *Blue Shadow* problems while the second, shorter, sortie on 5th January was completed to a 'revised' plan.

Blue Shadow was at its best mapping coastlines, as this output shows. The aircraft was flying on an east-west track at 30,000ft, passing over the south coast of the Isle of Wight.
via Ron Henry

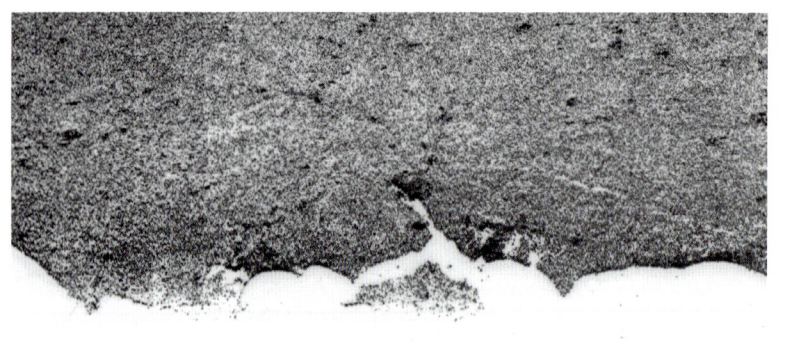

On return to Watton WH670 underwent an investigation into apparent G4B gyro-compass errors and had its navigation fit further improved with the installation of a modified Air Position Indicator and a Periscopic Sextant, the latter providing the Canberra's navigator with a very useful astro-navigation capability. In theory astro-navigation was possible from the Canberra using a hand-held sextant looking through the pilot's canopy but in practice the curvature of the canopy introduced errors and it was in any case very difficult for the navigator to move forward to take star shots. The Periscopic Sextant solved both these problems: it was fitted to the navigator's hatch with its head protruding above the hatch in the raised position. The drawback of the system was its limited field of view, requiring the navigator to pre-plan his shots to ensure the required stars could be quickly and accurately located.

Two further Border sorties were flown by the Washingtons from Watton during January: a 5 hour night flight (WZ967) on 21st January and a 5½ hour day flight (WZ968) on 28th January. Both were successful, with Decca working as expected.

During 1953 the LSIB had raised a new airborne Elint task known as 'Local Flight', designed to intercept communications transmissions which could not be intercepted by ground stations, and requiring a continuous series of sorties in the same geographical area over a prolonged period. During late 1953 it was decided to carry out a trial with one Washington to evaluate the 'Local Flight' task and assess how many aircraft would be needed to meet the required scale of effort. The trial was expected to last about a year, running from the spring of 1954 to the spring of 1955.

Since the 'Local Flight' task involved communications intercept it required a change of Washington Elint equipment as the aircraft's standard fit was optimised for radar intercept and D/F. Accordingly in November 1953 the Air Ministry asked 90 Group to investigate fitting a Washington with HF/VHF communications monitoring sets at all six Special Operator positions, with the proviso that the aircraft should easily be refitted with the standard Elint fit when required. By December the CSE put together a design for a Washington communications intercept installation (designated the No.2 Standard Fit) comprising two HF positions, using BC-348 receivers, and four VHF positions using the ARR-5 receiver, with all six stations fitted with an ANQ-1 wire recorder to record intercepted traffic. Changeover between the normal Washington fit and the communications fit would take 12 hours. Approval for the fit was received in January 1954 and conversion took place during January/February.

Following the conversion Washington WZ966 was detached to Wünstorf at the end of February for a series of ten sorties under Operation *Catarrh*. In all, nine daylight sorties each of 5 hours duration, were flown over two weeks, presumably monitoring Soviet Bloc voice transmissions along the East German border.

The RPF programme authorised in August 1953 was concluded in early March 1954. In a subsequent report to the Prime Minister the Secretary of State for Air noted: 'These flights have taken place and have obtained valuable information for us. The Russian Control and Reporting System has reacted on each occasion, but the Russian Air Force has not, so far as we can discover, attempted interceptions. No unusual incidents have been reported and there is still no indication that the Russians have fighters equipped with AI'.

In the same letter the Secretary of State requested approval for a further series of eight RPF sorties in the period 24th March to 6th June 1954. The aircraft would visit the same areas as the previous programme and operate under the conditions previously laid down by the Cabinet. The Foreign Secretary would be consulted before each flight. Winston Churchill approved the programme the following day.

Fitting the Second Canberra

The second Canberra B.2 (WH698) had been delivered to 192 Squadron in February 1953. As shown above, there had originally been considerable uncertainty regarding the optimal Canberra Elint fit and the installation in the second Canberra had been deferred pending the outcome of operational trials in the first aircraft (WH670). Thus WH698 was initially used for crew continuation training during the first half of 1953. Unfortunately the installation, and subsequent trials, of the provisional fit in WH670 took longer than expected and by mid-year that aircraft was still not fully operational.

Despite this, by July 1953 the CSE had apparently gleaned sufficient information from the working-up and calibration trials to submit proposals to 90 Group for the second Canberra Elint fit. By the end of July the Air Ministry had agreed that work could start on the second Canberra with the proviso that it would only be approved once the operational performance of

Green Satin used a twin-beam X-band Doppler radar to measure ground-speed and drift. The twin beams (one forward, one aft) compensated for aircraft pitch. On each scan the two beams scanned diagonally across the aircraft's heading, to compute drift. The mechanically-scanned aerial was installed under the port wing.

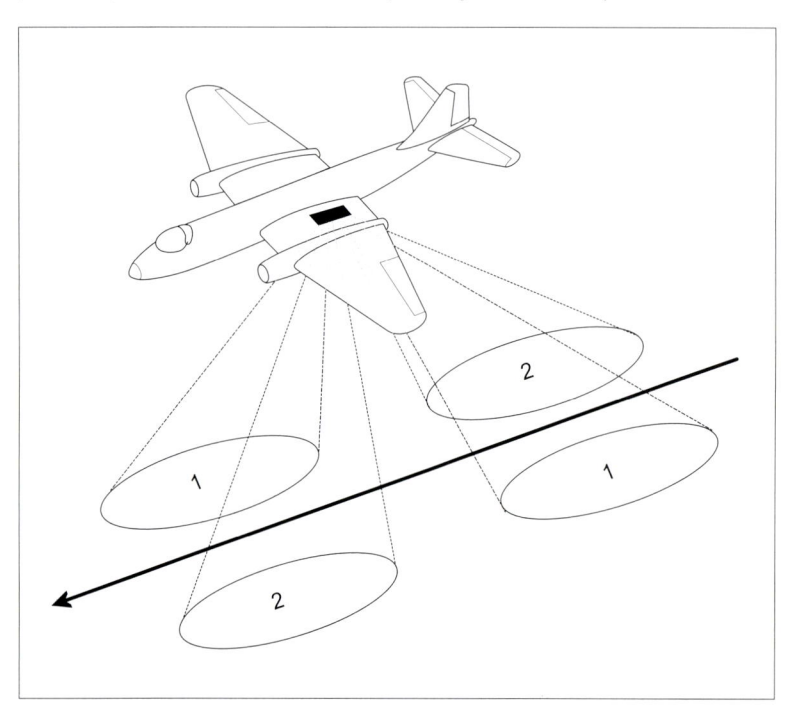

the installation in the first Canberra had proved satisfactory. It was also agreed that the fit would also be considered provisional since 'several untried facilities are to be incorporated'. WH698 went to the CSE Installation Flight at the end of June 1953 where an Elint fit similar to that applied to WH670, comprising ARI18021 *Flange*, APR-9 and APA-11, was installed. The *Flange* installation in the tail comprised two S-band aerials and two X-band aerials, arranged at an angle of 45° to allow reception of both horizontally and vertically-polarised signals. The nose installation comprised two horizontally-polarised and two vertically-polarised aerials (either X-band or S-band) installed on a motorised turntable. A *Flange* X.15 frequency selector and squint control unit were also fitted: the former allowing the frequency range of the *Flange* set to be restricted and the latter allowing the D/F accuracy of the system to be adjusted. An APR-9 receiver was fitted, along with fixed X and S-band aerials either side of the nose. The required band had to be selected before flight and the appropriate Tuning Unit fitted. Other equipment comprised an APA-11 pulse analyser, an ANQ-1A wire recorder and a modified ARI 5593 *Blonde* receiver/photographic recorder.

The initial installation in WH698 was completed on 31st October 1953 and air-tested in November. These flights uncovered a number of snags including an inevitable electrical interference problem which was eventually traced to the aircraft's VHF radio. A rewire of the VHF during December partially cured this problem and in January 1954 further modifications took place including repositioning parts of the *Flange* installation to give the Special Operator (SO) a better view of the APA-11 analyser, modifying the *Flange* rear aerials, and making provision for fitting an ARR-5 VHF search receiver and ANQ-1A recorder as an alternative to APA-11. The provision for the ARR-5 (included at request of Air Ministry, following enquiry as to feasibility in November 1953) gave the Canberra a voice intercept capability. At that time the Soviet air defence system was in the process of moving its GCI communications from HF to VHF and the ARR-5 would allow the Canberra to intercept and record VHF traffic between Soviet GCI stations and fighters. The navigation fit applied to WH698 was the same as that applied to WH670: principally *Blue Shadow* SLAR and Radio Compass. The installation in WH698 was finally completed in February 1954 and the aircraft returned to the squadron at the end of the month.

The entry into service of WH698 allowed Canberra WH670 to be withdrawn from service in March 1954, initially for an inspection and then to have its interim Elint fit brought up to the same standard as WH698.

Fixing the Navigation Problem: *Green Satin*

The RPFs in Canberra WH670 during late 1953/early 1954 had highlighted the shortcomings of the aircraft's navigation suite. Although WH670 had a better navigation fit than most Bomber Command aircraft, it was still rather limited, lacking a really effective self-contained navigation aid. The *Blue Shadow* radar only looked out to one side of the aircraft (thus restricting the direction of flight on coastal sorties), was of limited use over featureless terrain, and could, in theory, reveal the track of the aircraft to a suitably-equipped air defence organisation. This lack of effective navigation aids was of particular concern when operating in areas such as the Black Sea and Caspian Sea, where navigation was difficult and the consequences of a mistake could be severe. Accurate navigation was also of course necessary for the accurate plotting of Soviet radar stations. The CSE issued a paper in October 1953 describing the limitations of the Elint Canberra navigation suite and the consequent accuracy that might be expected when flying on Radio Proving Flights. The paper recommended that the navigation suite be improved by the addition of a Periscopic Sextant (for astro-navigation) and, most importantly, by the installation of *Green Satin* Doppler navigator.

Green Satin was a self-contained radar navigation device, providing an accurate measurement of an aircraft's ground-speed and drift by bouncing a radar signal off the ground and measuring the Doppler shift of the return. The output of *Green Satin* was fed into the associated Ground Position Indicator (GPI) Mk.4 computer which produced a continuous indication of aircraft position. Work on the *Green Satin*/GPI Mk.4 system had started in the late 1940s and by the end of 1953 the equipment was nearing the end of its development programme. *Green Satin* would in theory allow a Canberra to navigate over any terrain with a high degree of accuracy. Although the production *Green Satin* set (ARI 5851) was not likely to be available until the beginning of 1955, authorisation had been given for the manufacture of a small number of 'crash' production models (ARI 5871) in readiness for the introduction into service of the new Valiant V-bomber in mid-1954.

A party from the CSE visited TRE Defford towards the end of October 1953 to inspect *Green Satin* installed in a trials Canberra and to assess the practicality of fitting it into an Elint Canberra. The project appeared feasible and in December 1953 the Air Ministry Directorate of Operations suggested that two of the pre-production ARI 5871 *Green Satin* sets be diverted from the Valiant equipment programme and fitted to the two 192 Squadron Canberras.

Although fitting *Green Satin* into the Canberra B.2 was possible, it was not without its complications, requiring structural modifications to accommodate a downward-looking aerial in the aircraft's wing. Investigation then revealed that structural provision for the *Green Satin* antenna had already been made in the Canberra PR.3. Looking for a short-cut, the Air Ministry Operations branch suggested that 192 Squadron might want to consider swapping its two Canberra B.2s for two Canberra PR.3s. This suggestion was quickly dismissed as unrealistic, with various parties pointing out that the PR.3 only had two crew whereas the Elint task required three, that there were considerable differences between the pre-production and production models of *Green Satin*, thus complicating the PR.3 installation, and that fitting *Blue Shadow* into a PR.3 would probably take as long as fitting *Green Satin* into a B.2. Despite a number of objections, the Elint programme had sufficient importance that the proposal to fit the two Canberras of 192 Squadron with ARI 5871 *Green Satin* sets was eventually accepted.

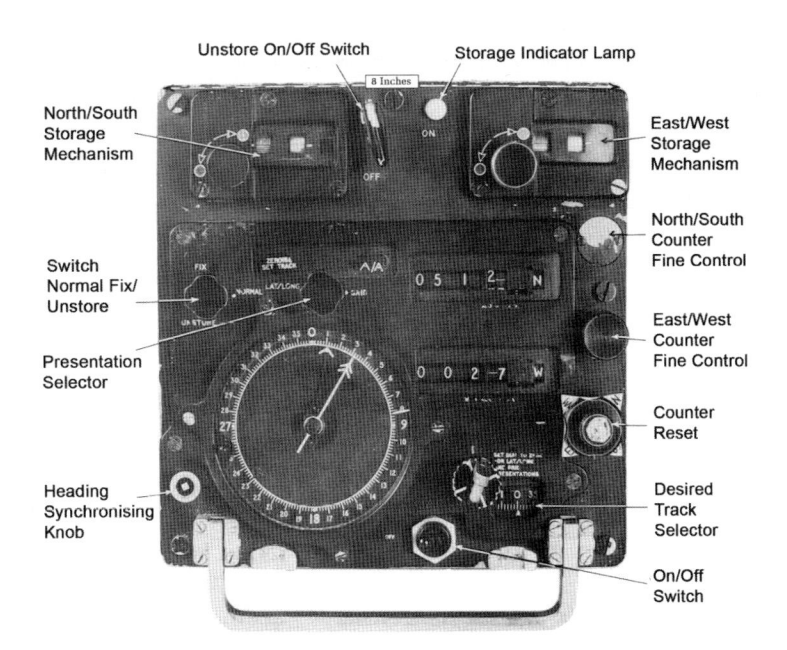

Spring 1954 RPF Programme

The return of Canberra WH698 to 192 Squadron, complete with an updated Elint fit, coincided with the start of the next RPF programme.

In early March authorisation had been obtained for a further series of eight RPFs over the Baltic, Black Sea and Caspian Sea. These would be flown between 24th March and 6th June and would be subject to the same conditions as before regarding conduct, nearest approach and approvals.

The first sorties in the new programme (and the first operational sorties by Canberra WH698) were a series of four RPFs from Habbaniya at the end of March/early April during the dark new moon/quarter moon period. Prior to the detachment a number of test flights were carried out in WH698 from Watton to check the Elint and navigational equipment in the Canberra. The latter included an improved API and a Periscopic Sextant, both of which worked well. The Canberra was detached to Habbaniya on 20th March, night-stopping at Luqa and arriving at Habbaniya the following day. The aircraft's Elint fit comprised *Flange* for centimetric D/F, APR-9 for centimetric search, an ARR-5 VHF communications receiver, a *Blonde* receiver/recorder and an ANQ-1A wire-recorder. Alternative *Flange* antennae and APR-9 tuning units were taken along on the detachment, allowing the aircraft to be fitted for either S-band or X-band search as required and an APA-11 signal analyser was also provided as an alternative fit to the ARR-5. The VHF receiver, fitted to intercept Soviet GCI transmissions, was a new

The output from *Green Satin* was fed into a Ground Position Indicator (GPI) Mk.4 electro-mechanical computer, which provided the navigator with a continuously-updated position display. This was leading-edge technology in 1954. *Author's collection*

addition to the Canberra Elint suite and the squadron Special Signals Leader was instructed to ensure that: '… the Special Operator using the ARR-5 understands the use of both his receiving and recording facilities and has practice in the making and annotation of recordings'. A GNQ wire-recorder playback machine was also taken on the detachment, allowing ANQ-1A recordings made in the air to be reviewed post-flight.

The first sortie (Operation *Natal*/Air Ministry Operation No.77) on 24th March was a 3 hour night flight over the Caspian Sea; a second, slightly longer, sortie (Operation *Revivor*) was flown on the 26th March. This did not go entirely to plan with the aircraft diverting slightly from its planned route due to difficulties in accurately fixing a position on the coast. Although the diversions from the planned track were minor, the Caspian Sea was an extremely sensitive area and the Air Ministry subsequently demanded an explanation for the deviation. A third night sortie (Operation *Lebanon*/Air Ministry Operation No.75) was flown on 31st March.

The sorties from Habbaniya were intended to investigate the Soviet air defence system on its southern borders. One of the major goals of the UK's 1953/54 Elint programme had been to determine if Soviet night fighters were fitted with AI radar and on 4th April the last RPF of the series (Operation *Minority*) flown by WH698 from Habbaniya picked up the very first indications that the Soviet Air Force was using AI radar.

The evidence of Soviet AI was tentative and, in an effort to gain confirmation, a follow-up operation over the Baltic was arranged in the last week of April. This appears to have been something of a rush job: previously-planned training detachments and operational sorties were cancelled, and Canberra WH698 was readied for detachment at 'short-notice'. The operation (Operation *Breach*) was planned as a joint Washington/Canberra RPF, being notable as both the first joint Washington/Canberra sortie and the first use of the Washington in the Baltic. Both aircraft were deployed to Wünstorf on 25th April and the operation was flown on the night of 26th April with both aircraft fitted for S-band radar and VHF voice search. The slower Washington departed Wünstorf at 1850hrs, followed by the Canberra at 1925hrs, the aircraft presumably rendezvousing in the operational area. It seems likely that the Canberra 'trailed its coat' along the coast of Germany, Poland and the Baltic States while the Washington stood off at a safe distance, monitoring any attempted intercepts. In the event no evidence of AI was uncovered. Following the earlier problems on Operation *Revivor*, the navigation of both aircraft came under particular scrutiny but this time everything was satisfactory, the Canberra navigation being described as 'excellent' and that of the Washington as 'quite good'.

Since the results of the Baltic flight had proved negative, plans were made for further investigations over the Caspian Sea, where the AI signal had first been detected. The question of Soviet AI was considered so important that the embargo on Washington flights over the Caspian was relaxed and a series of three joint Washington/Canberra RPFs was scheduled in the new moon period between 28th May and 6th June.

Unfortunately political considerations prevented the operations from taking place. An international conference on the future of Korea and Indochina had been convened in late April 1954 in Geneva, attended by representatives from the United Kingdom, the United States, the Soviet Union, France and the People's Republic of China. The flights over the Caspian were vetoed by the Foreign Office at the end of May due to worries about provoking an international incident during the Geneva conference, which was then at a critical stage. As a result the series was rescheduled for the next operational period.

The cautious approach taken by the Foreign Office and Air Ministry regarding the Caspian RPFs was doubtless informed by the RB-47E incident earlier in the month. On 8th May a USAF Boeing RB-47E photographic reconnaissance aircraft, operating from RAF Fairford, Gloucestershire, had flown a daylight penetration sortie over the Kola Peninsula, overflying Soviet territory in order to photograph airfields and other military facilities. The aims of the sortie were to gather intelligence on the suspected deployment of the new Mikoyan-Gurevich MiG-17 *Fresco* fighter, and to investigate the presence of long-range bombers in the area. The sortie unambiguously confirmed the deployment of MiG-17s: the RB-47E was intercepted and attacked by a number of these aircraft and was very lucky not to be shot down. The Soviet Air Defence system would almost certainly be on heightened alert following this incident, increasing the risk of the proposed Caspian RPF programme.

The deferral of the Caspian sorties marked the end of the March-June 1954 RPF programme. The results of this were summarised by the Air Ministry as follows: 'The most important result was the interception of suspected Russian AI signals in the Caspian areas. We have also obtained new information and confirmatory evidence on the Russian order of battle'.

The RB-47E photographic reconnaissance version of the Boeing B-47 Stratojet bomber. The RB-47 carried up to eleven cameras in an extended nose and formed the backbone of the USAF reconnaissance capability in the late 1950s and early 1960s.
T Panopalis Collection

Soviet AI

The nature of the signal picked up by Canberra WH698 over the Caspian in early April is still not entirely clear. By 1954 the Soviet Air Force had two airborne radar sets in service, both fitted in variants of the MiG-17. The first was the S-band SRD-1 (*Scan Fix*) range-finder, carried by the MiG-17F (*Fresco-C*). This was a very simple non-scanning radar, providing an indication of range to a target dead-ahead of the aircraft via a number of coloured lights mounted alongside the gunsight. The second equipment was the X-band RP-1 (*Scan Odd*) search and track radar, installed in the MiG-17P (*Fresco-B*): although of limited range, the RP-1 was a true AI radar. The fact that the Canberra was fitted with S-band search equipment for the follow-up investigation over the Baltic suggests it may have intercepted signals from the SRD-1 over the Caspian rather than the RP-1.

It may seem surprising that it had taken the Soviet Union up until 1954 to introduce AI radar equipment into service. In fact work had started on the development of AI radar in the late 1940s but development was complicated by the need to produce a system capable of installation in a small single-seat fighter with a nose engine intake. Work initially concentrated on a radar known as *Thorium*: a centimetric set with a sin-

The MiG-17P *Fresco-B* was the PVO's first all-weather fighter, equipped with the X-band *Scan Odd* AI radar.
via Yefim Gordon

67

LISTENING IN

The *Kite* radar proved to be unreliable, lacked range and was difficult for the pilot to operate, so its development was abandoned in favour of *Scan Odd* radar set that was subsequently fitted to the *Fresco-B*. *Author's collection*

The *Scan Odd* radar installed in a MiG-17P *Fresco-B* showing both aerials. The twin-dish search aerial (at top with back-to-back dishes) rotated in the vertical plane while the tracking aerial (centre) had a conical scan. *Author's collection*

gle dish, providing search and lock-follow facilities. Flight trials were carried out in a modified MiG-15 in 1949/50, with the radar installed in a large bulge above the engine intake, but unfortunately showed the radar was unreliable, lacked range and was extremely difficult for the pilot to operate, especially when changing from search to track. An improved, smaller, version of the radar, designated *Kite Hawk* exhibited the same failings and by 1950/51 attention had turned to a simpler AI radar known as *Emerald*, apparently developed as an insurance policy against the failure of *Thorium*. The *Emerald* radar used two aerials, with a rotating search aerial in a lip above the engine intake and a tracking aerial in a bullet fairing in the centre of the intake. More importantly, the set was easy to operate, providing automatic change-over from search to track once the aircraft had been positioned behind the target. Although lacking in range, *Emerald* was deemed superior to *Thorium* and was introduced into service as the RP-1 (NATO reporting name *Scan Odd*) in a new all-weather variant of the MiG-17.

Summer 1954 RPF programme

In mid-June 1954 Canberra WH670 was returned to 192 Squadron from the CSE Installation Flight following its Elint refit, allowing Canberra WH698 to be taken out of service for the previously-planned installation of the *Green Satin* Doppler navigation aid. The installation of *Green Satin* in a Canberra B.2 required, amongst other things, the fitting of a large plate aerial aperture under the wing. This was a difficult and expensive undertaking, outside the capabilities of the CSE, and the task was allotted to Boulton Paul (an EECo sub-contractor). WH698 was thus despatched from Watton to the Boulton Paul workshops at Defford in June 1954.

Canberra WH670 now became the operational aircraft and was readied for a new series of RPFs in late June. It appears that the Air Ministry had already made plans for another attempt at the Caspian Sea AI investigation held over from late May. By then however further intelligence seems to have been obtained on Soviet AI, probably from US and UK Elint ground stations, and the investigation had switched to the X-band. As part of the preparations for the Caspian Sea sortie, the Elint installation in WH670 was calibrated against a Meteor NF.14 night fighter fitted with an X-band AI.21 radar.

In the event the planned operation over the Caspian did not take place because in mid-June 1954 the RPF operations planned for the month were cancelled by the Air Ministry. This seems to have been due to a misunderstanding regarding political authorisation for the flight, the Lord De L'Isle and Dudley writing to CAS at the beginning of June: 'I am very glad to understand that you have instructed the Staff not to undertake any further flights of this kind in the area in question without your direct authority'. In reply CAS confirmed that he would not authorise any further flights without a resubmission of the plans to the Secretary of State.

Since plans had already been made for a Middle East detachment, it was decided to proceed instead with an alternative series of flights requested by HQ MEAF during the period 25th-30th June. The targets of these sorties were almost certainly Egypt, Israel and Syria. Canberra WH670 and Washington WZ967 were detached to Nicosia on the 24th June, accompanied by the support Washington (WW346) carrying spares and crew, and Varsity WJ940 (recently received as a support aircraft) carrying ground-crew. The Canberra was fitted for X-band radar search and the Washington for metric, S-band and X-band search.

The first operations were scheduled for 26th June but Washington WZ967 suffered an engine failure before take-off and did not fly. The Canberra took off and flew a 2 hour 30 minute day sortie from Nicosia, most probably along the northern and eastern borders of Syria, landing at Habbaniya in Iraq, carrying out a similar sortie on the return flight from Habbaniya to Nicosia later in the day. A second return trip to Habbaniya was attempted the next day but was aborted due to 'equipment failure', the Canberra returning to Nicosia after 1 hour 45 minutes. A final attempt the following day was only partially successful, the Canberra once again returning to Nicosia after 2 hours 40 minutes.

During the period of Canberra operations, Washington WZ967 had remained unserviceable at Nicosia due to engine problems. The support Washington, WW346, was despatched back to the UK and returned with a new engine on 28th June. After the defective engine was replaced the Washington carried out two day RPFs (*Brigand 1* and *Brigand 2*) in the first week of July. Both were relatively short sorties (3 hours 10 minutes and 2 hours 40 minutes respectively) and most probably went along the coasts of Israel and/or Syria. The Washington returned to the UK on 5th July.

Submission of the Summer Programme

At the end of July 1954 the Air Ministry proposed a further RPF programme including the Caspian flights held over from May/June. There is evidence to suggest these may have included incursions into Soviet airspace: so-called 'Penetration' sorties. In August the Secretary of State for Air (The Lord De L'Isle and Dudley), commenting on the proposed RPF programme, wrote: 'I presume that aircraft flying over the Caspian will be fitted with the additional navigation device which you mentioned in your last minute on this subject. If this has been done and the navigational accuracy of the aircraft so fitted has been satisfactorily tested – I agree to an approach to the Foreign Office'. The navigation device referred to was of course the *Green Satin* Doppler navigator, which was then in the process of being fitted to Canberra WH698.

By the end of August 1954 Foreign Office approval had been obtained for a series of nine RPFs in the Baltic, Black Sea and along the Greek side of the Bulgarian and Yugoslav borders while the more sensitive Caspian flights appear to have been held in abeyance pending the completion of the *Green Satin* installation in the Canberra. Permission was obtained from the Turkish authorities for the Black Sea sorties, but it is not clear whether the Greeks were also approached, and it is possible there may have been diplomatic problems since the sortie over Greece was eventually cancelled. Indecision regarding the programme had delayed authorisation and the Air Ministry were anxious to crack on 'Since we are now well into the period proposed for the first two flights'. The programme was submitted to the PM on 31st August, approval was obtained on 1st September, and Washington WZ967, accompanied by WW346 in support, left Watton for Nicosia the same day.

Unfortunately the detachment did not go well. Washington WW346 developed engine problems on the journey out and both aircraft diverted to Luqa. An engine change was required and WZ967 had to make a return trip to the UK to collect a new engine, then proceeding to Nicosia, leaving a ground team to carry out the engine change on WW346. Two Washington RPFs were flown from Nicosia over the Black Sea: a 7 hour night sortie on 5th September (Operation *Brigham*) and a 6 hour sortie the following night (Operation *Brindled*). No navigational problems were reported but the second sortie was 'not planned as ordered by Air Ministry', apparently due to corruption of a signal sent to the detachment. WZ967 also suffered an engine failure on the second sortie, returning to Nicosia on three engines. Inspection showed an engine change was necessary and WW346 (now repaired) was flown back to the UK to pick up a new power unit. Unbelievably, WW346 also suffered engine problems on the return trip and arrived at Nicosia on three engines. Washington WZ968 flew out with another engine and all three aircraft were back at base by 16th September.

Problems with the Wright R-3350 Duplex-Cyclone engines in the Washington had been an ongoing cause for concern at the CSE. The USAF rated the engines at a realistic, but short, 250 hours between major overhaul while the Air Ministry used a somewhat optimistic figure of 800 hours. Bomber Command and CSE experience suggested the US figure was the more accurate but Air Ministry were reluctant to reduce their life figure since it would 'overload

LISTENING IN

repair facilities'. The CSE continued to press for a more realistic figure, noting that: 'the nature of No.192 Squadron's operations make it imperative that no risks should be undertaken'.

During 1954, in addition to flying on the planned RPF programmes, 192 Squadron held a Washington in readiness for ad hoc Elint operations against Soviet naval units under Operation *Claret* (the successor to the earlier Operation *Reason*). Operation *Claret* was actually only one component of a larger naval shadowing exercise known as Operation *Grape*. This comprised three separate sub-operations: the previously-mentioned *Claret* (Elint operations by 192 Squadron), Operation *Moselle* (shadowing operations by Coastal Command aircraft), and Operation *Sherry* (photographic reconnaissance by Bomber Command aircraft). However, the Soviet Navy did not venture far from home port during the mid-1950s and the squadron had to wait until August for the first *Claret* operation of the year. Two sorties were flown over the North Sea, the first terminating early due to an engine problem, while the second lasted a marathon 16 hours.

In the second week of September 1954, HQ 90 Group submitted flight plans for the next two sorties in the summer 1954 RPF programme to the Air Ministry. These also appear to have been Black Sea sorties, the cover story being that they were investigations into the anomalous propagation of radio waves over Turkey and the Western Black Sea. By September however, the Air Ministry (probably at the behest of the Secretary of State for Air) had decreed that, due to doubts regarding the navigational accuracy of the aircraft, no further Canberra RPFs could be flown until *Green Satin* had been installed and, because of the restrictions on Washington operations, this effectively put the 1954 RPF programme on hold.

Green Satin was installed in Canberra B.2 WH698 by Boulton Paul at Defford between June and August 1954 using a similar scheme to that used in the Canberra PR.3. The fit involved the installation of the main *Green Satin* units in the rear fuselage, aft of the bomb-bay, the incorporation of an aerial unit under the starboard wing, and the installation of the *Green Satin*/GPI Mk.4 units in the rear cabin. Additional AC power for the system was provided via a new inverter mounted in the forward bomb-bay. The installation was inspected and cleared at a meeting at Boulton Paul on 19th August 1954 and, after some delay due to unserviceability, proving flights started at Watton at the end of October. In November a training flight was carried out to Idris, Libya for trials over tropical and desert terrain and by the beginning of December the preliminary *Green Satin* trials had been completed. The results showed that the combination of *Green Satin* and *Blue Shadow* (the latter providing fixes to correct the former) had a navigation error in azimuth of approximately 3% of distance flown, half the error obtained using *Blue Shadow* on its own and a valuable improvement in accuracy.

Unfortunately, shortly before the *Green Satin* trials were completed, the Air Ministry then decided to cancel operations for 'a variety of reasons other than the need for proved satisfactory operation of Green Satin'. The principal reason for deferring the resumption of RPFs seems to have been a proposal to transfer operational control of 192 Squadron to Bomber Command (see Chapter Seven).

The extension of the embargo on RPFs in December 1954 resulted in the *Green Satin* proving trials in WH698 being extended into 1955. A further overseas sortie was carried out to Khartoum in January, followed by a number of simulated RPFs to assess accuracy under operational conditions. The final report, delivered in February 1955, confirmed the accuracy figure of 3% in azimuth, using a technique in which a *Blue Shadow* fix was used to reset the GPI every 20 minutes.

Washington WZ966 in flight showing the under fuselage radomes and lack of gun turrets/sighting blisters. Although this is a monochrome image, the dark coloured areas seen on the fin and wingtips would have been red, carried over from the aircraft's USAF service.
via Paul Stancliffe

7 Transfer of Control

'This is a highly important method of obtaining intelligence. Last year's flights were especially fruitful in adding to our knowledge of Russian AI'
Lord De L'Isle and Dudley, May 1955

The transfer of operational control of 192 Squadron from 90 Group to Bomber Command had first been proposed in 1950 and then again in mid-1952 but met almost universal opposition from the various intelligence organisations. The subject of control of 192 Squadron was re-examined again in the second half of 1954, by which time the intelligence the squadron was collecting on the Soviet air defence system was seen as being of vital importance to the new V-bombers and therefore the future UK nuclear deterrent. As VCAS ACM Sir Ronald Ivelaw-Chapman noted to CAS: 'I am sure there is no other way in which we can obtain this crucial information, on which the success of our "V"-bomber sorties must so largely depend. It is our sole means of measuring the Russians' ground radar, their airborne radar, their VHF and UHF controls. We will be badly hampered in our RCM, unless we have the information that only these flights can produce'. Following representations from Bomber Command, another Working Party was convened by the Air Ministry to investigate the matter and visits were carried out to the CSE and various headquarters. At the beginning of December the Air Ministry announced that no RPFs (Canberra or Washington), other than Border flights, were to be planned or carried out until the question of the operational control of the squadron had been settled.

The linkage between Elint and the V-Force seems to have persuaded all concerned that it now made sense to transfer control of the squadron to Bomber Command. However, it was also clear that 192 Squadron remained heavily dependent on the CSE for technical support, and needed to maintain a close relationship with the scientific and installation units at Watton. In late January 1955 a compromise was reached whereby operational control of 192 Squadron was transferred to Bomber Command, but the squadron remained based at Watton.

Following the transfer of the squadron to Bomber Command, 90 Group drew up plans for a new series of RPFs and agreement on a provisional programme for May-December 1955 was obtained from the Air Ministry in March. However, the authorisation and conduct of RPFs now came under scrutiny and, as a result, the Air Ministry decided to delay the resumption of RPFs until that issue had been settled. Proposals were made in April by CAS for a modification of the way in which RPFs were planned and approved in order to ensure the 'success and safety' of the flights. The suggestion was for a two phase approvals process. In the first phase the general RPF programme, drawn up through 'normal Intelligence channels', would be approved by the Secretary of State for Air, then the Foreign Secretary and finally by the Prime Minister. The programme would be passed to Bomber Command where detailed planning of individual flights would take place. Once this had been completed, each flight would be subject to a similar chain of approval. Discussion of the RPF planning and approvals process resulted in the resumption of operations being pushed back until June 1955.

Border and Claret

The Washingtons continued to be flown on weekly Border sorties in the ten-month embargo period between August 1954 and June 1955. Flown from Watton, these normally comprised a 6 hour day sortie along the East German border. A change in the weekly Border flights occurred in January/February 1955 when both the US authorities and the Air Ministry agreed to a 90 Group proposal to extend them south into the American Sector and this revised track was flown from February onwards.

A further *Claret* operation was also carried out in April 1955, and a detailed account serves to illustrate the difficulty of these operations. Intelligence was received on 7th April that the

LISTENING IN

Valiant, Vulcan and Victor – Britain's V-Force carried the nuclear deterrent and had, by the time they entered service, become increasingly dependent on the intelligence supplied by the RAF's Elint operations.
via Terry Panopalis

Soviet Auxiliary *Neva*, proceeding from the Black Sea to the Baltic, had cleared Cape St Vincent, off the southern tip of Portugal, and was heading north. 192 Squadron was notified the same day that a *Claret* RPF should be carried out against the ship as soon as it came within range. Political approval for the sortie was obtained the next day (8th April), by which time the ship was half way up the Portuguese coast. A Washington sortie took place in the afternoon but unfortunately the aircraft had engine problems and was forced to return to base without making contact with the target. A shortage of Washington crew (it was Friday night) prevented a replacement aircraft being despatched and it was agreed to fly a second sortie the next day. It then transpired that new political approval would be required for a second sortie, which was rescheduled for the Monday. Accordingly a Washington departed Watton at 1717hrs on Monday 11th April for a second attempt at intercepting the *Neva*, which had now passed through the English Channel and entered the North Sea. This attempt was marred by a breakdown in communications between Naval Intelligence and the Air Ministry. The *Neva* was a Soviet Auxiliary ship and the Air Ministry had been referred to the appropriate page number in the standard naval ship identification book. Unfortunately no copies of the book were held by the Air Ministry department concerned so the Washington crew set out believing their target was a destroyer-class warship rather than a depot ship. A grey merchant ship was spotted, but was ignored as not fitting the bill. Only one fleeting S-band signal was picked up. The Washington finally landed at 0042hrs on 12th April after a fruitless search. The misunderstanding regarding the target was then discovered and details of the *Neva* rushed to Watton. A second sortie was flown later on 12th April over the North Sea and this time the ship was positively identified (and determined not to be the ship spotted the previous day) but no radar signals were picked up. Time was now running out, as the *Neva* was thought to be heading into the Baltic and *Claret* operations were prohibited once the ship had entered the Skagerrak. Luckily, as it turned out, the ship was actually heading for northern waters and a fourth and final sortie was authorised for 15th April, by which time the *Neva* was on passage northwards midway up the Norwegian coast. The crew of Washington WZ967 (Flt Lt G Wellum and crew) were warned not to make visual contact until the required Elint had been obtained. This time metric (572MHz, PRF 400Hz, 3.5μs pulse-width, 4rpm sweep rate) and S-band (3.05GHz, PRF 830Hz, 1μs pulse-width, 7.5rpm sweep rate) signals were intercepted during the 13 hour sortie but no visual confirmation of the target was possible due to heavy cloud.

A second *Claret* operation followed in mid-May 1955 against a Soviet naval force of two *Sverdlov*-Class cruisers and four destroyers on transit between the Skagerrak and the Lofoten Islands, en-route to the North Cape. Under Operation *Moselle* the ships were shadowed by Lockheed Neptunes of Coastal Command, the aircraft using their AN/APS-20 radar to maintain contact. Three *Claret* sorties were carried out by 192 Squadron Washingtons from Watton with the final flight in the series being a mammoth 20 hour sortie causing Washington WZ968 (Flt Lt D Comer) to land back at Kinloss low on fuel. This was a very successful operation, intercepting a 'considerable volume of Soviet signal transmissions'.

Improving the Elint Fit

The inaccuracy of direction finding (D/F) bearings taken by both Canberra and Washington aircraft was a constant cause for concern to 192 Squadron and so in the late summer of 1954, taking advantage of the lull in operational flying, serious attempts were begun to try to improve the D/F accuracy of the Washington. The first step was a partial redesign of the ARI 18021 S-band aerial, to improve coverage, and the fitting of bias boxes, presumably to correct known errors in the D/F accuracy. These modifications were applied to Washington WZ967 in the autumn of 1954 and resulted in a general operational improvement. As a second step improved APA-17B D/F sets were installed in all three Washingtons during late 1954. After some problems with missing parts, requiring the manufacture of replacements by REU Henlow, the installation was completed in early 1955. The modifications were followed by an intensive calibration programme in early 1955 during which 67 hours were flown calibrating ARI 18021 at S-band in one aircraft. Comparative centimetric D/F trials were also conducted using APA-17B and ARI 18021 to evaluate their respective accuracy. The results of these trials made it possible for the first time to effectively assess the Washington as a D/F platform. One outcome was a decision to use APA-17B on all future operations where accurate D/F was required, relegating the ARI 18021 (*Flange*) to the intercept role. Another outcome of the trials was the introduction of new operating techniques designed to maximise D/F accuracy. At the same time extra effort was devoted to Special Operator (SO) training and by mid-1955 the combination of equipment improvements, new techniques and intensive training were producing significant improvements during 'routine RPFs' (that is, Border flights over West Germany). Following the success of the S-band work, a similar programme to improve range and D/F at X-band was conducted in late 1955.

The Elint fit in the Canberras was also subject to improvement during the RPF embargo period. During November/December 1954, while the *Green Satin* trials were in progress, the Elint fit in WH698 was revised. A QRC-7 wide-open VHF receiver was installed and the ANQ-1A wire-recorder supplemented by a more sensitive EMI magnetic tape-recorder. The QRC-7 was a US wideband receiver covering the entire Soviet GCI VHF band. Because of its wide coverage the receiver could not be monitored in the air but instead had its output fed directly to an ANQ-1A recorder for subsequent analysis on the ground. The new EMI tape-recorder was installed specifically to record AI signals, as the ANQ-1A was not sensitive enough to satisfactorily record these weaker signals. A modified *Flange* tail aerial array was also fitted which gave a much wider angle of detection, providing better warning of interception by Soviet fighters.

In February 1955 the aircraft was detached to Wünstorf for Exercise *Wild Goose* in conjunction with Fighter Command. This was a series of simulated RPF exercises flown over the North Sea with three sorties flown in the period 22nd-25th February. The exercise apparently provided much useful information to both 192 Squadron and Fighter Command with the fighter crews reporting that the Canberra was a very difficult target and only two crews claimed a possible kill. A similar exercise, Exercise *Wild Duck*, involving the Canberra and two of the squadron's Washingtons, was carried out in April. These exercises also involved simulated RPF sorties, flown by WH698 with the Washingtons monitoring attempted interceptions by fighters and were clearly rehearsals for future RPFs. The new fit suggests the Canberras were now targeting Soviet night-fighter AI transmissions. Presumably the intent was that VHF traffic between GCI and fighters picked up on the QRC-7 could be correlated with any AI signals recorded on the sensitive new tape-recorder. The exercises may also indicate new thinking on the conduct of RPFs since, prior to 1955, most Canberra RPFs were flown unaccompanied while during and after 1955 they were more often than not flown in conjunction with a Washington.

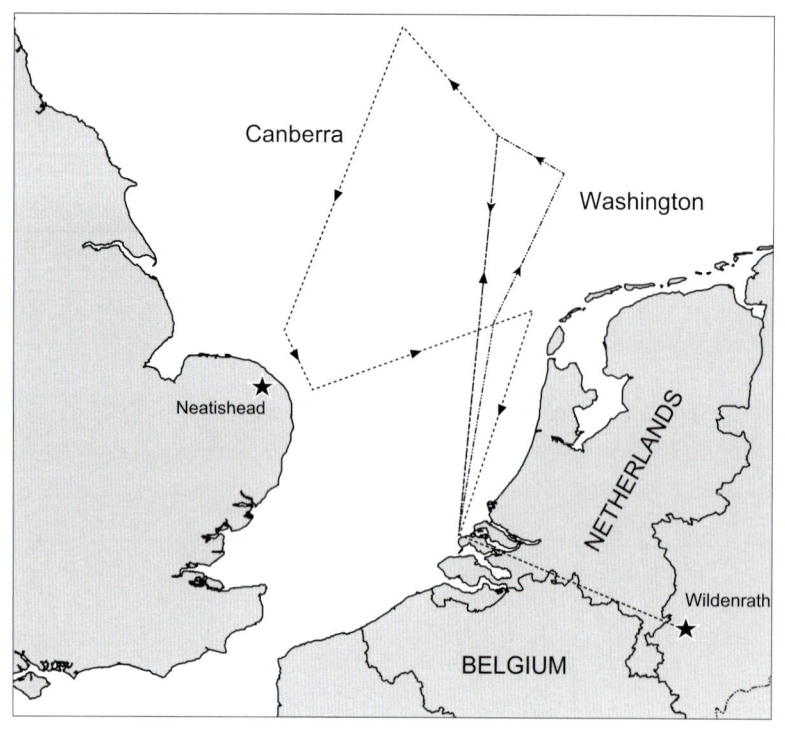

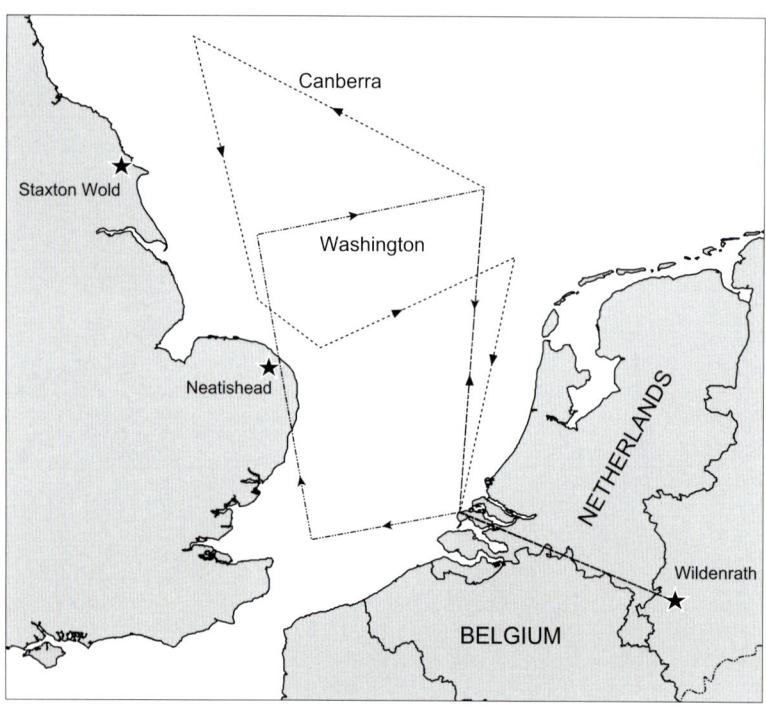

LISTENING IN

and US Navy'. In his submission Lord De L'Isle noted: 'This is a highly important method of obtaining intelligence. Last year's flights were especially fruitful in adding to our knowledge of Russian AI'.

The relatively large number of flights in the new programme caused some anxiety in government, with the Prime Minister concerned that there should be no increase over the previous year. In response the Secretary of State pointed out that the previous year's programme, although not completed, had comprised some 27 flights, and also stressed the importance of the Elint programme as a whole. The flights were: 'our prime source of accurate and up-to-date information about (i) the organisation and efficiency of Soviet air defences, (ii) the order of battle of the Soviet Air Force, (ii) Soviet radar counter measures, (iv) development in the technical field, particularly with guided weapons'. The major objectives of the programme were listed as: the investigation and assessment of Soviet AI; the investigation of the Soviet C&R system in order to assess day, night and all-weather fighter defence capability; the interception of VHF and HF point-to-point ground communications on which the Soviet Air Force depended for its organisation and operations; the investigation of Soviet RCM; the detection and assessment of new forms of Soviet ground radar; the investigation of new Soviet radar interception equipment on the Austrian border; and the investigation of the Soviet guided weapons programme. As regards the latter objective, the Secretary of State noted that: 'The whole field of Soviet guided missile activity is clothed in mystery. One of the best methods of gaining Intelligence is by means of Radio Proving Flights. Areas of activity are known within certain limits and many are within the range of aircraft flying beyond danger from the Soviet frontier. Scientific investigation of the Soviet guided missile programme can only be advanced by the interception of range procedures, control frequencies and radar transmissions from these areas'. The Air Ministry's reassurances were accepted and the Prime Minister approved the programme in July.

The first operation in the new programme was flown on 14th July 1955 when two Washingtons (WZ967, WZ968) were detached to Luqa, Malta, carrying out a daylight RPF over the Adriatic en-route. The primary aim of the 7 hour 30 minute sortie was an investigation of Albanian radar defences. A night operation was flown up the Adriatic three days later on the return flight to Watton, possibly including a detour along the Bulgarian border.

Operations Resume: the 1955 RPF Programme

In May 1955 the Secretary of State for Air requested authorisation for a resumption of RPF operations, following the transfer of control of 192 Squadron to Bomber Command. A programme of 14 operations was put forward, comprising 22 flights between June and December 1955, covering the Baltic, Adriatic and Black Sea, 'co-ordinated with similar flights by the USAF

74

Two Washingtons of 192 Sqn in flight over East Anglia, photographed from a third Washington. The radomes arrayed under the fuselage can just be seen on these aircraft. via Paul Stancliffe

In August two Washingtons (WZ967, WZ968) and Canberra WH698 were detached to Nicosia for operations over the Black Sea. On 22nd August Washington WZ967 and Canberra WH698 carried out Operation *Calomel*, investigating Soviet radar defences and 'stimulating reaction from ground stations' in the Eastern Black Sea. Some caution was exercised with the Canberra only approaching to within 55nm (100km) of the Soviet coastline while the Washington loitered 70nm (130km) offshore. This was the first Canberra operation since June 1954 and apparently went well with performance of the *Green Satin* Doppler in the Canberra declared to be 'most successful'. A similar joint Washington/Canberra operation was flown the following night but on that occasion, turbulence made astro-navigation from the Washington difficult, with the navigators unable to obtain a fix on the Georgian coastline, and the aircraft was forced to return slightly earlier than planned.

September saw two joint Canberra/Washington night RPFs into the Baltic, investigating radar defences along the Soviet coast and looking for indications of Soviet AI. Dates for the sorties were chosen to take advantage of 'no moon' conditions and avoid clashing with similar US operations in the Baltic. The Washington operated from Watton while the Canberra (B.2 WH698), used Gütersloh as a forward base, Wünstorf no longer being considered suitable. As this was the first RPF in the Baltic since April 1954 the Canberra, fitted for X-band search, approached no nearer than 60nm (111km) to the Soviet coast, while the Washington was employed in the 'stand-off' role and remained well clear of Soviet territory. Although a special equipment failure resulted in an early return by the Washington, the Canberra successfully completed its part of the operation. After a delay due to poor weather both aircraft (WZ967, WH698) successfully flew a second operation the following week, the Canberra once again operating from a forward base at Gütersloh.

In October two Washingtons (WZ966, WZ967) and Canberra WH698 returned to Nicosia for another series of RPFs over the Black Sea. The detachment did not proceed smoothly and operations were disrupted by all-too-familiar Washington engine problems, WZ967 being declared unserviceable shortly after arrival at Nicosia for an engine change. An attempt was subsequently made at the first RPF on 9th October using WZ966 and Canberra B.2 WH698 but the Washington had to abort the sortie 15 minutes after take-off due to an engine fire and the operation was abandoned. Arrangements were made for a Hastings to ferry two replacement engines from the UK. Meanwhile Canberra WH698 flew the first RPF on 13th October on its own. Washington WZ967 became serviceable the following day and flew a second night RPF in conjunction with the Canberra the same night. The Canberra approached to within 45nm (83km) of the coast while the Washington stood-off 75nm (139km).

November saw another detachment from Gütersloh for two operations over the Baltic during which WH698 was flown to within 30nm (56km) of the Soviet coast. Previous operations had skirted Danish airspace but for this (and future) operations the Foreign Office arranged 'informal' overflight clearance with the Royal Danish Air Force, making navigation easier.

Canberra B.6s

The last operation in the 1955 RPF programme was a dual Canberra operation from Nicosia over the Black Sea using Canberra B.2 WH698 and a new Canberra B.6 (BS) WT301. Before

Opposite page: Exercise Wild Duck involved practice RPF sorties by a Canberra and Washington over the North Sea to develop techniques for future RPFs. Fighter Command aircraft made interception attempts.

During Exercise Wild Duck 2 the Canberra was allowed to take evasive action to avoid interception by night fighters.

describing this last operation the introduction of the Canberra B.6 into 192 Squadron needs to be described.

During 1954 a plan had emerged to replace the three Washingtons and two Canberra B.2s of 192 Squadron with four Shackleton MR.1s and three Comet 2Rs (see Chapter Eight for the full story). By the autumn of 1954 however, financial restrictions and the Comet 1 accidents had led to the Comet order being reduced to a single aircraft. There were also concerns that the Comet might not be able to replace the Canberra on all operations as one of 192 Squadron's tasks was to carry out 'Penetration' RPFs, intruding into enemy airspace. As a matter of fact no such sorties had ever been authorised and the task was really a wartime commitment. Nonetheless the task, however notional, existed and there were doubts regarding the ability of the Comet to carry it out. In October 1954 the Air Ministry commented: 'There is, however, considerable doubt as to whether the Comet is satisfactory for penetration of "enemy" territory. If not, the Canberras will need to be retained'. By November agreement seems to have been reached that the Canberras would be retained. Unfortunately, the Canberra B.2 was not ideal for incursions into hostile airspace either, being somewhat short on range. As the same Air Ministry official noted: '[the Canberra's] duties entail penetration of 'enemy' territory to a greater range than is available from the Canberra B.2. Attempts to overcome this shortcoming included the use of wingtip fuel tanks and the fitting of special bomb bay fuel tanks. The former restrict the aircraft performance and the latter are impracticable'. The short range of the Canberra B.2 resulted in a proposal in late 1954 to replace it in 192 Squadron service with the longer-range Canberra B.6. In fact the Canberra B.6 had a couple of advantages over the B.2: it carried more internal fuel and thus had longer range, and it had a cabin air conditioning system, useful for operations in hot climates.

The proposal was not without complications however. The Canberra B.6 had the same navigation limitations as the B.2 and could not be used on RPFs without first being fitted with the *Blue Shadow* and *Green Satin* navigation aids. The problem was that, due to the rear cabin layout, *Blue Shadow* could not be retrofitted into a standard Canberra B.6, and could only go into aircraft specially modified on the production line. The RAF had ordered a small number of B.6 (*Blue Shadow*) aircraft to supplement the standard Canberra B.6 and the 192 Squadron B.6 aircraft would have to come from this *Blue Shadow*-specific production. Despite these complications, and after remarkably little argument, the Air Ministry concurred and plans were made to replace the two Canberra B.2s with two Canberra B.6 (BS). Agreement was also reached that one of the B.2s could be retained as a training aircraft.

As a result of the above negotiations, two Canberra B.6s, WJ775 and WT301, were delivered to 192 Squadron in December 1954. The intention was to fit the two B.6s to the same navigational and Elint standard as the 'operational' B.2, WH698, that is, *Blue Shadow/Green Satin* and APR-9, ARI 18021 and APA-11. Plans had also been made to fit both B.6 aircraft with an improved nose radome, based on a Canberra T.4 nose, to provide a better view for the ARI 18021 *Flange* D/F aerials. However, because the first aircraft (WT301) was urgently wanted for operations it was decided to fit it with the same 'short' nose radome fitted to the B.2s, in order to expedite its entry into service. Since the aircraft special equipment was in short supply the plan was to transfer the radio, navigation and Elint equipment from the B.2s to the B.6s. However, the squadron needed to retain an operational Canberra during the changeover to B.6s. As a result it was decided that the first B.6 (WT301) would be fitted with equipment removed from B.2 WH670, which would then be disposed of. The second B.6 would be fitted out with new equipment. This would leave Canberra B.2 WH698 providing an operational capability during the transition. Following the refit of both B.6s, WH698 would be retained for training and as an operational backup.

WT301 left Watton for Boulton Paul at Defford at the end of December 1954 for the installation of *Green Satin* as the first step in its conversion programme. The second B.6, WJ775, was fitted out at Watton with a temporary installation of Gee Mk.3, Radio Compass and Periscopic Sextant for long-range overseas flights, to acclimatise Canberra crews to long-range work at extreme altitude.

Canberra WT301 returned to Watton at the beginning of March 1955 after *the Green Satin* installation, ready for the main Elint installation programme under SRIM 1729. The task involved the installation of Twin TR1934 VHF with whip aerial, a QRC-7 wide-open VHF receiver with whip aerial, Gee Mk.3, *Blue Shadow*, ARI 18021 with forward and rear aerial systems, Squint control box, and Frequency Selector Unit Type 15, APR-9 receiver, APA-11 analyser, with provision for R.1949 as alternative fit, ANQ-1A wire recorder and *Blonde Recorder*. The Canberra was also fitted with a small nose radome, replacing its Perspex nose. All the above equipment (excluding *Green Satin*) was transferred from WH670. The conversion was completed at the end of August 1955 and, after a few adjustments, WT301 was finally returned to 192 Squadron in late October 1955. The second B.6 (WJ775) then went to Boulton Paul for installation of *Green Satin* and the T.4-style 'long nose' radome in June 1955.

The first B.6 RPF in WT301 took place in December 1955 from Nicosia when, together with the B.2 WH698, the aircraft carried out an investigation into radar defences in the Eastern Black Sea and along the north eastern Turco-Soviet border areas, testing the Soviet reaction to high-level flights in the area over two nights. The sorties had originally been scheduled to take place from Habbaniya, Iraq, but that base had been transferred to Iraqi control in May 1955 and, although Habbaniya was still used by the RAF, the sensitivity of the operations may have led to the operation being relocated to Nicosia. Both Canberras were fitted for X-band search with the B.6 also carrying QRC-7 to record VHF traffic. This operation was noteworthy in being the first conducted by two Canberras. These two RPFs were also the last of the August-December 1955 RPF programme.

In a subsequent summary of the programme the Secretary of State noted that: 'There were no incidents, territorial infringements or indications of any attempt to intercept our aircraft. Preliminary results show that a considerable amount of information, which could not be obtained by other means, has been acquired'.

Although the B.6 had been declared operational, there were some concerns about the performance of its navigation and Elint equipment and attempts at improving the performance of both were to continue throughout the year.

RPFs January-June 1956

At the beginning of 1956 192 Squadron had three Washingtons, a Canberra B.2 (WH698), and a Canberra B.6 (WT301) available for operations, and a second Canberra B.6 (WJ775) was under modification for the Elint role.

The proposal for the next RPF programme, covering January-June 1956 and comprising 13 operations and 36 sorties, was submitted for approval in late December 1955. The increase in the number of sorties was due to a number of the proposed operations being flown with two Washingtons and one Canberra. Previous joint operations had involved a single Washington and a single Canberra. Although the Washington had a longer range than the Canberra it was much slower. Therefore the size of the operational area that could be covered on a joint sortie was limited by the Washington's speed and the Canberra's endurance. Using two Washingtons in conjunction with a single Canberra would allow the operational area to be extended, each Washington covering one half of the area.

The first two operations of the new series took place in mid-January 1956 and involved two Washington 'working-transit' RPFs enroute to and from Malta. The first sortie (*Shingle 1*) took place on 16th January and went along the Czechoslovakian border with Germany, then out over the Adriatic, running down the Albanian coast, before landing at Luqa. The planned operation from Malta (*Shingle 2*) was cancelled. The slightly longer return trip to Watton (*Shingle 3*) on 18th January was another working transit, flying along the Greek borders of Bulgaria, Yugoslavia and Albania. A great deal of the return flight was carried out in cloud, without visual contact, and was described as 'less interesting' than the trip out. The second January operation took place on the 16th when Canberras B.2 WH698 and B.6 WT301 flew a co-ordinated joint night RPF (*Frigid 1* and *2*) from Gütersloh into the Baltic looking for AI and X-band radars.

February 1956 saw the very first 192 Squadron RPF into the Barents Sea examining Soviet defences in the far north. The principle targets were the the Kola Peninsula, including the important naval base of Murmansk, and the island of Novaya Zemlya and its weapons testing ground. Preparations for the sortie had begun back in November 1955 when the navigators received training in Grid Navigation techniques appropriate for high northern latitudes. Arrangements were also made to use the Norwegian airfield of Bodø as an emergency diversion/refuelling stop in case of problems. At the end of January Wg Cdr Dixon, OC 192 Squadron, visited the airfield in Washington WW346 to check on ATC and other facilities. The general impression was quite good and the suitability of the airfield for use as a diversion was confirmed. The crew seemed to have been impressed with the catering arrangements, commenting that: 'A lunch stop proved that life within the Arctic Circle could be as advanced as in temperate regions'.

On 9th February Washington WZ966 (Wg Cdr W M Dixon), fully-loaded with fuel, took off from the long runway at Marham on the first Barents Sea RPF (*Gentry 1*). The aircraft flew up the coast of Norway and around the North Cape into the Barents Sea, flying along the coast of the Kola Peninsula. Extreme caution was exercised and the aircraft stood off 90nm (167km) from the Soviet Coast. The Washington successfully returned to Watton, without stopping at Bodø, completing the sortie in a marathon 18 hours 10 minutes, 13 hours 30 minutes of which were at night. The loss of one engine and the failure of the aircraft's heating system in Arctic conditions made the trip especially arduous. The squadron OC remarked 'This flight represents a big advancement in 192 Squadron operations'. The sortie was very successful, obtaining the first evidence of an early warning radar in Novaya Zemlya and new information on navigation beacons on the Kola Peninsula. A signal was subsequently received from Senior Air Staff Officer, (SASO) Bomber Command, Air Vice-Marshal Sydney Bufton, personally commending the crew for the results obtained.

One intention of the operation seems to have been to prove that Washington sorties into the Barents Sea were possible from the UK, and would not necessarily depend on the use of a Norwegian base and thus approval by the Norwegian government.

Also on 9th February a joint Washington/Canberra night RPF (*Gentry 2-4*) was carried out over the Baltic using two Canberras (WH698 and WT301) in company with Washington WZ968. The two Canberras, fitted for X-band radar and VHF communications search, positioned at Gütersloh prior to the sortie, supported by Varsity WJ940. The Varsity had spent nearly all of 1955 being refitted as a dual role transport/mobile radio workshop aircraft and this was its first outing following the modification work. The Washington (fitted for HF, VHF, metric radar, S-band and X-band search) operated from Watton. In common with the majority of the joint Washington/Canberra RPFs, the role of the Canberras was 'mainly to stimulate reaction' while the Washington stood off and recorded the subsequent radar and radio traffic. Both the Washington and Canberra B.6 returned early, apparently due to a power failure in one aircraft affecting navigation equipment.

On 16th February a further joint Washington/Canberra sortie (*Gentry 5/6*) was flown over the Baltic using Washington WZ968 and Canberra B.6 WT301. Once again the Canberra was fitted for X-band and VHF search while the Washington carried a comprehensive search and D/F fit.

March saw a detachment of both Canberras (B.2 WH698 and B.6 WT301) and two Washingtons (WZ966 and WZ967), supported by Varsity WJ940, to Nicosia for two night operations over Black Sea. The initial attempt at the first operation, two RPFs over the Eastern Black Sea involving a single Washington and a Washington and two Canberras on 2nd March, was abortive due to unserviceable navigation equip-

Washington WW346 at Bodø in northern Norway during a 1956 detachment. The aircraft operated in the support role for most of 192 Sqn overseas operations. The crews of 192 Sqn, and successors, were assured of a warm welcome from the Royal Norwegian Air Force at Bodø. *via Paul Stancliffe*

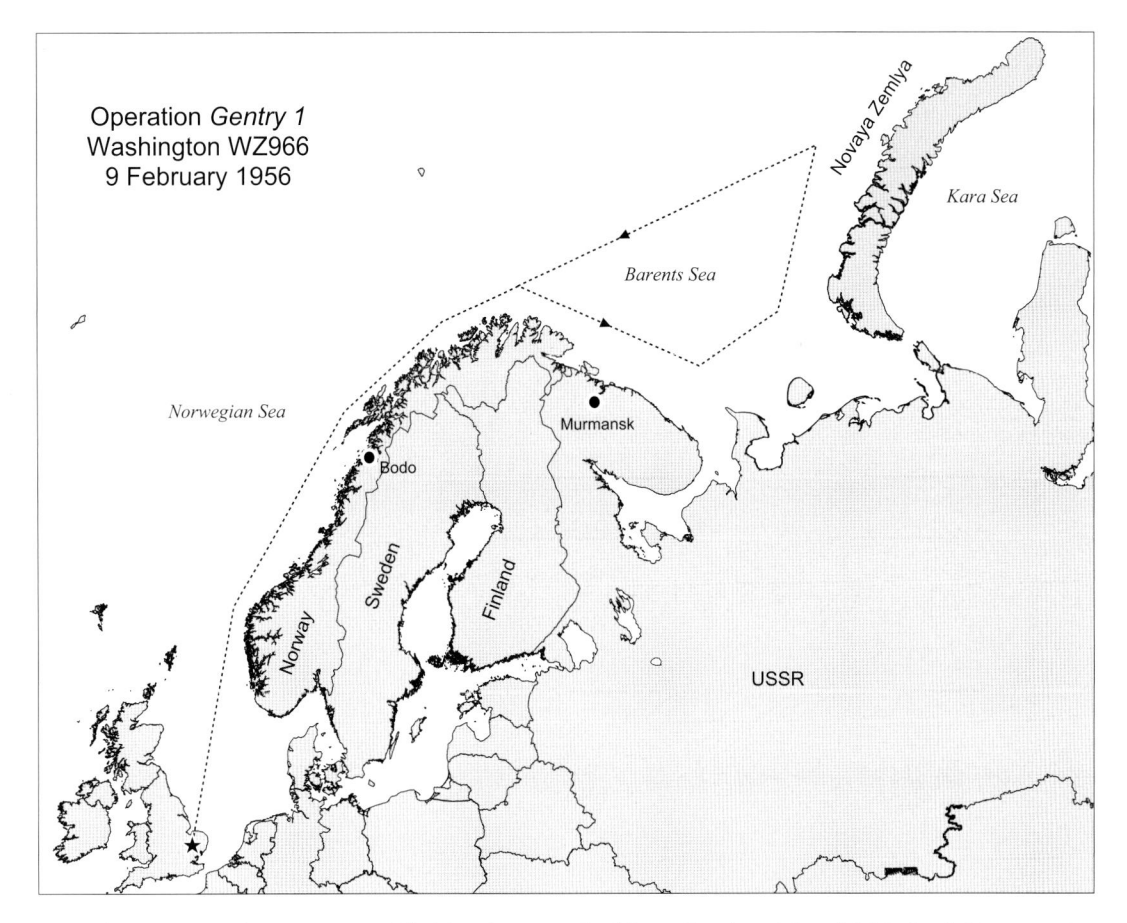

Operation *Gentry 1*
Washington WZ966
9 February 1956

The approximate track of the first sortie into the Barents Sea by an RAF Elint aircraft. The flight covered military bases on the Kola Peninsula and the island of Novaya Zemlya.

ment but a second attempt on 4th March was more successful with all four aircraft completing their sorties. During the second operation, carried out the next day, Canberra B.6 WT301 and the two Washingtons operated over the Central Black Sea, the Canberras approaching no closer than 30nm (56km) to the Soviet coast while the Washingtons stood-off at least 70nm (129km). During the RPF there was evidence of what appeared to be an unsuccessful attempt at intercepting one of the Canberras, although it was remarked that 'the fighter was positioned badly because of inaccurate height reporting and poor ground control technique'.

April 1956 saw a goodwill visit to the UK by the Soviet Premier, Nikolai Bulganin and the leader of the Soviet Communist Party, Nikita Khrushchev. No RPFs along Soviet borders were scheduled for April, probably in deference to the Soviet visit. The Foreign Office normally embargoed RPFs during the periods of official visits to minimise the chances of embarrassing incidents. Operations were however flown against Tupolev Tu-104 *Camel* airliner flights bringing elements of the Soviet delegation to the UK in April and a *Claret* seems to have been carried out against the *Sverdlov*-class cruiser *Ordzhonikidze* bringing Bulganin and Khrushchev to the UK in mid-April. The visit did not however pass off without incident. On the night of 19th April Commander Lionel Crabb, a famous ex-RN diver, entered the waters of Portsmouth harbour in an MI6-sponsored attempt to inspect the hull of the Soviet cruiser berthed there. He never returned. The UK government was forced to reveal the disappearance of Commander Crabb later in the month and a subsequent Soviet statement that a diver had been seen near the *Ordzhonikidze* on 19th April led to much speculation in the press as to Commander Crabb's fate, and considerable embarrassment for the government.

The last flights in the January-June 1956 programme were carried out in May and consisted of three joint Washington and Canberra operations from Nicosia into the Eastern Black Sea, along the Soviet Caucus border with Turkey and Iran, and along the Iranian coastline of the Caspian Sea to 50°East. These were the first flights in the Caspian Sea area since April 1954, reflecting the sensitivity of the region, and the consequent difficulty in obtaining political approval. The flights were also difficult to clear diplomatically, as agreement had to be obtained from both the Turkish and Iranian governments to allow overflights. The detachment (Washington WZ966, Canberra B.2 WH698 and B.6 WT301) supported by Washington WW346, positioned at Nicosia at the end of

LISTENING IN

One of the 192 Sqn Elint Washingtons taxying in at Watton after a long Radio Proving Flight. These operations took a heavy toll on the aircraft, particularly the engines, and led to a review of the overhaul schedule. *via Paul Stancliffe*

April. An attempt at the first operation, involving all three aircraft, was made on 2nd May but was aborted due to problems with the APQ-13 radar in the Washington. A second attempt the following day was more successful, with all three aircraft completing their tasks. A singleton Canberra RPF was flown in WT301 on 6th May, followed by a joint Washington/Canberra RPF on 9th May using WZ966 and WT301. On both these latter sorties the Canberra landed to refuel at Habbaniya before returning to Nicosia, highlighting the distance between Cyprus and the operational area. The final operation carried out by the detachment was a 6 hours and 20 minute day sortie by WZ966 on 11th May, possibly off the coasts of Egypt and Syria.

At least one of the Canberra sorties in this series went out over the western Caspian Sea. The aircraft was tracked both by Soviet radar stations and, indirectly, by 276 SU at Habbaniya. The Soviet Union maintained a chain of mobile Early Warning radar stations along the Caspian coast, controlled by a smaller number of sector control stations. As the Canberra progressed up the Caspian Sea the radar stations reported the aircraft's position, encoded in five-letter code groups, to their sector control station via HF W/T. The position reports were intercepted by the 'Russian Radar' team of SOs at 276 SU, each operator monitoring three radar stations. The Soviet position code had previously been broken by GCHQ, allowing the SOs to decode each report and maintain a real-time plot of the aircraft's location. One ex-SO recalls the track of the Canberra being plotted on a large map in the intercept room. The accuracy of navigation came under particular scrutiny during these sorties, due to the sensitivity of the area, but the standard was subsequently declared to be 'eminently satisfactory'.

The Commander Crabb affair in April had caused the government considerable embarrassment and measures were taken to avoid any similar incidents. In May the Prime Minister (Anthony Eden) minuted the Secretary of State for Air, noting: 'Since the frogman episode, we must take all possible precautions to avoid the recurrence of any further incidents of a similar kind. I should

The Yakovlev Yak-25M *Flashlight*, the Soviet Union's first true all-weather fighter. The large radome housed a *Scan Three* AI radar and with a second crew member to operate the radar, the *Flashlight* was more effective than the *Fresco-B*. *via Yefim Gordon*

therefore be grateful if you would look at the radio-proving flights again. The essential is to make sure that the aircraft keep well away from Soviet territory so that there is no risk of an incident, particularly at the present time'. The Black Sea RPF operation planned for June was cancelled at the request of the Prime Minister and Foreign Secretary while a review of intelligence operations was carried out and as a result only Border sorties were flown that month.

Following the conclusion of the January-June 1956 RPF programme the Air Ministry reported to the Secretary of State for Air that: 'The Soviet early warning and raid reporting organisations have reacted to all flights and fighter activity has been greater than on previous occasions. No "incidents" have occurred. Considerable information on Soviet communications and air defence radar has been obtained. In addition to the valuable intelligence gained by the aircraft, the increase in communication activity occasioned by the flights has provided our ground intercept stations with material which has considerably improved our knowledge of the Soviet air warning organisation'. Amongst the specific items of intelligence obtained were:

- first evidence of an early warning radar on Novaya Zemlya and new information on navigation beacons on the Kola Peninsula.
- new information on links between raid reporting organisations in Poland and the Soviet Baltic areas.
- identification of a loss of track continuity to the Soviet air defences north west of Gotland.
- clarification of certain raid reporting elements in the Crimea and changes in reporting procedures in the Tiflis (Georgia) area.
- identification of inaccuracies in height reporting and poor quality of ground control in the Black Sea affecting their ability to carry out interceptions.
- indications that a sudden loss of height will result in loss of track and that Soviet height-finding is poor.
- evidence that the Soviets have fighters equipped with limited-search AI in all areas visited.

By the end of the December 1955-June 1956 programme the squadron's aircraft were regularly picking up AI radar signals during their sorties, and there was evidence of one unsuccessful attempted interception during March. However, despite the existence of AI-equipped fighters, the Canberra remained pretty much safe from interception as most of the fighters were single-seat MiG-17P *Fresco-B* fighters fitted with the rather limited-range *Scan Odd* AI. Intelligence suggested the Soviets had also developed a true night-fighter, the two seat Yakovlev Yak-25M *Flashlight,* fitted with a longer-range AI radar (*Scan Three*), but there was no evidence that it had yet entered squadron service. The RPF sorties were always flown in total darkness forcing any interceptor to rely on GCI control to get him close enough to the Canberra to acquire it on his AI set. Since the Canberra presented a smallish, high-level and high-speed target this posed considerable problems for the Soviet air defence system, which still had difficulty tracking targets above 35,000ft (10,668m).

In fact, the Air Ministry had such confidence in the Canberra's ability to evade interception that it suggested in May 1956 that the Canberra be allowed to approach to within 15nm (28km) of Soviet satellite borders when over friendly territory, rather than the 30nm (56km) limit then in force. A further two requests were made at the same time: the first being to introduce more flexibility in the six-monthly RPF programmes by allowing the Intelligence Staff to change operational areas in response to changing requirements, while the second was to allow operations in conditions up to first quarter moon. RPF operations were normally only flown in the ten-day no moon period of each month. Since the Soviets were aware of this scheduling restriction it raised the possibility of radar systems being hidden by being switched off for the few days of no moon or, even worse, of spoofing operations being run during the same period. Extending RPF flights to the dark period from last to first quarter (covering 15-20 days each month) would make these sort of deceptions much more difficult to achieve. The Operational Research Branch of Bomber Command had investigated the probability of detection of the Canberra at night in early 1956, examining all available data on the *Scan Odd* AI radar, current exercise and post-war operational information on interceptions, and data on the visibility of the Canberra under various conditions. The results were two probability curves for detection by AI and visual means. These showed there was a 50% chance of detection by Soviet AI at 3½ miles, falling to 10% at 6 miles. In respect of visual detection there was virtually no chance of the Canberra being detected at ranges of a mile or over under starlight-only (no moon) but a 10% chance under full moon conditions. Contrails were however visible over quarter moon conditions. The conclusion of the memorandum was that

LISTENING IN

'it is therefore essential to fly under less than quarter moon conditions if one wishes to keep the risks down to an absolute minimum'. In the end the furore regarding the Crabb incident resulted in the proposals being shelved for consideration at a later date.

RPFs July-December 1956

Proposals for the next RPF programme, covering July to December 1956, were submitted by the Air Ministry in July. However there still appears to have been some fall-out from the Crabb affair, with the Foreign Office objecting and suggesting a six-month moratorium. The Air Ministry protested, pointing out that RPF operations had only just recovered from the ten-month gap in 1954-55 and that 'the present is a particularly important time for us to gain maximum intelligence on Russian radar defence systems and so on, when the V-Force is just beginning to build up'.

While the subject was under discussion, an alternative series of operations was scheduled for July, covering the less-controversial areas of the Adriatic and Middle East. At the beginning of July 1956 Washington WZ966 was deployed to Nicosia, carrying out an RPF en-route over the

Washington RPFs from Cyprus during and immediately after the Suez Crisis. Track 1 was flown prior to the invasion. Track 2 was flown post-Suez to monitor Syria and Egypt.

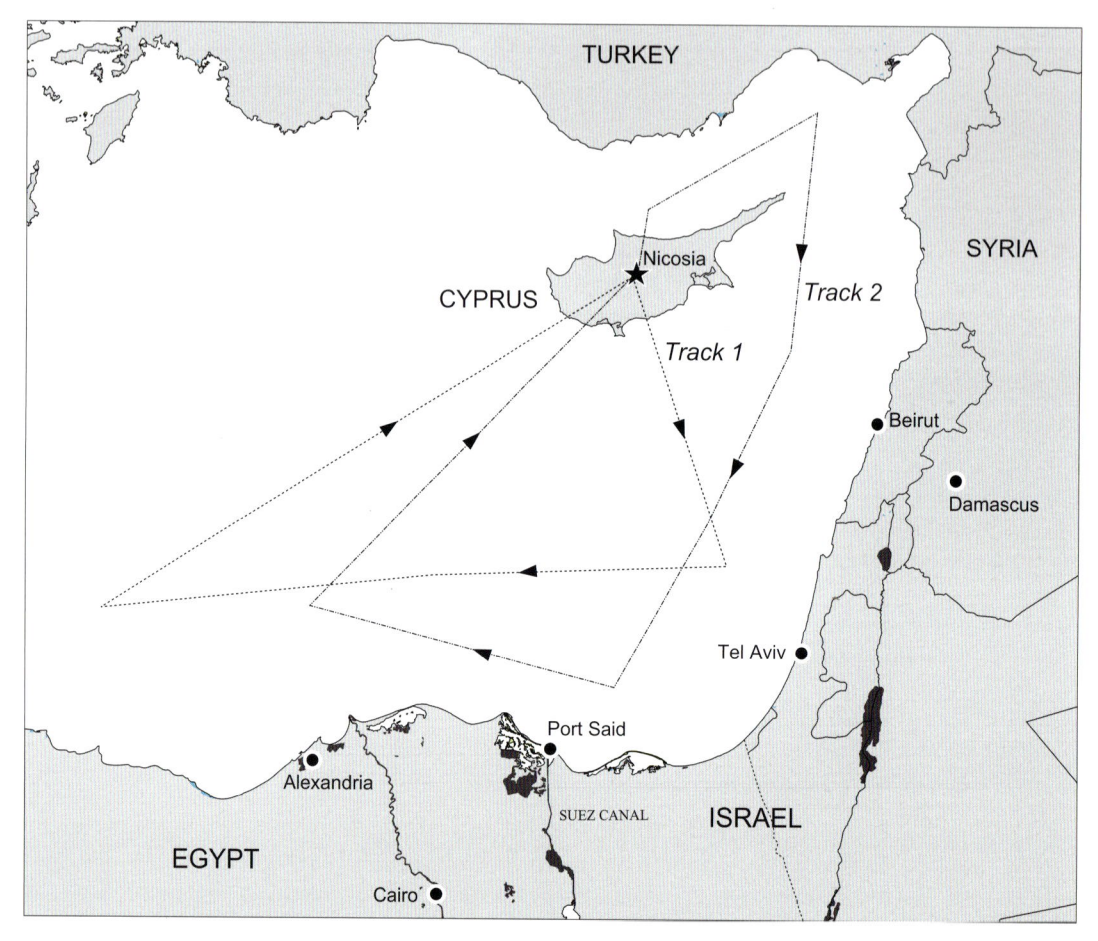

RPFs flown by Canberras and Washingtons in the period January 1953 to July 1956, excluding Border sorties. This illustrates the the limited scope of the UK airborne Elint programme during the first half of the 1950s. Note the lack of Canberra sorties in the period May-July 1955 as a result of *Green Satin* installation programme and discussions regarding administrative control of 192 Squadron.

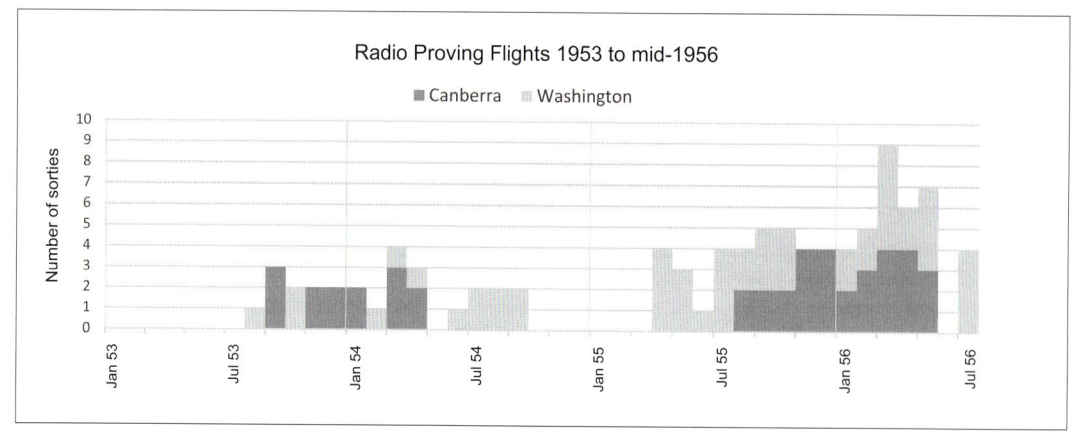

Adriatic during the 9 hour transit flight. Two further RPFs, this time 'in the Middle East area' were flown, one day, one a night return trip to Amman, Jordan, from Nicosia. These were almost certainly RPFs against Egypt and Syria, triggered by deteriorating political relations with those countries. Both Egypt and Syria were recipients of Soviet military assistance and the UK was keen to compile an electronic order-of-battle for both countries as a precautionary measure.

The 'Suez Crisis' proper began at the end of July 1956 when Nasser seized control of the Suez Canal, nationalising British and French interests. On 9th August Washington WZ968 was despatched to Nicosia for an indefinite period for operations against Egypt. This was a tactical Elint task, supporting the Anglo-French air, ground and naval forces, and the Washington was placed under operational control of HQ, Middle East Air Force for the duration of the detachment.

After a couple of weeks inactivity due to political indecision, the first RPF from Nicosia, a 3 hour 25 minute day sortie off the Egyptian coast, was flown by Flt Lt D Comer and crew on 26th August. From 27th August the 192 Squadron detachment was put on a 2 hour standby for Elint operations and a further five RPFs were flown during August and September. The last sortie listened in on VHF traffic during an Egyptian Air Defence exercise in the Canal Zone, the Washington standing off 80nm (148km) from the coast. Operations continued through October at a similar rate: a sortie on 20th was apparently flown in conjunction with a Bomber Command Canberra, the latter acting as bait to stimulate the Egyptian air defence system.

Although the RAF Nicosia authorities were copied on signals authorising the Washington flights, they were not kept fully in the loop for security reasons. This led to complaints from Nicosia to HQ MEAF that their ability to operate an efficient air defence and air sea rescue service was being compromised by the lack of co-ordination regarding the Washington flights. The Nicosia detachment came to an end on 5th November, following the end of hostilities against Egypt, when WZ966 and crew returned to the UK.

The work of the squadron seems to have been appreciated by higher authorities. At the end of October the Foreign Secretary (Selwyn Lloyd) sent a message of thanks to the unit: 'I would be grateful if you would convey to those concerned my congratulations on the excellence of the results, and my thanks for the great efforts they are making'.

Despite the end of the Suez operation, the UK continued to keep an eye on the dispositions of the Egyptian and Syrian armed forces. On 13th November, Nicosia was informed that a Washington would be arriving on 15th November for a detachment of at most 10 days, carrying out VHF listening sorties off the Syrian coast. Washington WZ966 arrived early in the morning of 15th November and sorties were flown on 16th, 17th and 19th November, none approaching closer than 50nm to the Syrian coast. The sortie on 19th November initially flew north east from Cyprus towards Turkey (presumably to conceal the origin of the flight), then following the Turkish, Syrian, Lebanese, Israeli and Egyptian coasts before returning to Cyprus. A final sortie was flown on 26th November. After the return of the detachment to Watton, a Washington was maintained on 48 hours' notice for further operations in the Middle East as a contingency measure.

RPFs October 1956-March 1957

The Foreign Office had objected to the new RPF programme proposed in July 1956 on the grounds that it was too soon after the Commander Crabb Affair. A slightly modified version of the RPF programme was resubmitted by the Air Ministry in September 1956 and this time it was approved.

The first operation in the new programme (and the first operation against the Soviet Union for five months) comprised Washington and Canberra RPFs into the Barents Sea, investigating Soviet radar development and deployment along the Kola Peninsula. This was the squadron's second visit to the area. The first operation, using a Washington, had operated from the UK but the shorter-range Canberra was forced to use a forward base in Norway, and was thus subject to restrictions imposed by the Norwegian government. Although Norway was a member of NATO and conducted its own ground-based Elint operations against the Soviet Union, it had to balance co-operation with the UK against the maintenance of a working relationship with its larger neighbour. The Norwegians thus insisted that any overland reconnaissance flights originating from their territory kept approximately 300km (162nm) from the Norwegian/Soviet border. As a result of the overland restriction the Canberra sortie was flown out over international waters, going up and around the North Cape and into the Barents Sea. The operation appears to have been particularly successful, producing the most complete recording of the Soviet air defence sys-

LISTENING IN

tem's reaction to an unidentified aircraft that had then been obtained as well as identifying the characteristics of the RP-6 *Scan Three* AI set in the Yak-25M *Flashlight* night fighter. It also obtained recordings of VHF traffic between GCI stations, a hitherto unknown practice. The Washington sortie into the Barents, operating from the Marham, was not so successful. The first attempt, using WZ967, was aborted 4 hours into the sortie due to APQ-13 radar problems; a second attempt a few days later suffered the same fate. The sortie was subsequently postponed pending a full analysis of the Canberra RPF results, and of data gathered by a USAF RB-47H Elint aircraft in the same area.

The final operation of October was a joint Washington/Canberra operation over the Baltic on 31st October, Washington WZ968 operating from Watton while B.6 WT301 flew from Gütersloh. The Baltic was a narrow sea and for that reason was still considered a relatively risky area for Washington operations, since it was difficult for the aircraft to keep the prescribed 90nm (167km) from the Soviet coast without infringing Swedish airspace. On this sortie, the aircraft operated between Denmark and the island of Bornholm, standing-off 30nm (56km) from the East German coast and 75nm (139km) from the nearest Polish territory.

The 'Long-Nose' Canberra

Plans had been made to fit the second Canberra B.6 (WJ755) with a T.4-type 'long-nose' radome. The Canberra went to Boulton Paul in June 1955 for fitting with the new nose and a *Green Satin* Doppler navigator, returning to the CSE in December 1955 for an almost identical Elint fit to that applied to WT301. Work on WJ775 appears to have been delayed by a shortage of airframe fitters, and also by difficulties in obtaining Elint equipment and the installation was not cleared until May 1956. After further adjustments and calibrations the Canberra was returned to 192 Squadron in August 1956 and finally declared operational in December 1956, nearly 18 months after it went for conversion. The squadron now had two operational Canberra B.6s. The new B.6 flew two night RPFs over the Baltic during December, the first (*Folium 2*) from Gütersloh, the second from Watton. Both operations were successful with the new radome apparently producing excellent results.

During December the squadron also began training for a new series of joint Canberra/Washington RPFs in the Barents Sea. The plan was to use the new B.6 on the operation and the crew of Canberra WJ775 carried out a grid navigation exercise early in the month in preparation. Unfortunately the neoprene covering on WJ775's radome started to deteriorate and the aircraft had to be withdrawn from service for a few months for repairs and the other B.6 had to be substituted. Operations began on 24th January when WT301 flew a 3½ hour night sortie from Bodø into the Barents Sea, investigating Soviet radar on the Kola Peninsula and along the coast of Novaya Zemlya, followed by a joint Canberra/Washington RPF, operating from the same base, on 30th January. The Washington had originally been scheduled to fly from the UK but the problems experienced during October seem to have persuaded the Air Ministry that it was more practical to use a forward Norwegian base for these operations. Unfortunately both aircraft returned early due to navigation equipment problems.

In February two RPFs were carried out over the Baltic. The first, on the night of 21st February, involved Washington WZ966 operating from Watton in conjunction with Canberras B.6 WT301 and B.2 WH698 from Gütersloh. A second operation was flown by Washington WZ966 and Canberra WH698 on the night of 25th February, again with the Washington operating from Watton and the Canberra from Gütersloh.

The last major operation of the October 1956-March 1957 programme was a series of three joint Washington/Canberra RPFs from Nicosia over the Black Sea and Caspian Sea in

Canberra B.6(BS) with the so-called 'long nose' radome, modified from that of a Canberra T.4

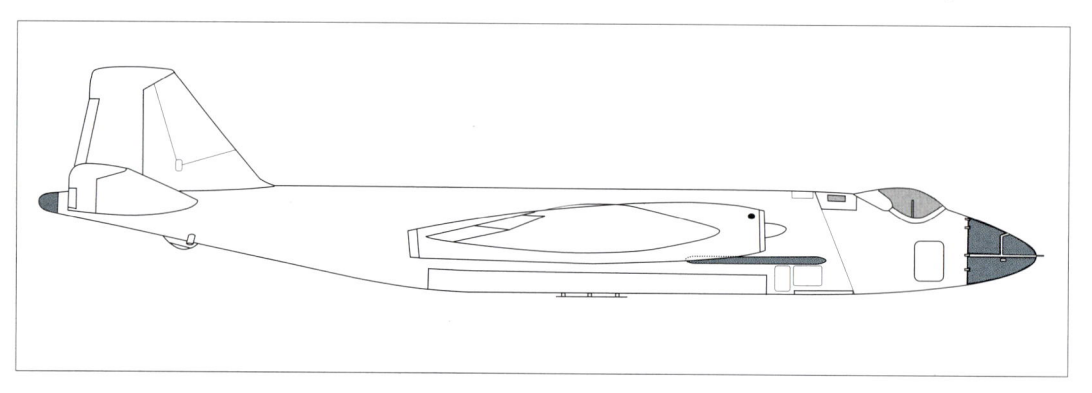

84

Canberra B.6 WJ775 circa 1956 with the new 'long nose' radome, modified from a T.4 nose. The metal skin has been replaced with fibre-glass dielectric panels. *via Peter Green*

March 1957. This had originally been planned for November 1956 but had been put back due to the Suez Crisis. The first RPF took place on 24th March using both Washington WZ966 and Canberra WT301, and was followed by a second on 28th March using the same two aircraft. This second operation appears to have flown over Turkey and Iran, along the Soviet Trans-Caucasus border (Armenia, Azerbaijan) and then into the Caspian Sea. Reports suggest that the Washington ran into adverse headwinds and, after an 11 hour sortie, was forced to divert to Habbaniya to refuel. The aircraft landed at 0155hrs and, after a quick turnaround, departed for Akrotiri 2 hours later. The final RPF, on 31st March, was less successful, the Washington suffering engine problems while the Canberra had navigation equipment problems.

The subsequent summary of the October 1956-March 1957 RPF programme noted that, although the Yak-25 *Flashlight* night fighter had entered service, the Soviet air defence system still had major problems in effecting ground controlled interceptions above 35,000ft (10,668m). There was thus little risk of the Canberra being intercepted as long as it operated at between 40,000 and 45,000ft (12,192-13,716m). The Washington was more vulnerable but was always flown in the 'stand-off' role some 60nm (111km) or more off the coast, well away from harm.

Operations April-September 1957

A new six month RPF programme for April-September 1957 covering the Baltic and Black Sea was proposed in February. At the same time the Air Ministry resurrected proposals for a more flexible planning process, and for a relaxation in the moon conditions under which sorties could be flown. Following the agreement of the Foreign Secretary, the proposals were submitted to the Prime Minister in April and subsequently approved.

The first operation, using WT301, WH698 and a Washington, was carried out over the Baltic on 26th April, the Washington operating from Watton while the Canberras operated from Jever. A similar operation was repeated a few days later on 30th April. In both cases the Washington sortie was around 7 hours and Canberras around 3 hours. Operations in May comprised two Washington RPFs from Nicosia along the coasts of Syria, Israel and Egypt, investigating the radar defences of those countries. Both sorties stood off 50nm (93km) from the coast and were flown during daylight because the target radars were normally switched off at night. The return of Washington WZ967 was delayed by a taxying accident at Nicosia on 31st May when the port inboard propeller struck a fire extinguisher. Spares were ferried out by WW346 on 1st June and WZ967 finally returned to Watton, via Luqa, on 3rd June.

A heavy programme was planned for June/July with four joint Canberra/Washington operations from Nicosia covering the Eastern, Western, Central and Western, and Southern areas of the Black Sea. By June the nose radome of the second B.6, WJ775, had been repaired, allowing the use of both B.6s for this intensive programme.

Information was received in mid-June that the Soviet authorities intended to send two naval detachments (firstly, a *Sverdlov* cruiser accompanied by two destroyers, and secondly a flotilla of three destroyers) through the Bosporus into the Mediterranean. Arrangements were made to shadow the ships using

LISTENING IN

Coastal Command Shackleton aircraft and carry out Elint sorties using a 192 Squadron Washington. The latter task was carried out as part of Operation *Chianti*, Elint operations against Soviet Naval Forces in the Mediterranean. 192 Squadron was placed at readiness on 18th June and on 21st June Washington WZ967 was detached to Luqa. Two *Chianti* sorties were flown against the Soviet naval units, one on the morning of 22nd June and a second during the night of 23rd June. Following the completion of the *Chianti* task WZ967 positioned at Nicosia for the month's major RPF.

The Washington was joined at Nicosia by Canberra B.6s WT301 and WJ775, supported by Washington WW346, at the end of June and a joint Washington/Canberra RPF was flown by WZ967 and the two Canberras on 29th June but WJ775 was forced to make an early return due to *Blue Shadow* problems. A second joint Washington/Canberra operation (WZ966, WJ775) was flown on 30th June. The Washington had to return to base early due to adverse weather but the Canberra completed its task. Operations continued through into July with a joint RPF on 2nd July using Washington WZ967 and Canberras WT301 and WJ775. A further sortie on 3rd July was a dual Washington sortie, using the support Washington WW346 as the second aircraft. Unfortunately the operation was abandoned when WZ967 became unserviceable. The operation was successfully flown the next day using the same two aircraft. This concluded the major RPF operation for the month.

Washington WZ967 was retained at Nicosia for a further few days to fly three 'special' sorties under the operational control of MEAF. These sorties (flown on 16th, 17th and 18th July) were doubtless operations against Egypt and Syria, keeping an eye on the military state of those countries. Washington WZ967 returned to Watton on 20th July on completion of these tasks.

Operations in August were to have comprised eight sorties. Unfortunately on 19th August the nosewheel of WT301 collapsed during an engine start at Watton, rendering the aircraft Cat.3. As a result the number of sorties was reduced to four. The first operation was a joint Washington/Canberra sortie from Nicosia using WZ967 and WJ775 on the 22nd August; the second was a similar operation on 24th August. The sortie by WZ967 on the latter operation was abortive due to engine failure. Further flights in support of MEAF were made at the end of the month. Washington WZ967 flew three successful sorties on 29, 31st August and 2nd September. The final operations in the series were three joint Washington/Canberra RPFs using WJ775 and WH698 (substituted for WT301) into the Baltic from Ahlhorn at the end of September/beginning of October.

In a memo to the Secretary of State at the end of July, VCAS (ACM Sir Ronald Ivelaw-Chapman) summarised the achievements of the programme as follows: 'The flights undertaken so far in the present series have continued to produce valuable intelligence on the Soviet air defence system. In particular, the technical analysis and operational evaluation of the new Soviet height-finding radar has been completed almost entirely on the basis of information gathered during proving flights in the Baltic area. In addition, recordings of at least one new night fighter radar have been obtained'. He went on to report that the Soviets were improving in their ability to control GCI interceptions at altitude and that as a result the minimum operating altitude of the Canberra in risk areas would be increased to 40,000ft (12,192m). Careful tactical routing and timing using more than one aircraft would continue to be employed as this appeared to confuse Soviet radar tracking.

Improving the D/F Performance of the B.6s

The results obtained by the Elint suite in Canberra WT301 during early 1956 were initially disappointing and led to an investigation into possible improvements, similar to that carried out on the Washingtons. An initial series of X-band calibration flights flown during mid-1956 confirmed that the polar diagram of the Canberra's ARI 18021 D/F aerial was up to specification, and showed the intercept performance of the receiver was acceptable, but also highlighted the poor accuracy of D/F obtained from the system. Further trials against high-looking X-band radars in early 1957 revealed there were inherent problems with ARI 18021 and that consequently there was little chance of getting accurate D/F with the equipment as fitted. Comparative trials using the ARI 18021 in the longer nose radome of Canberra WJ775 produced similarly poor D/F results, confirming the original diagnosis. Trials were then carried out in WJ775 using ARI 18058 *Breton* (see below) and APA-17B as alternative D/F systems in place of ARI 18021. Both produced encouraging results with the ARI 18058 in particular showing a remarkable improvement in accuracy. As a result of the investigation by mid-1957 WJ775 was being flown operationally using either ARI 18058 (X-band) or AN/APA-17 (X and S-band) for D/F.

The Third Canberra

In July 1955 Bomber Command decided that the planned squadron establishment of two B.6s was insufficient to meet future tasks and agreement was subsequently reached with the Director of Forward Planning, Air Ministry that the new establishment would be revised to three B.6s and a single B.2. The addition of a third B.6 came as an unexpected surprise to the CSE, commenting that it 'would undoubtedly prove beneficial'. The celebrations were however short-lived. In November 1955 the Air Ministry withdrew their agreement for a third B.6, 'unless it could be shown that there were the most compelling operational reasons for an increase in establishment'. Bomber Command did not give up, and negotiations for a third Canberra seem to have continued through 1956. Their persistence was finally rewarded when a third Canberra (WT305) was allotted to 192 Squadron in May 1957. By the time the third Canberra arrived there was some indecision regarding the future Canberra Elint fit and as a result, WT305 remained in its original unmodified state for over a year. (The eventual modification of WT305 is described in Chapter Nine).

Operations October 1957-March 1958

The programme for the six months beginning October 1957 was approved early that month. The first operations were two joint Washington/Canberra RPFs into the Barents Sea, the Canberra (WJ775 on both occasions) operating from Bodø as before. November 1957 saw two joint Washington/Canberra RPFs along the Soviet-Iran border from Nicosia, this time with WT301.

The plan for December had called for three joint Washington/Canberra RPFs into the Baltic from Watton or Jever but shortly before the first of these flights the Washingtons were finally withdrawn from operations. The serviceability of these aircraft had been in a constant decline and by December, had reached a stage where they could not be expected to complete a sortie without some serious unserviceability arising. This marked the end of Washington in RAF service.

As a result of the Washington's retirement, a single Canberra flew the three Baltic RPFs from Watton during December. Operations in January 1958 were then cancelled while the squadron waited for the Washington's replacement, the Comet, to enter service.

All four Washingtons were disposed of during February 1958. The first to go was WW346, which was flown to 23 MU at Aldergrove on 3rd February. More ceremony attended the disposal of the three operational aircraft when on 6th February the Commandant; CSE (Air Cdre B D Nicholas) inspected a parade of Washington Flight air and ground crew. Later the same day the three aircraft, whose crews included OC

The three Elint Washingtons, accompanied by a support Washington, perform a flypast at Watton to mark their retirement on 6th February 1958. The aircraft were subsequently scrapped. via Paul Stancliffe

A Washington makes a flypast at Watton on the occasion of the type's retirement, accompanied by a CSE Development Squadron Meteor. The CSE Washingtons were the last of their type operated by the RAF. J Cornwall via Paul Stancliffe

LISTENING IN

Diagram of the ARI 18058, the direction finding and rough frequency analysis element of *Breton*. It was designed to be used with a superheterodyne receiver for fine frequency determination and an analyser for pulse analysis.

192 Squadron and the Commandant, CSE carried out a flypast over Watton and 90 (Signals) Group HQ, Medmenham en-route to Aldergrove for final disposal.

Breton Development 1952-58

Before covering the introduction into service of the Comet the development of the *Breton* Elint system needs to be described.

As previously described, scientists at the CSE had started work on a new centimetric Elint suite, covering the 2-10GHz band, in 1949 (see Chapter Three). The intention was to produce a high-probability-of-interception system with direction-finding and signal analysis facilities. The basic concept was to use four units to intercept, D/F and analyse signals. The first unit was a wideband DV guard receiver (working on the same principles as *Flange*): this would provide a basic signal intercept and D/F capability on any signal in a 2GHz range. The second unit would comprise a number of narrow-band DV receivers, covering the same frequency range in a number of discrete steps, allowing the received signal frequency to be determined to within 400MHz. (The initial design had envisaged the narrow-band receivers covering 200MHz but this was later amended to reduce the number of receivers required). Signals from the first two stages would be fed into a superheterodyne receiver incorporating a 400MHz spectrum analyzer. The receiver would be tuned to the 400MHz band indicated by the narrow-band DV receivers and spectrum analyzer would provide an exact frequency for a signal. The final unit was a signal analysis unit which could determine PRF and pulse-width. The new Elint system was intended to cover a wide frequency range and was designed to be modular in nature, using a range of RF units feeding a common set of analysis and display units. Initial work concentrated on the 2-4GHz S-band, since this was considered one of the more important bands of interest and was technically less challenging than the higher-frequency X-band.

During 1951 the DRPC confirmed the basic approach taken by the CSE, suggesting that the UK concentrate on relatively simple wide-open receivers, leaving sophisticated search receivers to the Americans and by January 1952 the official requirement for the Elint system had been formalised via the issue of OR.3537. The wideband/narrowband DV receiver system was designated the X.17, the superheterodyne receiver the X.38, and the analyser the CRT-2. The complete system later received the codename *Breton*.

By the first quarter of 1952 the CSE had produced two prototype laboratory models of the S-band (2-4GHz) DV receiver (each incorporating two D/F channels and nine 400MHz-wide frequency-determining channels) and a laboratory model of the wideband superheterodyne receiver, the latter incorporating a 400MHz RF spectroscope.

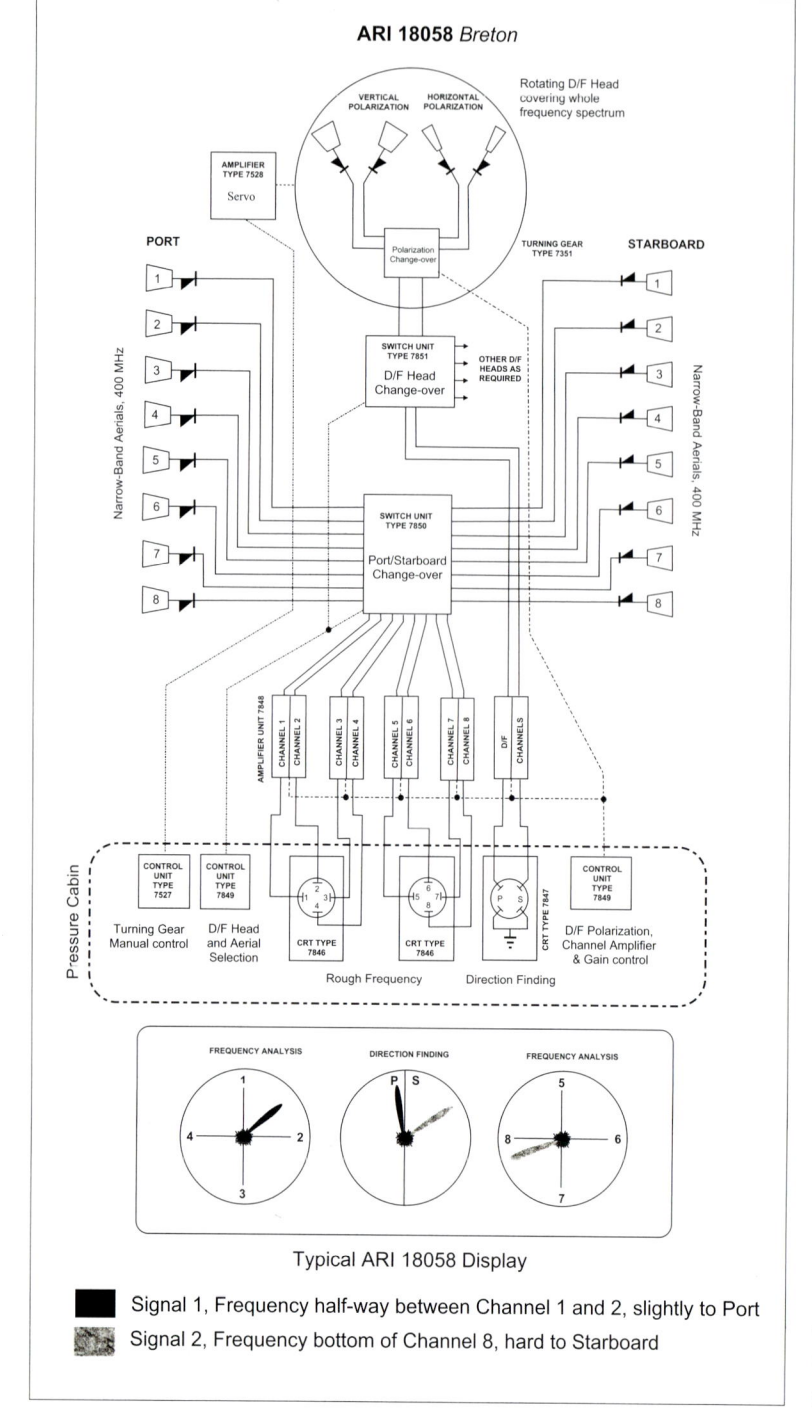

Flight trials of the system began in early 1952 when one of the S-band DV laboratory models was installed in Lincoln SX992 of the CSE Development Squadron. At that time it was not clear whether D/F was best provided via a series of fixed aerials, or via a rotatable aerial. The flight trials included an assessment of both aerial systems and suggested that, although the accuracy of a rotatable aerial was better, a combination of both rotating and fixed might be required on some aircraft. Flight trials of the superheterodyne receiver in the same Lincoln started in the second quarter of 1952 and by the end of the year the trials were examining operating methods, that is, how a Special Operator (or Special Operators) could use both systems to intercept, D/F and determine the frequency of a signal.

Once the basic design of the various components of the *Breton* system had been completed by the CSE (and proven by flight trials in the Lincoln), contracts for the further development were placed in industry. Bush Radio became the primary contractor for the DV receiver and McMichael for the superheterodyne receiver. The contractors (provided with full design details by the CSE) initially produced elementary experimental models of the units for bench and flight tests.

Parts of the first experimental model DV receiver were delivered to the CSE from the contractors for flight trials in first quarter of 1953. Following ground tests, the units were flown in the Lincoln as part of the CSE experimental DV system. A first model of the RF spectroscope was also received and flown in first quarter of 1953 and flight trials of all parts the first model DV receiver were completed in the autumn of 1953. The results of the trials at the CSE were fed back to the contractors, who then produced further, improved models for further assessment. The plan was to iterate through a small number of intermediate experimental models until the final requirements were clear and the design was finalised. The contractors would then produce a number of engineered models for final acceptance tests and service trials. Second experimental models of both the DV and superheterodyne receivers were delivered to the CSE during 1954.

In the first half of 1953 the CSE began to look at aircraft requirements for the later *Breton* development and service flight trials. The Lincoln (SX992) used for the initial flight trials was too cramped for trials of the final system, and too unrepresentative of the V.1000 for which the system was eventually intended. By mid-year it had been decided to replace the Lincoln with two aircraft: a Canberra for high-altitude trials of individual units and a larger aircraft for trials of the complete system. There was some indecision regarding the type of aircraft required for complete system trials and it was decided to continue with the Lincoln in the short-term until a suitable replacement could be found.

In June 1953 HQ 90 Group requested the allotment of a Canberra to the CSE for flight trials of the X.17 and X.38 receivers. In support of their request they explained that the design of the final engineered models of the receivers would be influenced by the types of aircraft in which the equipment would be installed for operational use. In particular, it would be necessary to ensure that the various elements of the system were suitable for operation in a restricted space at high altitude. A great deal of thought had been given to making the *Breton* system modular, allowing subsets of the system to be used, according to the search task. If the equipment could be accommodated in a Canberra then it was likely to work in any other aircraft under consideration. The request was quickly approved and a brand-new Canberra B.2 (WJ984) allotted to the CSE in September.

The Canberra went to EECo in November 1953 to be fitted with a new nose for the *Breton* trials. Working with the RRE, the company designed a modified T.4 nose section, replacing most of the metal skinning with fibreglass panels to produce a nose radome providing a suitable 'view' for the *Breton* aerials. (The same type of nose was later fitted to B.6 WJ775). Work on the new nose was completed in February 1954 and WJ984 was then returned to Watton for the main installation programme. The installation in the Canberra got under way in mid-July with a target date of October 1954. After stripping all bombing equipment, a number of *Breton* units were installed in the bomb-bay, and a new navigation fit of Gee Mk.3 and ARI 5428 provided. The installation programme was delayed by the need to enclose the *Breton* units in the bomb-bay (as a fire precaution) and

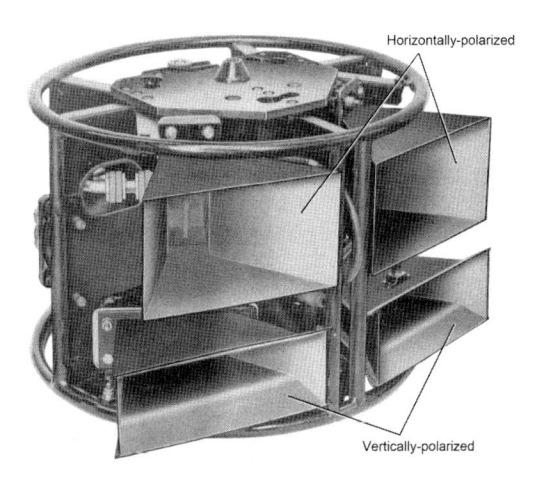

The ARI 18058 *Breton* S-band D/F aerial array comprised two horizontally-polarised and two vertically-polarised aerials mounted on an electrically-rotated turntable. It was designed to provide direction finding on any signal in the S-band.
Author's collection

LISTENING IN

Canberra B.2 WJ984 of the CSE Development Squadron was fitted with a T.4-style radome for *Breton* development. *via Peter Green*

was not completed at the end of February 1955, following which the aircraft underwent a major servicing and rectification programme. Canberra WJ984 flew its first *Breton* sortie in August 1955

By the beginning of 1954 a Varsity had been selected as the large *Breton* trials aircraft and Varsity T.1 WL687 was subsequently allotted to the CSE in March 1954. WL687 went immediately into the CSE Installation Flight where it was fitted with a full *Breton* installation, a new navigation suite, including *Blue Shadow* and *Green Satin*, and additional power supplies. This was a big task, requiring the assistance of RRE Defford and a Contractors Working Party, and was finally completed in May 1955.

In the second half of 1955 the CSE (somewhat optimistically) made an application for a Comet for use both as an additional *Breton* trials aircraft, and as a general flying laboratory aircraft. The request was turned down by the Air Ministry, on the grounds that a Comet could not be supplied before the conclusion of the *Breton* trials. Instead the Air Ministry suggested that the CSE should consider using one of the Comets scheduled for delivery to 192 Squadron. The CSE rejected the latter suggestion, as impracticable, given the expected high utilisation of the 192 Squadron Comets on operational Elint tasks.

During 1954-55, while the Canberra and Varsity installation were in progress, trials of *Breton* continued in Lincoln SX992. Flight trials were carried out in the first half of the year to obtain Air Ministry agreement on the display of frequency and D/F information. Towards the end of 1954 doubts emerged as to whether the rotatable *Breton* static-split D/F system (that is, using two static aerials for D/F) had any advantage over the 'spinner' D/F system used in US equipment. In order to investigate further an APA-17 system was installed in SX992 alongside the *Breton* D/F system in the autumn of 1954. The comparative trials showed that static-split had a better probability of intercept against a modern ground radar with a narrow rotating beam, but the system was prone to D/F errors due to airframe reflections, although this was dependent on where the aerial was sited. The APA-17 was not as prone to reflections and thus was generally more accurate. The conclusion was that both aerial systems might be needed: a static-split for high-probability of interception and a 'spinner' for D/F accuracy.

By the beginning of 1955 a representative *Breton* system (DV receiver, superheterodyne receiver and analyser) had been installed in a single rack assembly and demonstrated on the ground to a large number of interested parties. The rack was subsequently used to check the compatibility of the various *Breton* units when operating together and to establish an operating technique.

Airborne trials of a full *Breton* S-band installation in the Varsity began in mid-1955 and were supplemented by high-altitude Canberra trials of individual units from August 1955 onwards. By mid-1956 final models of the complete *Breton* system, covering S-band and X-band became available and from that point the airborne trials programme became a Combined Service/Development trials programme, with specially-selected and trained Special Operators from 192 Squadron flying in the Canberra and Varsity. Assessment of the superheterodyne receiver was completed at the end of 1956 but work on the DV receiver, including a comprehensive evaluation of D/F accuracy using static-split and spinner aerials, continued through until 1957. The D/F trials effectively confirmed the earlier Lincoln work, showing that D/F using static-split was accurate when the aerials were ideally located (for example, in the nose of a Canberra), but prone to reflections, especially at S-band, in other locations. The *Breton* service trials came to an end in mid-1957.

8 The Comet

'[it is]… not only opinion abroad, but opinion at home which may need to be satisfied about this rather curious operation by the RAF of non-military types of aircraft'. Sir Edward Bridges, HM Treasury, Spring 1954

Before describing Comet RPF operations, an examination of how and why the Comet was procured is required. By 1953 there were strident calls from various intelligence agencies for an increase in airborne Elint operations. At that time the Washington and Canberra had just entered service, but unfortunately neither aircraft was an ideal Elint platform: the Washington was too slow and vulnerable for some operational areas and the Canberra lacked the space to accommodate a comprehensive Elint fit. As described in Chapter Five, the long-term plan was to replace the Washington and Canberra with an Elint version of the V.1000 transport, but that aircraft was unlikely to enter service before the end of 1957 at the soonest. There were also concerns that it might not prove possible to keep the Washington in service up until the arrival of the V.1000 due to the former's poor maintainability. There was clearly a need for an interim aircraft to bridge the gap between the Washington and the V.1000.

Analysis of the Elint task showed that it could be broken down into a high-altitude/high-speed role (the 'regular' RPF task) and a low/medium-altitude role (maritime and communications intercept tasks). Both roles required an aircraft capable of accommodating at least six Special Operators and their equipment. Investigation by the Air Ministry, in the guise of ACAS(Signals), during 1953 revealed that the only aircraft with the required performance for the interim high-speed, high-altitude Elint role was the new Avon-powered Comet 2, which could operate at altitudes above 40,000ft (12,192m) and could accommodate nine Special Operators (SOs) and two supervisors. It was estimated that if an order was placed for Comet 2s immediately, then the aircraft could be delivered by the end of 1954 and enter service in the Elint

New and old, 1956. A pair of Swedish Air Force J.28 Vampires, approaching the end of their service, intercepts a CSE Vickers V.1000 on a Radio Proving Flight over the eastern Baltic. The *Flygvapnet* frequently intercepted RAF and USAF Elint aircraft. *Adrian Mann*

LISTENING IN

role in mid-1955. ACAS(Signals) thus recommended that three Comets 2s be procured to meet the high-level requirement, also suggesting that two shorter-range, Ghost-powered Comet 1s should be obtained to provide an immediate capability in advance of the Comet 2s. Although the Comet looked ideal for the high-altitude role, performance figures suggested the Comet 2 would have poor endurance when operating below 35,000ft (10,668m) and thus could not meet the medium/low altitude role requirements. A second type would thus be required, and ACAS(Signals) went on to suggest that the best candidate for the low/medium altitude role was the Shackleton MR.1 maritime reconnaissance aircraft, already in service with Coastal Command. The Shackleton MR.1 had a good low-altitude performance, had a ceiling just about sufficient for the medium-altitude role, could accommodate seven SOs and a supervisor, and was scheduled for replacement in the near future by the Shackleton MR.3. There would thus be a ready supply of aircraft for conversion. ACAS(Signals) proposed that four Shackleton MR.1s would be needed in addition to the Comets in order to meet airborne Elint requirements.

The recommendation made by ACAS(Signals) for the replacement of three Washingtons and two Canberras by three Comet 2s and four Shackleton MR.1s was endorsed by an Air Ministry committee in February 1954. Revised forecasts for the V.1000 indicated that delivery of the Elint version of that aircraft would not now begin until late 1958/early 1959. The Comet 2 aircraft would be available in mid-1955 and would take six months to convert to the Elint role. It thus appeared that the Comet 2 might enter service at the beginning of 1956 and see three years operational service before being replaced by the V.1000.

The recommendation for three Comet 2s had been based on the assumption that the aircraft would be required for both 'Boundary' RPFs, flying along the coastline and borders of the Soviet Bloc, and occasional 'Penetration' RPFs, overflying Soviet airspace. The meeting noted that political approval had not yet been obtained for 'Penetration' sorties, and that, in the meantime, two aircraft would be sufficient to meet the 'Boundary' RPF requirement. The proposal to acquire four Shackleton MR.1s for use at low/medium altitude was also approved but the proposal to procure two Comet 1s as short-term gap-fillers was rejected on the basis that production was already spoken for and it was thus extremely unlikely the aircraft would be available in time.

Work on the Comet 2 Elint project got under way in March 1954. The two immediate concerns were ensuring the procurement was handled in a secure manner, to avoid disclosing the role of the aircraft, and the need to draw up a Standard of Preparation for the aircraft. Designing an Elint fit for a large complex aircraft like the Comet was a demanding and time-consuming business, making it difficult to estimate the final costs of conversion.

The Air Ministry were also looking at acquiring a number of Comet 2s for use by Transport Command and in late March/early April the Air Council agreed that orders should be placed for eight Comet 2s for the RAF: six for Transport Command and two for Signals Command in the Elint role. The acquisition of Comets for Transport Command was welcome news to those concerned about the security aspects of the Elint Comet 2s. The hope was that the entry into service of the Comet 2 in the transport role would deflect attention away from the acquisition of the two Elint aircraft.

In mid-March the Air Ministry put the Elint Comet proposal to Sir Edward Bridges, then chairing a Treasury committee overseeing the costs of intelligence. The Air Ministry noted that the cost of procuring the two Comets would be

The Shackleton MR.1 originally designated GR.1, although far from ideal as an Elint platform, being noisy and uncomfortable, had an excellent low-altitude performance. *via Phil Butler*

around £1.3M, with subsequent running costs coming in at £200,000 per annum. The price for the Shackletons was, by comparison, negligible: the aircraft would be transferred from Coastal Command free of charge as surplus and the only costs would be for an Elint refit, estimated at £180,000 for all four aircraft. The Air Ministry suggested that the Chiefs of Staff should be solicited for their opinion on the Comet proposals. The plan was that, following Sir Edward's approval, the Air Ministry would then approach the Treasury for official approval for the order. The cover story for the two Comets would be that they were intended for the radar-calibration role, a subterfuge that would provide cover for the RAF's Elint types for the next four decades.

The Comet Disaster

Unfortunately by March 1954 the safety of the Comet 1 was coming under scrutiny following the unexplained loss of a BOAC aircraft (G-ALYP) on 10th January of that year, shortly after take-off from Rome. The loss of a second Comet 1 (G-ALYY) on 8th April 1954 led to the grounding of the fleet, leading the Secretary of State for Air (Lord De L'Isle and Dudley) to question the viability of both the proposed Comet Transport and Elint orders for the RAF.

Sir Edward Bridges replied to the Air Ministry proposals regarding the Elint Comets in mid-April, noting that the Comet accident meant that any employment of the aircraft by the RAF was *sub judice* until the cause had been established. He also expressed disquiet regarding the cost of the proposal, indicating that his recent review of intelligence spending had decided to cap expenditure on Sigint at £12M for the coming year and that £1.3M was a substantial chunk of the intelligence budget, cap or no cap. He therefore suggested that the Comet proposal should be examined by the Chiefs of Staff and the LSIB in order to confirm that it was an effective use of the Sigint budget. Sir Edward also expressed some concern regarding the cover story for the aircraft, noting that it was imperative that the Chiefs of Staff and the Foreign Office reach agreement on the cover plan, since it was '… not only opinion abroad, but opinion at home which may need to be satisfied about this rather curious operation by the RAF of non-military types of aircraft'.

In accordance with Sir Edward Bridges' wishes a paper on the Elint Comet 2 was circulated to the Chiefs of Staff Committee at the beginning of July 1954 and was considered at a subsequent meeting of the Chiefs of Staff on 21st July. In view of the high cost of the aircraft, and the need to make economies in intelligence spending, the Chiefs of Staff asked the Chairmen of the Joint Intelligence Committee (JIC) and LSIB, in conjunction with the Foreign Office, Treasury and Air Ministry, to examine the proposal in more detail, in order to determine 'how these aircraft could be purchased, if possible during the current financial year without involving disastrous economies to other intelligence services'. The conclusions of the investigation (chaired by Sir Edward Bridges) were examined at a second Chiefs of Staff meeting on 25th August. The Chiefs of Staff were not entirely convinced of the need for two Comet 2s (and by implication, airborne Elint operations), citing the 'political and operational uncertainty which must inevitably surround operations of this kind'. There was also concern regarding the future of the Comet and 'restrictions that might be placed on its operation' once the cause of the accidents had been determined. Because of the doubts surrounding both the aircraft and their employment, the committee decided that only one Comet should be approved (producing a notional saving of £0.75M), with procurement of further aircraft dependent on the experience gained with the first. Furthermore, the single Comet had to be purchased during the existing (1954-55) financial year to take advantage of funds resulting from a predicted underrun in spending.

Although the Chiefs of Staff had approved the procurement of a single Comet 2, there was still the problem of the Comet accident investigation. By August 1954 the RAE had determined that the cause of the Comet 1 accidents was metal fatigue: repeated cabin pressurisation cycles had led to fatigue and then failure at a fuselage weak point. Although the cause was known, more work was required to prove the fairly extensive strengthening modifications required to fix the problem. There was thus some reluctance for an approach to the Treasury regarding funding for the Elint Comet until the fix had been proven and the long-term future of the Comet assured. This was a major problem for the Air Ministry who needed to order and pay for their single authorised Comet in the 1954-55 financial year.

By September, investigation by the Air Ministry had produced a possible solution. The Comet 1 fatigue problem was directly related to the magnitude of the pressure differential applied to the cabin during flights: this was normally 8½ psi (58Kpa), required to keep the cabin altitude at a passenger-friendly 10,000ft (3,048m) when the aircraft was cruising at 40,000ft (12,192m). If the Air Ministry installed an oxygen system in their Comet 2 aircraft they could operate the aircraft at about half the regular pressure differential and significantly reduce the stress on the fuselage.

The expectation was that the Elint Comet 2 would have a limited operational life of only 2,000 hours, less than half the 5,000 hours of the Comet C.2 transport aircraft. The aircraft was a gap-filler until the arrival of the V.1000, and so would have a relatively short service life, and Elint aircraft had, in any case, a low utilisation rate. Calculations suggested the reduced cabin operating pressure, combined with some precautionary local strengthening to the aircraft windows, would eliminate the fatigue problem over the aircraft's limited operational life. Little difficulty was expected in obtaining an aircraft due to the Comet 1 accidents and subsequent suspension of orders for the Comet 2. The de Havilland Company was holding five Comet 2s, originally constructed for BOAC but now facing an uncertain future, and it seemed likely that the Air Ministry would be able to buy one of those aircraft. The purchase would also have the beneficial side-effect of giving some financial assistance to the company, to help it through the crisis.

The revised proposal was considered by the Chiefs of Staff in mid-October 1954 and this time it was approved. An approach was then made to the Treasury in early December 1954 and approval was obtained in mid-month. The projected cost of the aircraft was now £575,000, comprising £350,000 for the basic aircraft and the rest for modification to the required standard. In order to ensure the majority of the cost fell in the 1954/55 financial year the Air Ministry planned to formally 'take delivery' of the unmodified Comet prior to its conversion.

By late 1954 the Air Ministry plan was therefore to replace the two Canberras and three Washingtons of 192 Squadron with one Comet 2 and four Shackleton MR.1s. Unfortunately by mid-1954 further investigation had revealed that the Shackleton MR.1 was excessively cold and suffered from high levels of noise and vibration, making it unsuitable for the RPF role. Attempts were made to obtain the quieter Shackleton MR.3 but these were rebuffed by the Air Ministry (all available production was wanted for Coastal Command). With no alternative to the Shackleton MR.1 readily available the CSE reluctantly agreed to accept the aircraft.

Two Additional Comets: February 1955

The main justification for the procurement of the Shackleton MR.1 had been that the aircraft would provide a capability against naval radars, some of which, it was thought, could only be intercepted at low-altitude. However, a technical investigation into the polar diagrams of ship and submarine radars by 90 Group during late 1954/early 1955 revealed that not only could these radars be intercepted at high altitude, but in some cases, high-altitude was essential for interception. The requirement for a low-altitude intercept capability thus disappeared. At the same time, a reassessment of Comet performance suggested that it would have sufficient endurance to fulfil the medium altitude task.

The disappearance of the low-level task, and the ability of the Comet to operate at medium altitude, removed any remaining justification for accepting the otherwise unsuitable Shackleton MR.1. In January/February 1955 the Air Ministry suggested cancelling the four Shackletons and buying an additional two Comets in their place, taking advantage of savings thrown up by an underspend on the Supermarine Swift to help fund the procurement. The proposal would of course require Sir Edward Bridges and the Treasury to agree to the purchase of two additional Comets, at a cost of £1.3M, in addition to the one already authorised in December 1954. Given the struggle required to obtain approval for one Comet, this may at first appear to have been an ambitious goal. However, by February 1955 operational control of 192 Squadron had been transferred from 90 Group to Bomber Command. It seems likely that the transfer was accompanied by an increased recognition of the importance of Elint within the Air Ministry. It is also worth noting that, although expenditure on the proposed two additional Comets aircraft was classed as Intelligence spending, in reality their cost would be charged against the Air Vote, giving the Air Ministry an important say in the matter. The proposal was put to the Secretary of State for Air at the beginning of March, with a strong endorsement by the Chief of Air Staff, and by mid-month both the Minister and Sir Edward Bridges had agreed to the purchase of the additional two aircraft. Treasury approval was obtained at the beginning of April on condition that payment for the two aircraft was deferred until the 1956-57 financial year.

Procurement

The Air Staff issued a requirement for three Comet 2s for Elint duties on 28th February 1955 to Standard of Preparation C.63353/54/OR2. The Air Ministry agreed to accept the aircraft equipped to the original 1953 BOAC specification, that is, without the structural modifications being developed for the Comet C.2 version for Transport Command. An Instruction To Proceed (ITP) for the supply and modification of three Comet 2Rs for 'Radar Calibration' duties was issued to de Havilland in March 1955 under contract 6/Aircraft/11808/CB.10(a). The estimated price was £350,000 per basic aircraft, with an additional £185,000 for modification, giving a total cost for three modified aircraft of £1.65M. The aircraft were required to have a life of 2,000 hours, the same cruising speed as the civil Comet 2 when operating at similar weights, sufficient fuel capacity for a range of 3,060nm (5,667km) at normal take-off weight, and a take-off distance not exceeding 2,000 yards (1.8km).

The Comet 2Rs still needed some structural modifications for their new role, principally the conversion of the under-floor freight holds into equipment bays housing aerials covered by two large radomes, the addition of various other aerials on the airframe, a rework of the passenger and galley compartments to accommodate a number of SO stations, and a new Navigator's station. Other major systems modifications included a revised navigation fit and the provision of three complex intercommunication systems. The major modifications were outside the capabilities of the CSE and were farmed out to de Havilland, who in turn sub-contracted the structural work to Marshalls of Cambridge. The plan was to have de Havilland/Marshalls carry out the structural and system modifications, after which the aircraft would be delivered to the CSE for an Elint equipment installation.

The three airframes initially allocated against the Comet 2R contract were serial numbers 06027, 06035 and 06037 and the first of these (06027) was delivered to Marshalls in March 1955 for preparatory work prior to the start of airframe and systems modifications. Shortly afterwards, in June, two earlier-production aircraft (06023 and 06025) were substituted for 06035 and 06037 on the Comet 2R contract. In July 1955 the three aircraft (06023, 06025 and 06027) were allotted the serials XK655, XK659 and XK663.

An initiation meeting for the Comet 2R conversion was held at Marshalls on the 1st June 1955 when the intended fit of the aircraft was discussed. Accurate navigation was an important requirement for an Elint aircraft and the Comet 2R Standard of Preparation called for a completely revised navigation fit. A new twin-navigator navigation station was specified, installed opposite the starboard entrance door in the space previously occupied by the forward luggage and freight hold. The comprehensive navigation suite comprised twin *Blue Shadow* SLAR, *Green Satin* Doppler and GPI, Gee Mk.3, DME, Loran, Decca, twin AD7092B Radio Compass, Radio Altimeter Mk.6, CL2 gyro compass, AMU Mk.4, API Mk.1 and a Periscopic Sextant.

The two passenger cabins were converted into an SO's compartment, accommodating a number of forward-facing SO positions, either side of a central aisle. Each position comprised a double seat, table and an equipment rack mounting receivers, analysers and recorders. The double seating arrangements would allow a total of 26 personnel (including SOs) to be carried, providing a capability to transport ground-crew on overseas detachments. In order to make space for the SOs stations all unnecessary equipment associated with the passenger-carrying role was removed. The galley was stripped out, although a water tank and heated water urn were retained at the front of the compartment and one of the two toilets at the rear of the aircraft was also removed to provide general stowage space.

The Elint fit was complicated by the fact that the *Breton* system intended for the aircraft would not be ready in time for the Comet 2R's entry into service. It was decided to fit the Comet initially with an interim Phase 1 Elint installation using similar equipment to that installed in the Washington. The layout would be designed such that a changeover could be made to the Phase 2 *Breton* fit at a later date. The original plan for the Phase 1 fit provided ten Special Operator positions, comprising six radar intercept and direction finding (D/F) positions, two communications intercept positions, and two supervisor positions, with an option to convert two of the radar positions for communications intercept if required. In the Phase 2 fit, three of the radar positions would be fitted with *Breton* and the total number of SOs increased to thirteen. Communication between such a large team of SOs was important in order to effectively co-ordinate operations and so three intercom systems were specified for the Comet 2R, one for all the crew, one for the two

LISTENING IN

navigators and a third, multi-channel system, for the SOs.

The Comet 2R had been accepted to the original BOAC build standard, without additional fuselage strengthening. To eliminate the fatigue problem which had plagued the Comet 1, the 2R would be operated at a reduced cabin pressure differential of 4psi (27.6kPa), just under half that of the passenger aircraft. This would give a cabin altitude of 22,000ft (6,705m) at the maximum altitude of 50,000ft (15,240m), making the use of oxygen essential for all crew. The flight crew were equipped with a demand system using Mk.17 regulators while the Special Operators were supplied via a Mk.10A regulator and economiser system. Each dual-seat SO's position was equipped with two economisers, with Mk.5 walk-around oxygen sets provided for the two supervisors using plug-in points at each rack. An economiser was also provided in the single toilet compartment. In addition to operating at a reduced cabin pressure the cabin windows were strengthened, although, unlike the C.2s, the Comet 2R retained the original square windows.

One of the major challenges raised by the Comet 2R was the design of an aerial system for the aircraft which satisfied both the Air Ministry's requirements for D/F performance and coverage, and de Havilland's structural and handling requirements. After much discussion it was decided to fit four rotating D/F aerials underneath the aircraft (one forward, three aft), covered by two streamlined fibreglass radomes. Other aerials included a nose-cone aerial, a window-mounted aerial, an aerial in the tailcone, whip aerials and an HF wire aerial running from the fin to a mast on the rear fuselage. Additional DC power for the installation came from a 112V Rotax 505 generator on each engine, with AC power provided by a Rotax Inverter.

The Air Staff had hoped that the first aircraft would become available from Marshalls in July 1956 and might enter service around the beginning of 1957. Unfortunately the conversion of the aircraft was delayed, primarily due to changes in the Standard of Preparation, which went through three issues: 22nd March 1955, 4th July 1955 and 18th July 1956. Clearance of the modifications to the first aircraft finally took place at Marshalls over a couple of days at the beginning of October 1956 with no major amendments apart from some 'snagging'. Following completion of the modification programme at Marshalls, the aircraft went to de Havilland in December for a planned six weeks of flight trials. Unfortunately a number of defects had emerged in the Comet Mk.2 wing and flight trials were delayed by an inspection and rectification programme, pushing the delivery date of the aircraft out to April 1957.

Preparing for the Comet

During the spring of 1955 the CSE began to make arrangements for the entry into service of the Comet 2R. The two major concerns were the Elint fit of the aircraft and how to train crews for the new aircraft.

During April 1955 the CSE prepared a paper on the proposed Elint fit for the Comet. The initial specification for the Phase 1 fit called for a total of ten stations, comprising three centimetric (S, C and X-band) intercept positions, three metric (L, and P-band) intercept positions,

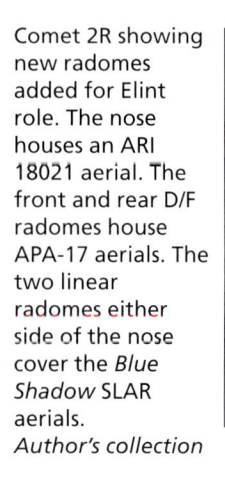

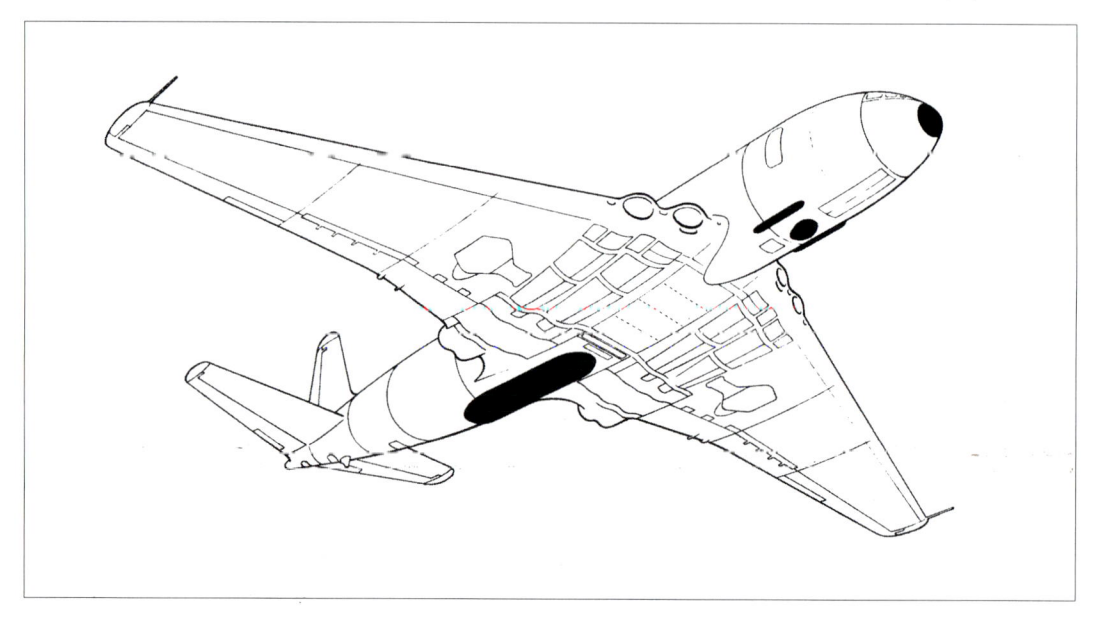

Comet 2R showing new radomes added for Elint role. The nose houses an ARI 18021 aerial. The front and rear D/F radomes house APA-17 aerials. The two linear radomes either side of the nose cover the *Blue Shadow* SLAR aerials.
Author's collection

two communications (HF and VHF) intercept positions and two supervisor positions. The stations were fitted with same equipment installed in the Washington that is, APR-4, APR-9, APA-17, APA-11 and ANQ-1A, with D/F provided via three centimetric and two metric APA-17B aerials. The major new items were APA-74 signal analysers fitted at the two supervisor position. These could display up to five signals simultaneously and would allow a supervisor to inspect signals received at any of the other eight stations.

It soon became clear that additional Elint equipment would need to be obtained from the US since the equipment due to be removed from the retiring Washingtons would be insufficient to equip the Comet. The Elint equipment supply position remained a concern throughout 1955-56 but in the end sufficient equipment was received from the US, found in store at MUs and recovered from the Washingtons, to equip the Comets.

The other big problem was training aircrew and ground crew for the Comet. It was not practical to detach existing flying personnel from 192 Squadron for training on the Comet because of the disruption it would cause to operations. The only solution was to form what was effectively a new unit with new crews. In August 1955 the Air Ministry approved the formation of C Flight, 192 Squadron as the new Comet Flight. This was established as a self-contained entity within the squadron with the intention that it eventually replace the existing Washington Flight when that disbanded.

Special Operator Training

In its final form the Comet would carry twelve Special Operators, twice the number carried in the Washington, which meant that a large number of new SOs would have to be recruited and trained in advance of the aircraft's arrival. The CSE's original intention was to carry out the airborne phase of the training programme in the Washingtons between operations, but by October 1955 it had become clear that the size of the training task, and the upcoming major inspections on two of the Washingtons, ruled this out. In the absence of the Washingtons the only way to train sufficient operators in time was to use a dedicated training aircraft. In October 1955 the CSE requested the allocation of an additional Varsity to 192 Squadron as a SO training platform. The intention was to use the aircraft initially for ab initio training for the new Comet SOs, later using it for on-going continuation training in order to minimise the training load on the Comets once they had entered service.

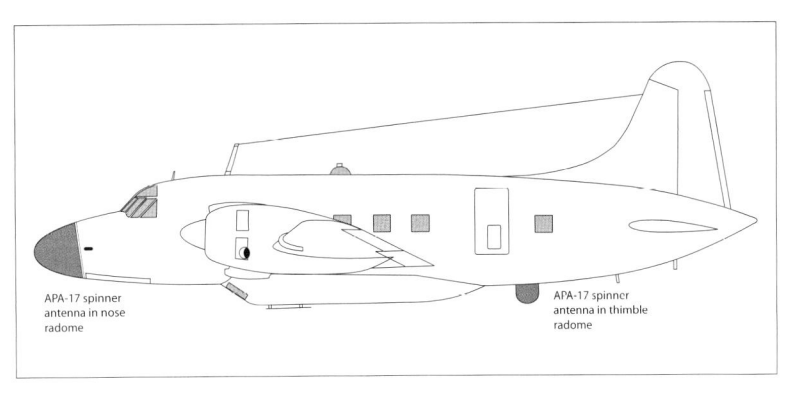

The CSE began modification work on Varsity WL686 in November 1955 to SRIM 1952, fitting five SO positions in the aircraft and relocating the navigator's position to provide the necessary space. The five training positions used Phase 1 equipment and comprised: Position 1 (metric intercept), Position 2 (centimetric intercept), Position 3 (metric D/F), Position 4 (centimetric intercept) and Position 5 (centimetric D/F). One APA-17B spinner aerial was mounted in the nose of aircraft, with the second under the rear fuselage, covered by a radome.

Varsity WL686 entered service as a SO training aircraft in mid-1956 by which time the plan was to provide some 60 hours of training per SO course. The CSE subsequently reported: 'The Varsity aircraft for special operator training is now in use and the addition of the aircraft to the training resources has enabled the back-log of air training for all new operators to be brought up to date ... Excellent results are being obtained from the special equipment installations and the aircraft availability has been exceptionally good'. By the time the Comet entered service in 1957 the training commitment had been fully met and sufficient SOs were available for the aircraft.

Training SOs was a lengthy and expensive business and it was a cause of some frustration in the CSE that these highly-trained personnel only served a three-year tour. The CSE subsequently convince the Air Ministry to allow an SO tour to be extended to five years in order to improve the return on the training investment.

Flight Crew Training

Members of the first 192 Squadron Comet crew arrived at Watton in December 1955 and were joined by their captain, Sqn Ldr Allen, in February 1956. The Comet conversion training programme began the same month with a detachment to de Havilland, Hatfield for a

This drawing shows the Varsity T.1 converted to Elint trainer with APA-17 aerials housed in the nose and in the thimble radome under the rear fuselage.

preparatory ground course. The original intention had been to convert the crew using the first production Comet T.2 (XK699) whilst the aircraft was at Boscombe Down. Unfortunately the airborne phase of the conversion programme was delayed due to the initial refusal by the Air Registration Board to grant the Comet T/C.2 a Certificate of Airworthiness. After some delay, flying training finally began in September when the crew was detached to Lyneham to take the 216 Squadron (Transport Command) Comet conversion course. Further flying was carried out in March 1957 during one of the regular 216 Squadron detachments to El Adem, Libya, taking advantage of the clear skies and good weather there. The 192 Squadron crew also flew the first Comet 2R on test flights at Hatfield during March 1957. Thus by the time the first Comet 2R was ready, in mid-April 1957, the squadron had a single trained crew.

The second 192 Squadron Comet crew also went through the Transport Command conversion programme at Lyneham, completing their basic conversion to type in October 1957. Unfortunately Transport Command announced in March 1958 that it would be unable to provide similar training to the third crew 'due to other commitments'. It pointed out however that it had three surplus crews on strength, all with 100 hours on the Comet, and suggested one of these crews be transferred to 192 Squadron. The CO of 192 Squadron declined the offer, citing problems of 'crew temperament' and subsequently made arrangements to have the third Comet crew trained at Watton. 192 Squadron aircrew were carefully selected due to the demanding nature of the squadron's task and the CO clearly did not want to take 'pot luck' on the new crew.

Delivery

The first Comet 2R conversion, XK663, was flown from Hatfield to Watton by a 192 Squadron crew on 18th April 1957, some 15 months later than the schedule originally agreed in March 1955. The aircraft was immediately fed in to the CSE Installation Flight for an Elint installation to SRIM 2113. This was described as the 'largest radio modification attempted in any aircraft in the Royal Air Force' and required the installation of one and a half tons of equipment. Some elements of the installation had been prefabricated in advance by the REU but it was still a mammoth task, eventually taking four months and around 700 man-hours, with the CSE installation team working under 'high pressure'.

As previously described, the Elint fit for the Comet had been designed as a two-phase process with the initial Phase 1, interim, fit involving the installation of US-provided Elint equipment similar to that already installed in the Washingtons. In Phase 2 the centimetric portion of the installation (the S, C and X-band positions) would be replaced by *Breton* equipment.

The specification for the Phase 1 Elint was refined and amended during 1955-56 while the aircraft were undergoing modification at Marshalls. By mid-1956 the, the number of racks had been increased from 10 to 12, the S and X-band positions were split into separate search and D/F positions, and the two supervisors had been moved to a shared position. Other changes included the removal of the *Blonde* automatic recorder and reducing the number of rotating D/F aerials to four.

Special Operator positions were installed in the cabin either side of the gangway. Each position comprised a double bench seat and an equipment rack, constructed from tubular aluminium and reaching to the cabin ceiling, carrying the various units of the Elint fit. A complex cable loom was installed in the cabin, running through a number of cable ducts on the cabin walls and ceilings, carrying power and signals to the equipment racks. Care was taken to separate the power and signals cable where possible to avoid interference problems. The under-floor equipment bay was fitted out with APA-17 spinner aerials, their driving gear, and those elements of the Elint fit that needed to be positioned in close proximity to the aerials. Despite the complexity of the task the installation appears to have gone reasonably well: it was completed early in July and approved at an acceptance meeting on 19th July 1957.

Working Up the First Comet

Installation of SRIM 2113 in XK663 was completed in July 1957 after which the aircraft was subjected to a couple of months of ground checks and minor modifications before being cleared for flight testing in September. The initial schedule for XK663 called for 130 hours calibration flying, with the aircraft becoming operational in December 1957. Some 50 hours were completed in September, by the end of which the expected in-service date had slipped back to January 1958.

The calibration sorties in XK663 uncovered a number of problems with the Elint installation

THE COMET

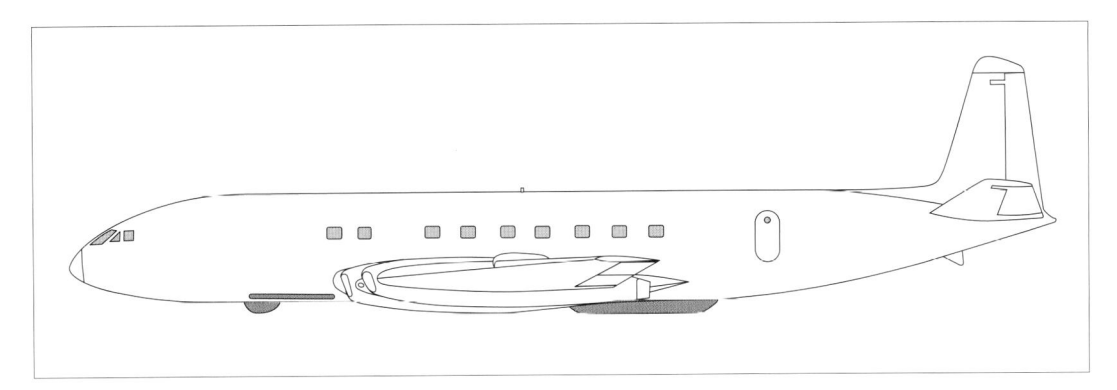

Diagram of the Comet 2R in its original configuration. Note the square windows and radomes covering direction finding aerials under fuselage.

in the aircraft, which was probably to be expected given the complexity of the fit. The most serious problem was electrical interference from the aircraft's DC power supply, the intensity of which varied from one item of equipment to another, particularly affecting signal recording. The problem was considered serious since it adversely affected the sensitivity of the Comet Elint installation. As the CSE noted: 'It is essential to clear this trouble before we can really find those signals which were too weak for the Washington to pick up but which we expected to receive on the Comets'. Other problems were experienced with mutual interference between the various items of Elint equipment and interference produced by the aircraft's W/T set. The DC interference problem was cured by the addition of smoothing filters to the DC supplies while the other interference problems were cured by improved earth bonding and other measures.

The calibration programme on XK663 also included a fairly exhaustive D/F calibration exercise, to measure and, where possible, correct D/F accuracy in both centimetric and metric bands. The calibration sorties showed that, in general, the initial Elint collection capability of the Comet was equal to, if not better than, the Washington, but problems were experienced in the metric and UHF/VHF bands, where the large number of signals picked up during test flights made direction-finding difficult. The increased number of signals was thought to be due to the better sensitivity of the installation and the higher operating height of the Comet compared to the Washington. Large X-band D/F errors were also encountered in the rear port quadrant and these were fixed by removing the IFF Mk.10 aerial. Eventually, after a considerable amount of work by the CSE, and over 130 hours of calibration flying, the first Comet 2R became available for operations in February 1958.

The Second Comet

The second Comet, XK659, arrived on 11th July 1957 and was initially used for much-needed flying training for a few weeks, completing a long-distance sortie to the Mediterranean and back, prior to entering the installation flight in August. By then SRIM 2113 had been slightly amended, replacing the APR-4 metric receivers with the more capable AN/ALR-5 receivers. These comprised an AN/APR-4Y receiver (a modernised version of the APR-4) coupled with an AN/ALA-2 panoramic adaptor. The installation in XK659 took a similar amount of time to the first Comet and was completed in late November 1957.

The second Comet XK659 at Nicosia in Cyprus during 1959. At this stage the aircraft had not been fitted with an HF aerial mast.
Glyn Ramsden

99

LISTENING IN

Flight trials of XK659 were initially delayed by various electrical problems and only got going in January 1958. In order to expedite the aircraft's entry into service it was decided to fly a minimal calibration programme, with only one D/F check on each position, effectively clearing the aircraft by analogy with the first Comet. The results from XK659 were cross-checked with those obtained from XK663 and further investigation only carried out in the case of discrepancies. An electrical interference problem soon became apparent, thought to be due to noise on the DC power supply, similar to that encountered on the first aircraft. This was a serious problem as a CSE report noted: 'any recordings taken ... are virtually useless because of the noise present'. Power supply suppressor kits were fitted but the noise persisted and, after some investigation, it was then discovered that the aircraft wiring diagrams were incorrect and that the interference was due to an electrical fault unique to XK659. By April the problem had been largely overcome. The effort required to fix the interference problem in XK659 highlights the complexity and uniqueness of the three Elint Comets.

Comet Operations Begin

The first operational Comet sortie took place on 3rd February 1958 when XK663, carrying the Commandant, CSE as an interested observer, flew a 4 hour 15 minute daylight Border sortie along the East German border.

Planning for the integration of the Comets into the ongoing RPF programme had begun in August 1957, when the Secretary of State for Air, The Hon. George Ward, had notified the Prime Minister, Harold Macmillan, that: 'I propose to substitute Comet aircraft for Washingtons in the provisional programme as and when the aircraft become available. The Comets will take part in the programme subject to the conditions laid down for these flights generally'.

The plan for January-March 1958 had originally comprised a number of Canberra/Washington sorties in the Barents Sea, Washington sorties against Syria, Israel and Egypt, and Canberra/Washington sorties in the Baltic. All three of these operations were subsequently re-arranged and rescheduled for the Comet. The first Comet night RPF took place on 11th February when XK663 flew a 4 hour 25 minute sortie over the Baltic in conjunction with Canberras WT301 and WJ775. A second night Baltic sortie took place on the 14th February, this time in conjunction with Canberra WT301 operating from Ahlhorn in Germany. Towards the end of February the second Comet XK659 was detached to Nicosia from where it carried out four day and one night sortie along the Mediterranean coasts of Egypt, Israel and Syria. Unfortunately problems with the navigation kit resulted in two flights being only partially successful. The final operation in the rescheduled RPF programme was the detachment of Comet XK663 and Canberra WJ775 to Bodø in March 1958 for two joint Comet/Canberra operations over the Barents Sea. This concluded the September 1957-March 1958 RPF programme.

The entry into service of the Comet seems to have gone well, the CSE commenting that: 'The results so far obtained on the Comet aircraft are as good as, or in some cases, better than those previously obtained on the Washington aircraft ... it is considered to be a noticeable step forward that no noticeable ground was lost in the changeover from old to new aircraft'.

A New RPF Programme

April 1958 saw the start of a new RPF programme using the two Comets (XK663, XK659) and the two operational Canberra B.6s (WJ775, WT301). The programme comprised operations in the Baltic (April), Black Sea and Mediterranean (May), Black Sea (June), and Black Sea and Mediterranean (September). Border sorties over Germany were slotted into the programme when the schedule and aircraft serviceability allowed. Most of the operations were joint affairs, with a Comet flying in conjunction with a Canberra (or sometimes two Canberras). All sorties were flown to the existing RPF rules, which normally mandated night sorties in international or friendly airspace, standing off a healthy distance from the target border. The exceptions to the night sortie rule were Border sorties over West Germany and operations over the Mediterranean off the coast of Egypt and Syria, both operational areas being assessed as low risk.

In early June the squadron was alerted for a possible maritime Elint operation against Soviet naval units and flew a simulated *Claret* operation in a Comet by way of rehearsal. Later in June one Soviet cruiser, two destroyers and four submarines left the Baltic and by 24th June had reached 61°N 03°W, steaming NNW at approximately 10 knots. The Air Ministry obtained approval for operations *Moselle* (Shadowing), *Claret* (Electronic Reconnaissance) and *Sherry* (Photographic Reconnaissance) against the force and two *Claret* sorties were flown by

THE COMET

Comet 2R XK659 on approach to Luqa showing the small forward radome. Malta was a valuable staging point and operational base for many 51 Sqn operations throughout the Cold War. *Victor Pisani*

Comet XK663 over the North Sea at the end of the month.

In the first half of July 1958 two Comets (XK663, XK659) and two Canberras (WJ775, WT301) were deployed to Nicosia for a planned series of twelve sorties into the Black Sea. The first RPF, a double joint Comet/Canberra operation utilising all four aircraft, was flown on the night of 15th July. Unfortunately, Middle Eastern politics then intervened when on 14th July a bloody military coup overthrew King Faisal II of Iraq, establishing an Iraqi republic and triggering widespread unrest in the region. In response to the coup, British troops were deployed to Jordan to support King Hussein (Operation *Fortitude*) and US troops landed in Lebanon. As a result of the upheaval the Comet/Canberra detachment was cut short, returning to the UK via Malta on 16th July.

Despite the early return from Cyprus, operations using the Comet and Canberra seem to have been generally successful. In August the CSE reported that: 'The work of [the] Squadron continues to progress well. The two operational Comets and the Canberras are producing good results and no major snags have occurred in either operating or maintaining their complex special installations'. Operations for the remainder of 1958 followed the usual pattern with night RPFs by Comets and Canberras over the Baltic, the Black Sea and Barents Sea.

The importance of Radio Proving Flights as part of the UK Intelligence-gathering effort was confirmed in September 1958. A JIC examination into the state of UK intelligence, conducted in conjunction with the LSIB, recommended five main measures to improve the situation, the third of which was the 'continuance of radio proving flights'.

In August 1958, at a ceremonial parade at Watton, 192 Squadron was renumbered as 51 Squadron in keeping with the policy of retaining the older squadron 'number plates'. The squadron had been initially formed in May 1916 and the previous incarnation of 51 Squadron, equipped with Avro Yorks in Transport Command, had disbanded in October 1950.

The Third Comet XK655

The third Comet, XK655, arrived at the CSE on 24th March 1958 and was initially used for flying training before going into the Installation Flight for the application of SRIM 2113 in mid-April. The Phase 1 Elint fit to SRIM 2113 had been intended as an interim installation pending the availability of the *Breton* system, and the intention was that the Phase 2 fit would replace the APR-9 receivers and APA-11 analysers at the three centimetric stations with *Breton*. By mid-1958 the first *Breton* units were becoming available and it was decided to fit XK655 with a partial *Breton* installation (S-band *Breton* in racks 1 and 2) to allow an operational evaluation of the system to be carried out prior to retrofitting all three Comets. The other positions in the XK655 were fitted to the existing Phase 1 standard. The *Breton* installation used the ARI 18058 DV receiver as a wideband 'guard' receiver for coarse frequency determination, the ARI 18050 superheterodyne receiver for fine frequency determination and the CRT-2 analyser for pulse analysis. Since the static-split D/F method originally intended for *Breton* had been found inferior to the spinner technique on large aircraft, D/F was provided by an ALA-6 spinner system (replacing the APA-17

LISTENING IN

installed in the first two aircraft). The installation in XK655 was designated Phase 1½, being half way between the original Phase 1 fit and the final Phase 2, full *Breton* fit, and was completed at the end of August 1958.

Ground testing of the fit in XK655 took three months and the aircraft became available for flight trials at the end of November 1958, subsequently flying its first operational sortie (a Border flight) at the beginning of February 1959. A number of equipment and operating snags emerged and calibration sorties and trials continued during the first few months of 1959, slotted in between operational sorties.

The installation of *Breton* in Comet XK655 raised a requirement for *Breton* training facilities and so in September 1958 work began on the construction of a ground *Breton* training rack, to allow Comet Special Operators to gain experience with the equipment. The initial cadre of *Breton* SOs received training on the system from the CSE Research Squadron, after which the ground rack was moved to the RCM School and instruction on *Breton* was introduced into the general Special Operator training syllabus. Arrangements were also made for airborne training and Varsity WL687 was modified as a *Breton* training aircraft to SRIM 2351 during 1958. The conversion work, which included fitting ARI 18058, ARI 18050, CRT-2 and an ALA-6 spinner aerial, was completed in January 1959.

Loss of Comet XK663

The completion of XK655 in February 1959 finally brought the newly renamed 51 Squadron Comet establishment up to the planned three aircraft, relieving some of the CSE concerns regarding available flying hours. Unfortunately this situation did not last very long.

In late April 1959 Comet XK663 went into No.4 hangar at Watton for airframe and Elint equipment modifications. The airframe modifications were carried out by a contractor's working party while the Elint modifications were carried out by CSE personnel. Both modification programmes were completed on 25th May, following which work began on a 600-hour scheduled service, prior to returning the aircraft to service. On 2nd June a number of tasks were being carried out on the aircraft as part of the service, one of which was the replacement of an air conditioning dehumidifier unit, located in the forward equipment bay beneath the navigator's compartment. Part of this job involved painting felt-lagged pipework with a hydraulic-fluid-resisting lacquer as a protective coating. The work was carried out by a Senior Technician, assisted by a Corporal, using a mains safety lamp to provide illumination in the dark and cramped equipment bay.

Due to an unfortunate misunderstanding, the safety lamp was left, still switched-on, in the equipment bay at the end of the working day.

The sad remains of Comet 2R XK663 following the fire in No.4 Hangar at Watton in February 1959. The relatively undamaged hangar is a testament to the efficiency of the Watton fire crews. *Author's collection*

By itself, leaving a lit safety lamp inside the aircraft should not have been particularly hazardous. Unfortunately the Senior Technician working on the dehumidifier had also left an unsealed tin of highly flammable lacquer with a low vapour point in the equipment bay. Furthermore, the live safety lamp was old and poorly maintained, with (it was later speculated) many cuts and abrasions to the electrical lead. Sometime in the early hours of 3rd June a spark from the defective lead is assumed to have ignited the vapour in the Comet's equipment bay. A member of the Security Patrol spotted smoke issuing from the hangar at 0405hrs and immediately raised the alarm. The station fire tender was on the scene in minutes and firemen began directing foam, then water onto the blazing Comet. Civilian fire appliances from surrounding towns also attended. Meanwhile the Commandant, CSE, had taken charge of the situation and organised the removal of a Lincoln and Varsity which were also occupying the hangar. Thanks to the combined efforts of the RAF and civilian fire crews the fire was extinguished without any major damage to the No.4 hangar structure or other aircraft. Comet XK663 was, however, a total write-off, with the fuselage completely consumed by fire.

The loss of XK663 was a heavy blow to 51 Squadron. Three aircraft were considered the minimum number to meet all operational requirements. Regular maintenance, flying training, calibration, overseas detachments and periodic refits meant that the remaining two Comets would have insufficient flying hours to meet all the planned operational tasks. A replacement aircraft was clearly required. The problem was, the Comet 2R was a very expensive aircraft and funding for a replacement was not a foregone conclusion. Before asking for a replacement aircraft, Signals Command established a Working Party to review future RPF requirements and determine the best aircraft for the job. The Working Party confirmed the need for RPFs: the information they obtained was highly regarded by ACAS(I) and, although GCHQ resented the high cost of airborne Elint, they recognised its importance in the exchange of intelligence with the US. In addition, an aircraft provided a 'Quick Reaction' capability for Elint investigations, able to deploy to areas of interest at short notice. The Working Party decided that the ideal future Elint platform was the larger, longer-range, VC10, with the Comet acceptable as an interim type.

While the Working Party report had confirmed the requirement for a replacement for XK663, obtaining a VC10 was clearly out of the question and so negotiations were opened to acquire a BOAC Comet 2, then being used for engine trials. Unfortunately by the end of 1959 the negotiations had stalled over the price. Doubts then emerged regarding the likely fatigue life of the Comet 2, which in turn raised questions as to the expected life of the remaining two Comet 2Rs in the Elint role. As a result of this uncertainty, plans to replace the burnt-out Comet were put on hold while the fatigue life question was investigated. Thus 51 Squadron had to struggle on with only two Comets during 1960.

Operations 1959

Operations during 1959 followed a similar pattern to 1958 with operations alternating between the Baltic and the Black Sea/Caspian Sea, plus two short detachments to Bodø (spring and winter) for forays into the Barents Sea. Up until 1959 all RPF operations against the Soviet Bloc were carried out on dark moonless nights in order to minimise the chance of interception. During the early/mid-1950s Soviet GCI radar had a relatively poor tracking capability against small high-altitude targets and Soviet AI radar was restricted in range. Thus operations in darkness provided a reasonable measure of protection, particularly for the small and fast Canberras of 51 Squadron. However, the downside of restricting operations to moonless nights was that it limited the number of sorties that could be flown each month, and made the dates on which they were flown predictable, reducing the chances of surprising Soviet defences.

By the late 1950s Soviet GCI and AI radar had improved to the point where night interceptions were no longer such a challenge to the Soviet air defence system and therefore RPF operations in the no moon period no longer offered the same protection. Thus, sometime in 1959, the UK authorities relaxed the 'no moon' rule and allowed RPFs in daylight. However, the rule was only relaxed for single aircraft sorties: multiple aircraft RPFs were considered more provocative (and thus more risky) and continued to be flown in darkness. The Comets and Canberras of 51 Squadron were regularly intercepted by Soviet fighters during daylight sorties, with close approaches often being made for identification purposes although in the main, interceptions were generally conducted according to the rules with no display of hostile intent.

Improving the Comets

Shortly after the Comets entered service work began to improve their Elint installations by the substitution of newer equipment. The first change was an upgrade of the VHF communications intercept receivers. In May 1958 the wartime-vintage ARR-5s in XK663 were replaced by the UK-designed R.216 receiver operating in the 20-155MHz band The R.216 had been developed in the early 1950s by the SRDE under War Office sponsorship to meet an inter-service requirement for a high-quality communications receiver for the ground intercept and D/F role. Due to difficulties obtaining suitable power supplies only three sets were initially fitted in the Comet, but more were fitted later as the necessary units were obtained and the R.216 was reported as being a great improvement over the ARR-5.

The second change was an upgrade of the recording facilities. The Comets were initially fitted with a single-track ANQ-1A magnetic wire recorder at each Special Operator's position to record voice intercepts, radar pulse modulation and Special Operator commentary. The problem was the ANQ-1A was obsolete, with poor reproduction quality, and a replacement was urgently required. The SRDE developed a single-track magnetic tape recorder to CSE specifications, two of which were fitted in Comet XK659 and used on operations in December 1958. Although the SRDE recorder proved superior to the ANQ-1A, CSE requirements had evolved and now called for a multi-track recording capability to allow the simultaneous recording of intercepted signals, operator commentary, a reference tone, timing information and other ancillary data. In early 1959 a contract was issued to the Data Recording Instrument Company (DRIC) for the design and manufacture of an airborne 14-track magnetic tape recorder. In the interim the CSE built a two-track, and two prototype four-track recorders themselves to obtain operational experience of multi-track recording. The first four-track unit was installed in a Comet in the second half of 1959 and by the end of 1960 the Comets were flying with a combination of the CSE four-track recorders, a number of single-track SRDE recorders and the original ANQ-1A wire recorders. The first DRIC 14-channel recorder, built to CSE specifications, arrived in the second half of 1960 but unfortunately proved unsuitable for airborne use and was eventually relegated to the ground recording role. Attention then turned to the Solartron 14-channel tape recorder, based on a US military recorder, and under development for the Canberras. A prototype was delivered in 1961 and the system appears to have been installed in the Comets and Canberras during 1962-63.

The third main change was the replacement of the APA-17 D/F system with the more modern ALA-6 system. Like the APA-17, the ALA-6 used a rotating 'spinner' aerial. The system appears to have been fitted in XK655 as part of the *Breton* 1½ fit in 1958 and later replaced APA-17 in XK659 in August 1961. The training Varsity was fitted with a single ALA-6 in mid-1957 (replacing one of the APA-17) to familiarise SOs with the system in advance of its introduction in the Comet.

Efforts were also made to improve the pulse analysis equipment in the Comet. The first two aircraft had been fitted with APA-11A analysers under Phase 1. However the long-term plan was to replace the APA-11 with the CRT-2 analyser developed as part of the *Breton* suite. Comet XK655 appears to have been fitted with CRT-2 during its *Breton* 1½ installation, and the system replaced APA-11 in four racks in Comet XK659 in the first half of 1959. At the same time the CSE introduced the X378 miniature pulse-width meter, which turned out to be superior to the CRT-2 for pulse-width measurement. Unfortunately CRT-2 did not prove particularly reliable in service and the units were eventually removed from the Comets and replaced by the APA-11A circa 1960. The failure of CRT-2, and the absence of suitable alternatives, eventually led to the development of the Integrated Display and Measurement System (IDMS) during the late 1960s and this system was installed in the Comet's successor, the Nimrod.

Work was also carried out to improve the *Breton* installation in XK655 as operations during the first half of 1959 using the *Breton* 1½ system had revealed a number of shortcomings. During the second half of 1959 the CSE made a number of changes and improvements to *Breton*, including the addition of a new optical system providing superimposition of analysis information on the D/F display, which allowed a single SO to carry out the complete task of search, D/F and signal analysis. Work was also carried out on the *Breton* receiver units to improve range and signal discrimination and to enhance pulse analysis. Flight trials of this improved system, designated *Breton* 1¾, were carried out in the period 1959-61 using Varsity WL687, the *Breton* training aircraft.

9 Elint in the 1960s and 1970s

'The aim of Radio Proving Flights is to provide electronic intelligence on the Soviet Air Defence System, which is then used in conjunction with information from other sources to design electronic counter measures which are incorporated in the weapons system of the 'V' Force. The flights are essential to maintain the effectiveness of the 'Nuclear Deterrent'.
Aide-memoire to Lord Mountbatten

At the beginning of 1960 the UK's operational Elint assets were two Comets (XK655, XK659), three Canberra B.6 (WJ775, WT301 and WT305, the latter still under conversion) and a Canberra B.2 (WH698). In addition to the operational aircraft the squadron also used a Varsity (WL686), maintained by the CSE Development Squadron, for Special Operator training.

The primary operational role of 51 Squadron continued to be the provision of intelligence on the air defences of the Soviet Union in support of the nuclear deterrent. By 1960 the Soviet Union had begun to reorganise its defences to counter the high-altitude bomber threat, deploying the new SA-2 *Guideline* surface-to-air missile, introducing new radars with improved range and height performance, and automating parts of their air defence control system. The intelligence provided by 51 Squadron was thus of vital importance to the V-Force, providing not only details of the existing defences, but also early warning of emerging threats.

The official view of the airborne Elint task at that time is best summed up by a contemporary aide-memoire to the Chief of the Defence Staff, Lord Mountbatten: 'The aim of Radio Proving Flights is to provide electronic intelligence on the Soviet Air Defence System, which is then used in conjunction with information from other sources to design electronic counter measures which are incorporated in the weapons system of the "V" Force. The flights are essential to maintain the effectiveness of the "Nuclear Deterrent".

The task comprises ascertaining the location and wave bands of all Soviet radars in the operating area, including those of Surface-to-Air Guided Weapons, by means of recording apparatus in the aircraft and in certain ground stations (Operation "Viking" and others) co-operating with them. By stimulating Soviet reaction the flights also make it possible to investigate Soviet systems of communication.

In addition to supplying information required to maintain an effective airborne jamming system the flights provide as a bonus information on Order of Battle of Soviet radar, anti-air Guided Weapons systems and communications systems.

Some of the information gathered on radar Order of Battle wavelengths and equipment employed by the Soviet Union is ultimately included in a publication made available to NATO countries. However, the various sources of information used to compile this volume, including Radio Proving Flights, are not divulged.

In conclusion, the information provided by Radio Proving Flights is vital if Bomber Command is to penetrate the Soviet defences effectively. The task is a continuous one since new equipment and techniques are constantly being introduced by the Soviets'.

U-2 and RB-47H Incidents

In late 1959 arrangements had been made for a 'Four Power' (UK, US, Soviet Union and France) summit conference, to be held in Paris in mid-1960, to discuss the status of Berlin. In line with standard practice, operational Elint flying was restricted during the first half of 1960, in the run up to the Paris Summit, to avoid the possibility of an accident causing a diplomatic incident with the Soviet Union. As a result the overseas detachment to Nicosia in January 1960 was used to fly sorties against Middle Eastern countries rather than the Soviet Union. Unfortunately the shooting down of the Lockheed U-2A flown by Gary Powers on 1st May

LISTENING IN

The Boeing RB-47H, introduced into squadron service during 1955, was a mainstay of the USAF Elint fleet until replaced by the Boeing RC-135C in 1967. *T Panopalis Collection*

1960, shortly before the start of the summit, effectively scuppered any chance of an agreement being reached with the Soviet Union. The loss of the U-2 also had a direct impact on UK Elint operations since the British government was anxious not to exacerbate an already difficult situation. As the 51 Squadron OC noted: 'Operational flying was limited during this month by decision of higher authority'. In fact all RPFs were cancelled and only daylight Border sorties were flown during May and June.

The Soviet response to the U-2 incident included threats by Khrushchev and Marshal Malinovsky (Soviet Defence Minister) in late May/early June that any further U-2 overflights of the Soviet Union would result in 'rocket' attacks on the bases from which they were made, and bases in the UK, France, Italy, Japan, Pakistan, Turkey and Norway were specifically mentioned. The Soviet threats had to be taken seriously and in June the JIC took the precaution of commissioning a report on their implications for the UK. The conclusion was that, although the Soviet Union did have the means to attack air bases, they were unlikely to do so for fear of provoking US retaliation. The assessment was that the statement was largely political in nature, intended to discourage the US from further flights, dissuade US allies from hosting such operations and to reassure the Soviet population. The report also noted that the threat had specifically referred to U-2 flights and had made no reference to other means of collecting intelligence. There was, however, a need for caution, and this was reinforced on 1st July when, exactly two months after the loss of the U-2, the Soviet Union shot down a USAF RB-47H over the Barents Sea.

During the second half of 1955 the USAF had re-equipped their main strategic Elint unit, the 55 SRW, based a Forbes AFB, Kansas, with the specially-modified RB-47H variant of the B-47 Stratojet, replacing their obsolescent Boeing RB-50Gs. The RB-47H carried three Special Operators in a pressurised bomb-bay capsule. The aircraft's primary Elint system was the APD-4 automatic receiver/analyser (see Chapter 11), supplemented by manually-operated APR-17 receivers, an ALA-6 direction finding (D/F) system and ALA-74 pulse analysers. The RB-47Hs of the 55 SRW made regular monthly detachments to the UK from where they flew Elint sorties over Europe.

On 1st July 1960 an RB-47H (53-4281) of the 55th SRW based at RAF Brize Norton, Oxfordshire took off on an Elint sortie into the Barents Sea. The aircraft's route took it up the coast of Norway, around the North Cape and then south east along the coastline of the Kola Peninsula, flying in international airspace some 50 miles (80km) off the Russian coast. As the RB-47H approached the end of the peninsula it was intercepted by a MiG-19 *Farmer* which flew in close formation with it. Some 50 miles (80km) off Cape Svyatoy Nos (Holy Nose Point), near the entrance to the White Sea, the RB-47H made a planned turn to the north-east, turning away from the coast. On the turn the MiG-19 broke away from the RB-47H before swinging back to attack with cannon fire. Two firing passes were made and the aircraft was shot down. Sadly only two of the crew of six survived and all three Special Operators were lost. The Soviet Union subsequently justified the attack by claiming that the aircraft had infringed Soviet territorial waters, approaching to a

ELINT IN THE 1960s AND 1970s

In-flight view of an RB-47H showing the bulged belly compartment housing three Elint operators. These aircraft were regular visitors to USAF bases in the UK, from where they flew their missions over western Europe. T Panopalis Collection

point 11.83 miles (19km) north of Svyatoy Nos.

The U-2 and RB-47H incidents prompted the Prime Minister to request that the JIC prepare a report on UK intelligence activities 'which, though not contrary to international law, might be disagreeable to the Russians'. The loss of the RB-47H also brought the activities of US units operating from the UK under greater scrutiny. It emerged that the UK was not given the route of US Elint sorties in advance, but were normally given the Navigator's log and intelligence results from each flight some time afterwards. As such the Air Ministry was only aware in general terms of the likely aircraft track of any sortie.

The Air Ministry were later passed the flight plan for the 1st July RB-47H sortie, along with ground intercepts of Soviet air defence activity, including tracking data. From this it appeared

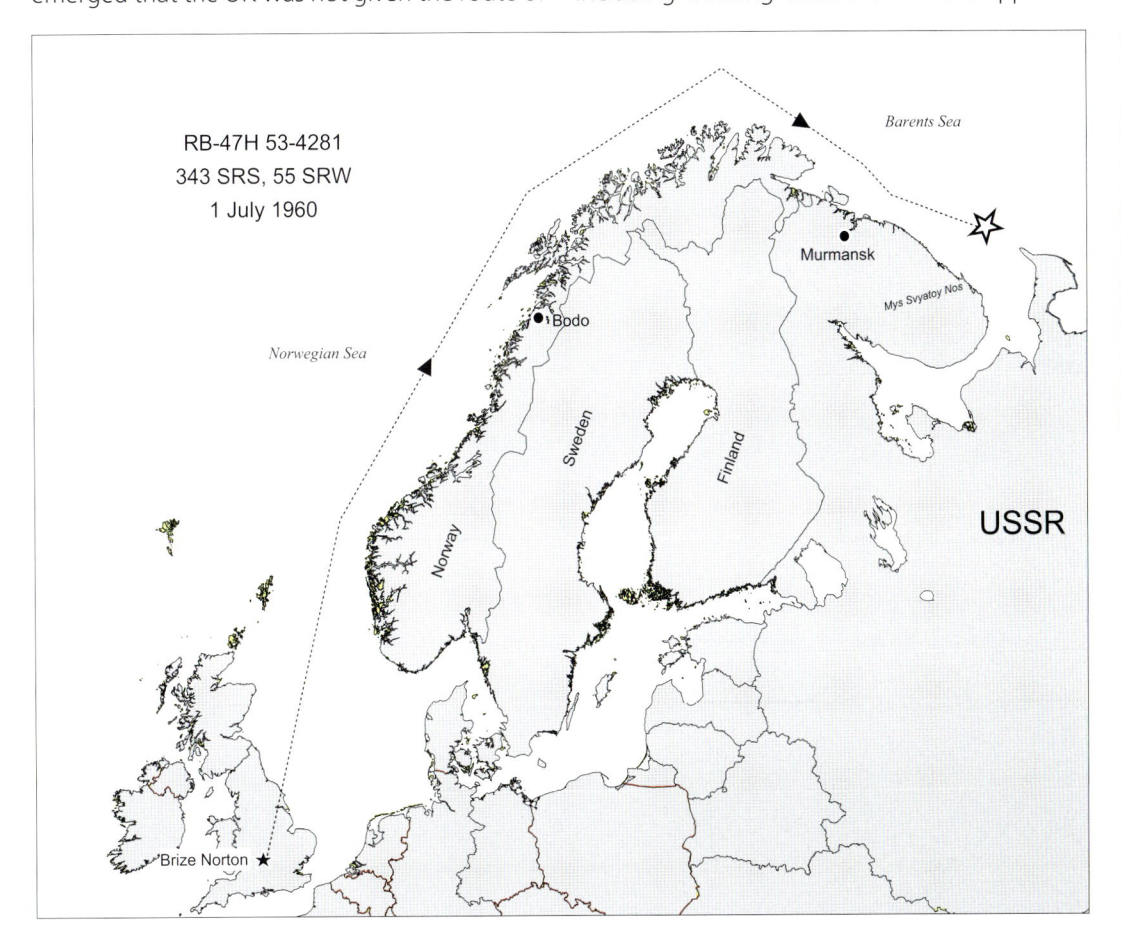

The ground track of RB-47H (53-4281) shows its progress up to the point at which it was shot down by a Soviet fighter. Only two out of the crew of six survived the incident. Following the shoot down USAF flights from the UK came under increased scrutiny.

RB-47H 53-4281
343 SRS, 55 SRW
1 July 1960

107

LISTENING IN

that the plan had been to keep 60nm (111km) from the Soviet coast, although intercept evidence suggested the RB-47H had actually closed to 28nm (52km) before resuming its planned track, then standing off some 66nm (122km). Tracking data was inconsistent, but appeared to show the aircraft may have closed the coast again before being shot down.

Following a JIC report on the matter, a new US-UK agreement governing the use of UK bases for reconnaissance purposes was proposed. This required the US to supply the UK government with full details of reconnaissance sorties intended to be flown from the UK by the 15th of the preceding month and effectively subjected US Elint flights to a similar approval regime to that applied to UK flights. The JIC report on UK intelligence activities was examined by the Prime Minister, the Foreign Secretary, the Defence Secretary and others at a meeting in mid-September. The report admitted there was a possibility that aircraft conducting RPFs over international waters might be attacked and shot down by Soviet defences. In mitigation, however, it noted that this was unlikely in the case of Baltic flights as the aircraft would be under surveillance by friendly and neutral radars, and the Soviets would find it hard to justify any aggressive action. It also pointed out that the UK had stopped flying RPFs into the Barents Sea as Norway no longer provided staging airfields (see Chapter 10). Finally, the report stressed the benefits that UK RPFs brought in the exchange of intelligence with the US, the benefit being 'out of all proportion to the effort involved'.

The Ministerial meeting appears to have generally accepted the main thrust of the JIC report: that RPF flights had great intelligence value and that, if prudently conducted, the risk was small. There also seems to have been a view that giving in to Soviet pressure on RPFs would simply encourage the Soviet Union to intensify its efforts against all intelligence gathering operations. Authorisation was thus given for the resumption of RPFs. However, the Prime Minister requested that the rules governing RPFs should be reviewed to ensure these flights respected Soviet claims regarding territorial waters, even when such claims were not admitted by the UK government. This would remove any possible Soviet objection to the sorties on the grounds that they infringed their territory. He also requested that the US should be consulted to ensure their Elint sorties from the UK followed the same rules, thus preventing arguments regarding the safety margins of individual sorties.

A subsequent review of the rules governing UK RPFs confirmed that the 30nm (55km) limit used on these sorties kept aircraft well outside any territorial waters claimed by the Soviet Union. The rules also appear to have been revised to reduce the risk of an adverse response from the Soviets, with sudden deviations in track toward Soviet territory during RPFs to stimulate a radar reaction, no longer allowed. Other aspects of the comprehensive rules governing the flights were also reviewed. Sorties were made under conditions of radio silence, but with all aircraft in W/T or R/T range of a controlling base, allowing them to be recalled if necessary. To this end the Comet was equipped as a radio relay aircraft, allowing W/T messages to be passed to Canberras via R/T. Radio silence could be broken at any time at the

Canberra B.6(BS) WT301 in early 1960s. Note the Canberra T.11 nose but absence of tail warning radome under tail.
via Peter Green

108

discretion of the RPF aircraft's captain, but the aircraft would then immediately abandon its sortie and return to base. In the event of an emergency the aircraft would be treated exactly as any other RAF aircraft in distress, except that special arrangements existed for search and rescue over the Black Sea. The rules also governed how many Elint aircraft could operate in the same area at the same time: one in daylight and four at night.

Curiously, given the long-standing US-UK co-operation in the Elint field, it appears that the two governments had not previously exchanged their respective rules on the conduct of Elint sorties. Thus, at the Prime Minister's request, the UK Ambassador in Washington was instructed to show a summary of the rules under which UK sorties operated to the relevant US authorities and to enquire whether theirs 'differ in any major respect'. The intention was to use the approach to try to get the US and UK rules to converge. If that could be achieved it would have obvious political benefits for the government, who would then be able to state that US reconnaissance flights conducted from the UK were carried out under similar rules to those governing UK sorties.

The loss of the U-2 and RB-47H had in both cases resulted in US airmen being captured and imprisoned by the Soviet Union, focussing attention on the lack of a cover story for UK Elint sorties. The standard UK government response to any questions on 'spy flights' was to refuse to comment on intelligence matters. However, there was a growing realisation that it would be impossible to hold that line if an RAF Elint aircraft was shot down and its crew captured. A cover story, even a 'weak one open to ridicule', might be helpful for a number of reasons. It might reduce the propaganda value of such an event, could be used as a cover when flying sorties from foreign bases, and it might even help smooth matters with the Soviets (by giving them a face-saving excuse not to take retaliatory action). A basic cover story was proposed, that RPF aircraft were engaged on technical flights for the calibration and testing of radar and navigational systems. Approaches were also made to the Turkish and Iranian governments to see if they would support localised versions of the cover story that is, that the RPF aircraft were testing Turkish or Iranian radars.

The Air Ministry continued to prepare for RPFs in the period May-September 1960, while the review was still in progress, presumably on the off chance that operations would be resumed. This resulted in operations being cancelled at the last moment when political clearance was withheld, and was a source of some frustration to 51 Squadron and the CSE, who complained: 'It should be explained that the bringing of the Comets and the Canberras to stand-by for operations invariably requires their grounding for three or four days before the operation so that the delicate special receiving equipment can be tuned to the highest peak of serviceability and kept that way. Last minute cancellations waste all the preparation time, which could be most usefully absorbed in training ...'.

The first UK RPFs following the U-2 and RB-47H incidents took place at the end of October 1960 when Canberra B.6s WJ775 and WT305 were detached to Jever, near Wilhelmshaven on the north German coast for an operation during the new moon period. After a few days delay due to poor weather, a joint RPF (*Ripper One/Two*) was flown into the Baltic on the night of the 28th October.

The US planned to restart Elint sorties from the UK at around the same time. The first submission under the new approvals regime was made on 14th October when the USAF requested authorisation for a series of four RB-47H flights during November. These comprised a sortie into the Adriatic and Aegean, two sorties over the Black Sea and a single sortie over the Barents Sea. A request for authorisation of an additional sortie, a flight along the Bulgarian border and Albanian coastline at the end of October, followed a few days later.

The USAF submission, comprising a short description of each sortie and an accompanying chart showing the proposed track, was first delivered to the Air Ministry, who copied it in turn to the Foreign Office. The Secretary of State for Air (George Ward) then had a summary of the proposal prepared for approval by the Prime Minister. Curiously, the initial summary submitted to the PM merely listed the sorties, dates, and areas of operations. Harold Macmillan was clearly taking no chances and requested the same level of detail as supplied for UK authorisations that is, whether the sorties would be flown by day or night, and the closest approach to Soviet/satellite territory.

In fact, the US were being understandably cautious: the Black Sea sorties stood off 85nm (157km) from the Soviet and Bulgarian coasts, while the Barents Sea sortie remained west of the Norwegian/Soviet border, approaching no closer than 150nm (277km) to Soviet territory. The rather less risky Adriatic and Aegean flights approached to 44nm (81km). All the sorties were to be flown at night or in twilight. The late

LISTENING IN

This map shows the routes of two USAF sorties into the Baltic submitted for approval by the UK government in late 1960. US RPF planning was still impacted by the loss of the RB-47H at this time and the sorties keep a healthy distance from the Soviet Bloc coast.

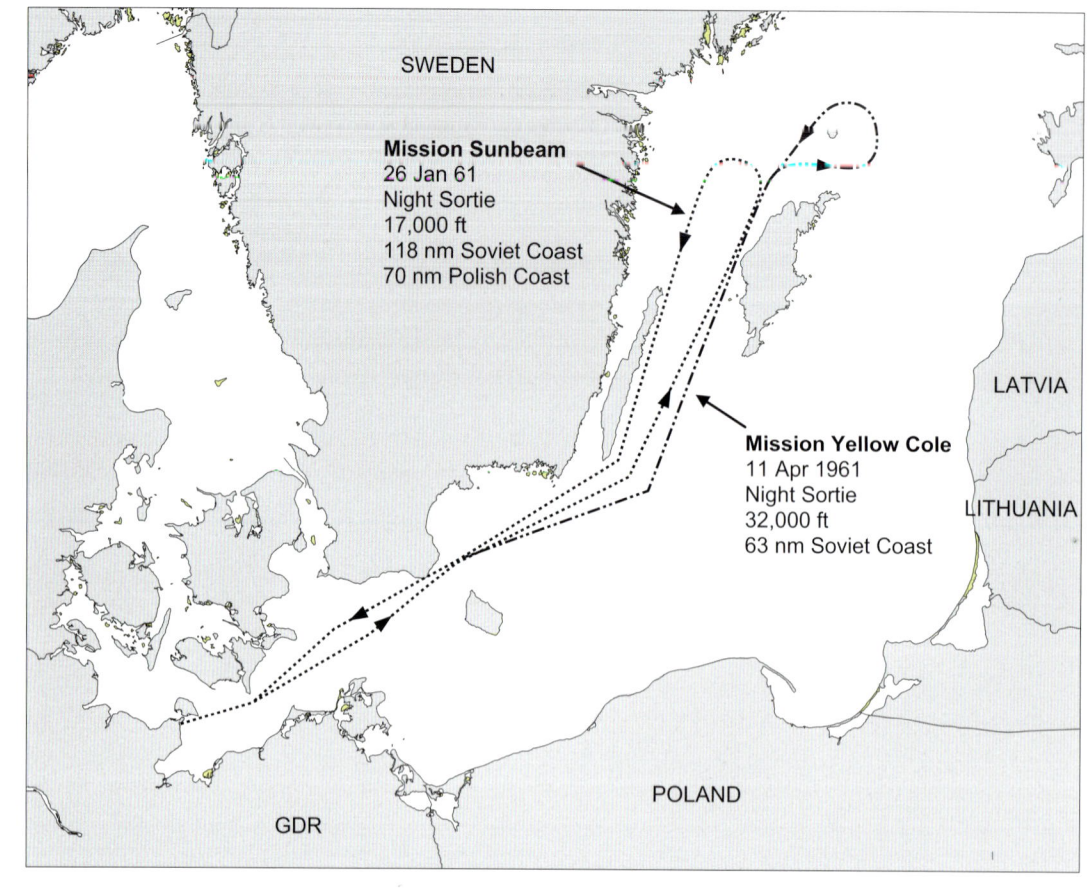

Mission Sunbeam
26 Jan 61
Night Sortie
17,000 ft
118 nm Soviet Coast
70 nm Polish Coast

Mission Yellow Cole
11 Apr 1961
Night Sortie
32,000 ft
63 nm Soviet Coast

USAF submission for the sortie at the end of October drew some criticism from the Secretary of State for Air. The request had been received on 19th October with a planned sortie date of 28th October, leaving little time for a scrutiny of the proposals. ACAS(I) defended the Americans, noting that: 'Unlike ourselves, who regard Proving Flights from a strategic viewpoint, the Americans look upon them as tactical operations and therefore are likely to initiate flights at short notice'. He also pointed out that the US had to ferry aircraft into the UK to fly the sorties, which further complicated planning. The late submission in October appeared to be a one-off and the evidence suggested the US were doing their best to comply with the terms of the new authorisation agreement. Finally ACAS(I) observed that: '... the US authorities have gone a long way to meet our requirements and I feel it would be unwise to press this matter too far'.

In the end the UK government's approval for the November sorties was academic. A Presidential election campaign was under way in the United States and on 1st November the USAF announced they were cancelling all SAC Elint sorties in the run-up to the election itself. On 8th November 1960, John F Kennedy (Democrat) beat Richard M Nixon (Republican), ushering in a change of government. USAF Elint sorties from the UK did not restart until mid-January 1961.

Upgrading the Canberras: the T.11 Nose, *Breton* and Its Successor

In 1958, when the Comet entered service, 192 Squadron had two operational Canberra B.6s on charge (WT301 and B.6 WJ775), a shorter-range Canberra B.2 WH698, mainly used for training purposes, and an unmodified Canberra B.6, WT305, awaiting a decision on its conversion to the Elint role.

The delay in converting WT305 was due to indecision regarding the future modification standard of all three B.6s. After much discussion the CSE decided in early 1958 to fit all three Canberra B.6s with a specially-tailored and adapted version of *Breton*. The opportunity was also taken to try to improve the D/F accuracy of the Canberras by fitting a larger nose radome. Operational experience with the existing two Canberra B.6s had shown that neither the short-nose radome on WT301, nor the longer T.4-style radome on WJ775 provided an adequate 'field of view' for the direction-finding

ELINT IN THE 1960s AND 1970s

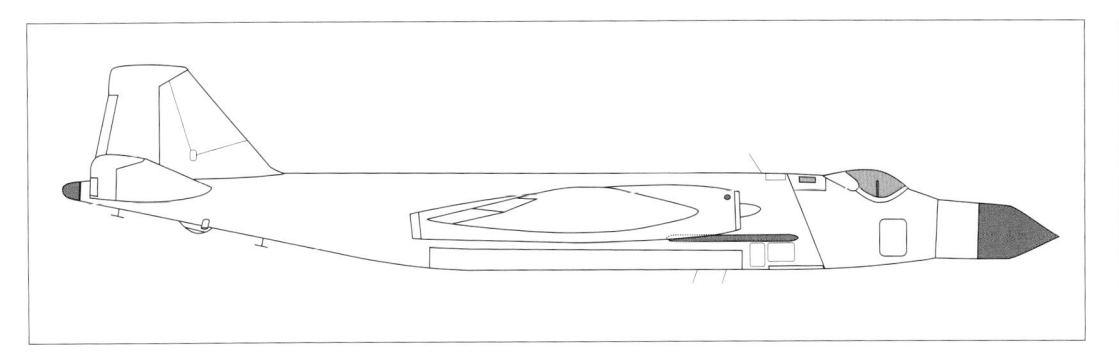

Canberra B.6(BS) with extended nose, based on the Canberra T.11. Unlike the T.11 AI radar trainer, the new nose section did not include a hydraulically-operated windbreak forward of the entrance door.

(D/F) aerials installed in the aircraft's nose. The plan was to use the extended nose and radome combination developed for the new Canberra T.11 radar-trainer as an off-the-shelf solution, thus avoiding the cost of new development. WT305 was selected as the initial T.11/*Breton* conversion, going to Boulton Paul in June 1958 for the installation of a modified T.11 nose.

While WT305 was at Boulton Paul the CSE started work on the design of the Canberra's S/X-band *Breton* installation. This was a major undertaking: the main challenges were to modify the system so that the search, D/F and analysis functions could be carried out by a single SO, and to rework the system layout so that it fitted into the cramped confines of the Canberra. By the time WT305 returned to the CSE in June 1959 most of the layout had been settled, the new *Breton* system boxes had been designed, and some prefabrication of parts had been carried out. Even so, the installation, to SRIM 2340, was a complex task, taking almost a year to complete. In parallel with the aircraft installation programme, the Canberra *Breton* equipment layout was reproduced in rack form for use in the Watton ECM School, as a training aid for the Canberra SOs.

The main installation in WT305 was finally completed in May 1960, following which an ARI 18147 receiver was added, supplementing *Breton* and extending the Canberra's frequency cover into the Q-band. The ARI 18147 was the result of a 2½-year development programme at the RAE and EMI to design a range of so-called 'guard' receivers working at frequencies above X-band. These frequencies were at that time pretty much unused, but needed to be monitored in order to detect any new Soviet activity. The development programme had started in 1957 and the first result was the ARI 18147, designed for installation in the tail of a Canberra. The system used an array of ten aerials, each feeding a separate crystal-video detector, providing approximately 180° of azimuth cover and 20° in elevation. The output of the system was displayed on a radial CRT D/F display, giving an accuracy of around 4°.

The opportunity was also taken to fit a new CSE-developed X.390 tail-warning system into WT305 for assessment trials. By 1960 the Canberras were being regularly intercepted by Soviet fighters and the existing rear-warning system, using ARI 18021 and a tail-mounted aerial, did not have the required performance to provide accurate and timely warnings. The X.390 had an improved range and angle of cover, providing visual and audio warning of the presence of an AI-equipped fighter closing from the rear, and positive indication of the AI radar switching from search to lock-on.

Following the installation of *Breton*, ARI 18147 and the X.390 tail-warner, WT305 went

Canberra B.6(BS) WT305 at Boulton Paul in 1959, after conversion with a T.11 nose and immediately prior to its Breton installation. Boulton Paul Association via Les Whitehouse

111

LISTENING IN

Canberra B.6 WT301 on a damp hardstand in the early 1960s. The T.11-style nose is obvious, but less so is the small radome fitted under the tail and the *Blue Shadow* antenna on the fuselage. *Author's collection*

for a major inspection, finally becoming available for acceptance flight trials in August 1960. These were primarily an evaluation of the *Breton* installation but also revealed severe interference problems between *Green Satin* and the X.390 rear-warner, resulting in the latter being removed. Following the conclusion of the acceptance trials, WT305 flew its first operational sortie in October 1960 when it carried out a night RPF into the Baltic from Jever in conjunction with Canberra WJ775.

The *Breton* fit in WT305 had not been an easy task, and the installation was something of a disappointment, the CSE reporting that: 'This was a difficult and expensive installation. Although this particular Breton installation was designed with the Canberra in mind it has proved rather bulky for a Canberra and its operation is rather complicated for the limited space available in the cockpit. It is not therefore proposed to continue with the fitting of Breton in the remaining Canberra aircraft of No.51 Squadron'. The CSE seems to have started investigating an alternative to *Breton* in the Canberra around May 1960 and, after discussions between the CSE and the Air Ministry, intelligence requirements were solicited and studied by the CSE Research Wing. By mid-1960 proposals for a standardised Canberra fit to meet the requirements had been submitted to the Air Ministry and GCHQ, and these were approved sometime in the second half of 1960. By December 1960 work had started on the design of the new system, and an old Canberra B.2 ground-instruction fuselage (ex-WF887) was refurbished by the CSE and used for a mock-up installation.

The second Canberra selected for conversion to the T.11/*Breton* standard was WT301, going to Boulton Paul for a T.11 nose fit in in June 1960, and returning to the CSE in November 1960. By then the *Breton* fit had been cancelled and the new replacement fit was still in the process of being designed. It is believed that, in the absence of a replacement fit, WT301 had its previous Elint equipment re-installed, possibly replacing the ARI 18021 with the DV front-end of *Breton*, along with the X.390 tail-warning receiver previously removed from WT305. Canberra WT301 became available for operations in March 1961, allowing WJ775 to be withdrawn in turn for its T.11 nose modification.

Canberra WJ775 was despatched to Boulton Paul for fitment of a T.11 nose in March 1961, returning to the CSE in late October. By the time WJ775 was returned to Watton, the CSE had completed the design of the new post-*Breton* Canberra fit, using a mock-up in the B.2 fuselage to prove the layout. Details of the new fit (to SRIM 2696) are sketchy but appear to have involved supplementing a basic search fit (possibly using DV elements of *Breton* in conjunction with the APR-9 and APA-11) with one or more of the new RAE-designed upper-band D/F guard receivers. Since a DV guard receiver only gave an extremely coarse frequency indication it was associated with an RAE-designed Instantaneous Frequency Indicator (IFI) which provided a much more accurate frequency measurement. Both the guard receiver and IFI were mounted in the aircraft's tail, and used the same radial CRT display to economise on space in the cabin. The installation also included a new X.448 tail-warning receiver, the transistorised replacement for the X.390. Installation of the new fit in WJ775 was a large task: work got under way in November 1961 and was not completed until September 1962. Rectification of the inevitable snags meant the aircraft was not returned to 51 Squadron until May 1963. The installation appears to have been successful, the squadron subsequently reporting that: 'The refitted Canberra WJ775 operated extremely well and yielded first class results'.

112

Operations 1961-1963

Operations during 1961 followed a similar pattern to those of previous years with regular detachments to Cyprus for sorties over the Black Sea and along the Turco-Soviet border, alternating with Baltic operations flown from the UK. Due to the Canberra upgrade programme the squadron had only two B.6s available for operations during the year, requiring the use of the shorter-ranged Canberra B.2 on occasion.

Early 1962 saw a major change in the pattern of RPF operations, with the first detachment to Sharjah in the Persian Gulf, from where RPFs were flown over Iran, into the Caspian Sea and along the eastern Iran-Soviet border. From 1962 onwards, approximately three detachments to Sharjah were carried out each year. Transit sorties from Akrotiri to Sharjah and return via Turkey were also used for operations against the Soviet Trans-Caucasus and Iraqi borders. 1962 also saw the first operations against Iraq operating from Bahrain, flying along the Iran-Iraq border.

On 15th October 1962 a USAF U-2 flight over Cuba revealed the construction of Soviet MR/IRBM sites on that island. The US government considered this a provocation and demanded that the Soviet Union dismantle the missile sites and withdraw all weapons from Cuba. A naval blockade of Cuba was also instituted in order to prevent further weapons being delivered. After initial Soviet protests the US and Soviet Union reached a mutually-acceptable (and secret) agreement under which the US withdrew IRBMs from Turkey as a *quid pro quo* for the Soviet withdrawal of their missiles from Cuba. The crisis officially ended on 28th October and the blockade was removed at the end of November. The Cuban Missile Crisis had a relatively minor effect on the RPF programme for October 1962, originally intended to comprise a number of single Comet and Canberra sorties over the Baltic and two Comet sorties into the Barents Sea. The Baltic sorties took place but, although the Comet was detached to Bodø, the two sorties into the Barents Sea were not flown.

There was a fear that the Crisis could affect the following month's programme (put forward for approval in late October), with the Foreign Secretary, Sir Alec Douglas-Home, commenting that: 'In normal circumstances these [flights] would appear to present little risk. Until Soviet reactions to the United States move on Cuba can be assessed, however, it is hard to say what the risk is likely to be by the time they are flown. In the circumstances I do not wish to raise objection in principle at this stage, in order that planning may proceed, but do so on the clear understanding that I may request their suspension at any time prior to their being flown, if the political situation warrants it'. In fact the November 1962 programme seems to have been completely unaffected and, comprising 17 sorties, was one of the largest programmes ever flown by 51 Squadron up to that time.

During early 1963 the Air Ministry decided to place all units of the UK Reconnaissance Force under the command and control of one headquarters. As a result the Central Reconnaissance Establishment (CRE), based at Wyton, was elevated to Group status and assumed command of 51 Squadron, 58 Squadron (Canberras) and 543 Squadron (Victors). As part of the re-organisation, 51 Squadron moved from Watton to Wyton in March 1963, joining the other elements of the CRE there. Despite the move, the CSE at Watton remained responsible for development and installation of equipment in 51 Squadron aircraft. However, the relocation of 51 Squadron led to CSE Watton being redesignated RAF Watton, with the CSE reduced to a resident unit.

1963 saw another extension of operations by 51 Squadron, with the first detachment of a Comet to the Far East in May, in connection with Indonesian opposition to the formation of the Federation of Malaysia. RPF operations were carried out against Indonesia over the Java Sea and Indian Ocean from bases in Singapore and Australia. Regular Comet detachments were carried out to Singapore for the duration of the Indonesian 'Confrontation' crisis, finally ending in 1967. Thus, by mid-1963, the main 51 Squadron operational areas comprised: the East German, Polish and Soviet Baltic Sea coast (from Wyton and Germany), the Soviet Barents Sea coast (from Norway), the Soviet Black Sea coast and Turco-Soviet border (from Akrotiri), the Soviet Caspian Sea coast and eastern Soviet-Iran border (from Sharjah), the Iraq-Iran border (from Sharjah and Bahrain) and the Indonesian coast (from Singapore).

Despite the extension in RPF operations, the UK government continued to keep a careful eye on their conduct, anxious to avoid anything that might cause political embarrassment or an international incident. Operations were often suspended during visits by UK ministers to the Soviet Union, during visits by senior members of the Soviet government to the UK and during important international conferences. USAF sor-

LISTENING IN

From 1963 onwards the RAF's reconnaissance assets were overseen by the Central Reconnaissance Establishment. Seen here are aircraft from the CRE's three flying units: 58 Sqn (Canberra PR.7), 543 Sqn (Victor SR.2) and 51 Sqn Canberra B.6).
RAF via T Panopalis

ties operating from the UK were expected to observe the same embargoes.

During 1963 the UK joined the US and Soviet Union in negotiating a Partial Nuclear Test Ban Treaty. As part of the negotiations Lord Hailsham, the UK representative, and W A Harriman, the US Under-Secretary of State for Political Affairs, visited Moscow on 14th July to discuss the proposals with the Soviet Foreign Minister. The visit to Moscow brought the Elint programme under particular scrutiny by the Foreign Secretary of the time, Sir Alex Douglas-Home. His initial reaction was to suspend all such flights for the duration to exclude the possibility of either an accident causing a political incident, or the Soviets exploiting the flights to provoke an incident. Although Sir Douglas could halt UK flights, this still left the problem (from the UK government point of view) of the Americans and their much larger Elint programme. Pressure could be brought to bear on the US regarding flights from UK bases but nothing could be done about flights from other bases, so the subject was raised by UK Prime Minister with the US President. For their part the US were not keen to halt the flights completely, noting that the majority carried little risk and suggesting that a complete break followed by resumption might be misinterpreted by the Soviets. They suggested that the programme should be subject to additional scrutiny to weed out any particularly sensitive sorties which might be considered provocative. As a result the UK authorities proposed that two Elint flights be cancelled. One (Operation *Massive*) was a planned high-level Canberra sortie from Norway into the Barents Sea, which was still considered a sensitive area as a result of the 1960 RB-47 incident. The other was a US night sortie, *Denver Thomas*, approaching to within 31nm (57km) of the Soviet coast. The US State Department had its own concerns, resulting in the cancellation of a further US operation and an increase in distance for some of the remaining high-north sorties. As it turned out, the conference was successful and the treaty was initialled on 25th July and formally signed in early August.

114

ELINT IN THE 1960s AND 1970s

Canberra B2 WJ640 used by 51 Sqn in the period 1958-1962 for jet continuation training. Interestingly, this aircraft carries the 51 Sqn 'Flying Goose' emblem on the fin.
via Peter Green

Training the Crews

The provision of continuation training for 51 Squadron flight crew, particularly the Comet crews, was a subject of much concern during the late 1950s/early 1960s as the heavy RPF schedule and regular Elint upgrades meant that there were relatively few hours left over for flying training in the Comet 2Rs. In addition, the CSE was reluctant to use expensively-modified aircraft, crammed full of delicate electronic equipment, for 'circuits and bumps'. What 51 Squadron really wanted was its own training Comet, to take the pressure off the operational aircraft, but a Comet was an expensive aircraft and an application in 1958 for a dedicated training aircraft was initially turned down. However, some limited resources were made available and in March 1958 the squadron was allotted Canberra B.2 WJ640 as a training aircraft, providing jet continuation training to both Canberra and Comet pilots. The B.2 was joined by Canberra T.4 WJ873 in September 1959 for Canberra pilot training. In the absence of a training Comet, specific-to-type continuation training was provided using the squadron's operational Comet 2R aircraft, supplemented by visits to RAF Lyneham to use the Comet simulator there, and by the occasional loan of Transport Command Comet C.2s.

After much negotiation, the squadron finally got its Comet training aircraft in May 1962 when Comet C.2 XK715, from 216 Squadron, was attached on temporary loan for training duties. The loan of the Comet rendered the Canberra B.2 WJ640 surplus to requirements, and that aircraft was disposed of shortly before the Comet arrived. A second Canberra, B.6 WT206, briefly used by the squadron in the period March-November 1962, was primarily a reserve aircraft for airborne sampling operations and was foisted on 51 Squadron for maintenance purposes.

In September 1962 the loaned Comet (XK715) was returned to 216 Squadron and replaced by Comet C.2 XK671 on long-term allocation. Thus by 1963 the squadron had a fairly reasonable training establishment of a Canberra T.4 and a Comet C.2. In addition to its training role the Comet C.2 was also flown occasionally as a transport aircraft, to support overseas detachments.

Canberra B.2 WH698, Canberra T.4 WJ873 and Canberra B.2 WJ640 lined up on the ramp at Watton circa 1959. Note the 'Flying Goose' emblem on tip tank.
via Peter Green

115

LISTENING IN

Replacing the Third Comet

The autumn of 1963 saw the long-awaited arrival of a third Comet (XK695), to replace the aircraft (XK663) lost in a hangar fire in June 1959. As previously noted, provision of a replacement Comet had originally been put on hold due to doubts regarding the fatigue life of the type, forcing 51 Squadron to soldier on as best it could with only two Comets through 1960 and into 1961. Transport Command helped out by loaning Comet C.2s for continuation training, but even so, some Elint operations were restricted for lack of a third Comet. As Brian Humphreys-Davies of the Air Ministry recounted to his opposite number at the Treasury: '… Signals Command have, under very considerable handicaps, continued to try to carry out their task with two aircraft only. This has given rise to real difficulties. … We have therefore had to restrict our operations, even though it has been possible to use Transport Command Comets to some extent for crew training. It is clear to us that three aircraft are needed and that there is a continuing need for them'.

The concerns regarding Comet Mk.2 fatigue life were eventually resolved in the first quarter of 1961, when tests by the Ministry of Aviation showed that, although the life of the Comet C.2 in Transport Command was expected to be quite short, the Comet 2Rs of 51 Squadron were likely to have an almost indefinite life, subject to some minor modifications. This difference in fatigue life was due to the fact that the Elint Comets operated at a much reduced cabin pressure, which produced less strain on the fuselage. The good news regarding fatigue life meant that the future of the Comet 2R in 51 Squadron service was assured and that negotiations for a replacement aircraft could be re-opened. The plan was to transfer a Comet C.2 from Transport Command, who were scheduled to re-equip with the Comet C.4, thus avoiding the need to buy a new aircraft. The majority of expenditure on the replacement aircraft would therefore be the cost of conversion to the Elint role, that is, installation of aerials, radomes, equipment racking and additional power supplies. This was expected to come in at around £0.25M. The Ministry of Aviation intended to place the contract with Marshalls of Cambridge, on sub-contract to de Havillands, and asked the company for a fixed-price quotation. A further, smaller, sum would be required to fit the aircraft out with Elint equipment. The US equipment could be provided from existing MDAP stocks at no additional cost but the UK-built equipment (principally *Breton*) would cost £54,000, of which £43,000 would have to be specially-provisioned. Thus the whole project was expected to cost around £300,000, with the installation and fitting work expected to take around 18 months. Given the timescales, the Air Ministry suggested that a start should be made as soon as possible, even if this left Transport Command temporarily short of a Comet 2.

The acquisition of the third Comet was supported by GCHQ with Leonard 'Joe' Hooper, then a high-ranking official at the agency, later to become Director, writing to the Treasury to support the Air Ministry case. In his letter, Joe Hooper admitted that Radio Proving Flights were an expensive form of Sigint compared to ground based-operations, but went on to stress that some of the results obtained were unique. He then noted that all aspects of Radio Proving Flights had been exhaustively examined in 1959, and again in 1960 following the U-2 and RB-47 incidents. A 'carefully controlled' programme of Radio Proving Flights remained national policy and three Comets were, in his view, 'the minimum with which an effective programme can be carried out'.

The Treasury were initially somewhat sceptical of the submissions, questioning the long-term future of airborne Sigint operations and wondering whether the Comet 2R was simply a by-blow from the government's rescue of de Havilland following the 1954 Comet disaster.

Comet 2R XK695 was the replacement Comet and being modified from a standard C.2, was distinguished by its round cabin windows. Note the longer forward ventral radome.

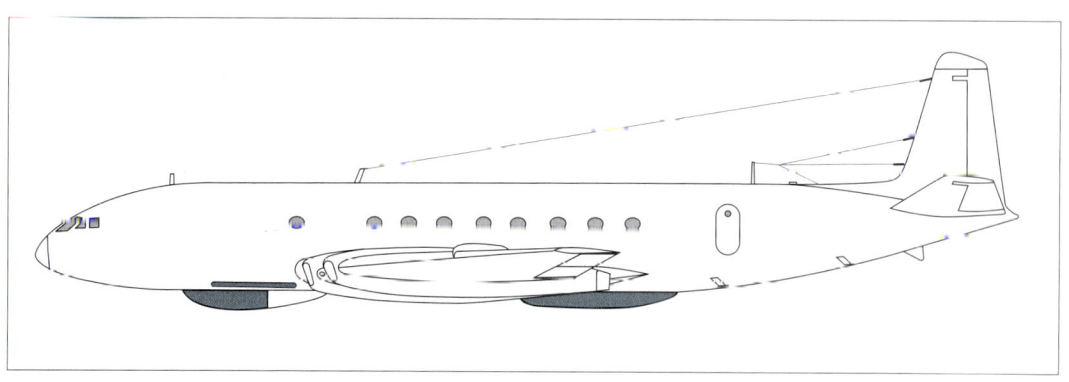

116

Comet 2R XK695, probably at Akrotiri in the later 1960s. The Elint Comets were regular visitors to Cyprus throughout their time in service. *T Panopalis Collection*

The matter was duly referred to the Ministry of Defence for their views. The MoD, in the person of the Deputy Chief of the Defence Staff, Air Chief Marshal Alfred Earle, were equally supportive of the case for the replacement Comet, noting that the Sigint task was a national one, portions of which could only be accomplished from the air, and that the rundown in Sigint ground stations meant the airborne collection task would actually become more important over time. As regards the aircraft itself, although the Comet was slightly lacking in range for the task, in practice it was adequate for 'most cases'. The only alternative aircraft was a VC10 (see Chapter 11) but one of those would come in at an eye-watering £3.5M after modification. Thus the conversion of a Transport Command Comet at around £0.3M was a cheap and convenient way to meet the requirement.

The allocation of a replacement Comet to 51 Squadron was duly approved by the Treasury and in July 1961 an ex-Transport Command Comet C.2 (XK695) was delivered to Marshalls for conversion to the Elint role. Perhaps not surprisingly, given the dynamic nature of the Elint task, XK695 was modified to an improved standard compared to the original three Comet 2Rs. As the CSE noted: '… there is considerable activity at CSE in preparation for [XK695's] special fit. The task of 51 Squadron is daily becoming more complex and diverse in step with rapid developments elsewhere and the new Comet must be fitted with equipment which will not only cater for the operational tasks as they exist today but, as far as possible, satisfy those which can be forecast for some years ahead. Some of this equipment does not exist at present, and a quite complex development programme is in hand at CSE to ensure the squadron aircraft are provided eventually with facilities capable of detecting the latest electronic techniques'.

The conversion of XK695 was a lengthy process and the aircraft was not completed until March 1963. Following the conversion at Marshalls, the aircraft was delivered to CSE Watton on 8th March 1963 for the embodiment of the Elint fit under SRIM 2604. Some of the equipment had already been prefabricated and work got under way a few days after delivery. The CSE were working to an aggressive programme and estimated that, using shift and overtime working, the task could be completed at the end of July and the aircraft made operational by mid-December.

The installation in XK695 was an improvement on that applied to the previous two aircraft. A new, larger, front radome had been fitted by de Havilland under the forward fuselage, providing more space for D/F aerials. The main elements of the installation in XK695 are thought to have comprised ARI 18050 (centimetric), ALR-5 (metric) and R.216 (VHF communications) receivers, ALA-6 for D/F, APA-11 and APA-74 analysers, and Solartron magnetic tape recorders. Some custom-built parts of the fit were manufactured by the CSE. The installation in XK695 was completed bang on target, a Ministry of Aviation acceptance meeting was held on 25th July and the installation was signed off the following day. Work then began on ground tests of the installation, followed by a flight trials programme. A series of eight flights began on 3rd September and successfully proved the installation. The aircraft was handed back to 51 Squadron on 30th September 1963, one day ahead of schedule.

Operations 1964-1966

Operations during 1964 followed the pattern set in 1963 with home-based sorties over the Baltic and regular detachments to Akrotiri, Sharjah, Singapore and Norway.

The January 1964 detachment to Sharjah had to be cancelled when Comet XK695, cruising at 37,000ft (11,277m) over the Mediter-

LISTENING IN

ranean en-route to Akrotiri, ran into severe clear air turbulence. The Comet diverted to El Adem, where its arrival and the subsequent security measures generated considerable speculation, with a popular theory being that the aircraft had been overstressed evading MiGs. An examination of the aircraft revealed cracks in the starboard centre section lower skin and a crack in the port rear fuselage. The squadron training Comet (XK671) ferried out a maintenance team and, after inspection and repairs, XK695 returned to the UK a few days later.

The UK continued to host USAF RB-47H operations, with occasional visits by specialist ERB-47H aircraft. The primary task of the RB-47H was to map and analyse the Soviet air defence system, compiling a Soviet electronic order of battle. The equipment fitted in the RB-47H (the APD-4 and ALD-4 automatic Elint systems, supplemented by APR-17, ALA-6 and ALA-74 manually-operated systems) provided facilities to intercept Soviet radar signals, measure their basic characteristics (frequency, PRF, pulse-width, scan-rate and so on) and plot the radar positions. However, there was also a requirement to examine the signals from some Soviet radars in fine detail, to fully determine their performance and capabilities, and to meet this technical investigation task the US converted three B-47s to ERB-47H standard during the early 1960s. The ERB-47H was equipped with a suite of extremely specialised receivers and analysers, operated by a crew of two Special Operators.

Approval of US Elint sorties from the UK was normally a fairly automatic affair but questions were occasionally asked. In mid-June 1964 the Prime Minister noted that one proposed US flight by an ERB-47H into the Baltic (*July Osbourne*) approached to within 25nm (40km) of the Soviet coast and asked that this be looked into. A subsequent investigation by VCAS noted that the flight was routed around the Baltic at the normal height and speed for the type, with the leg during which it made its closest approach to the Soviet coast being parallel to the coastline. The report noted that flights approaching to within a similar distance of the same stretch of coast had been mounted in January and February 1964, and that two of these had flown a more provocative profile than that planned for *July Osbourne*, making their closest approach at right-angles to the coast during a descent to low-level. Neither had provoked much of a reaction from Soviet air defences and on that basis there was no reason to expect any untoward reaction to *July Osbourne*. However,

measures were put in place to ensure that 48-hour review of the flight would be carried out with particular thoroughness.

Operations by 51 Squadron during 1965 were much the same as previous years, visiting the same operational areas as before. The major change during that year was a further relaxation on the rules governing daylight sorties. Previously these had been restricted to a single aircraft per operational area, but from the spring of 1965 onwards, multiple aircraft operations in the same area were permitted in daylight.

In December 1965 the squadron lost access to one of its most productive operational areas when Turkey withdrew permission for Elint operations over its territory following the loss of a USAF aircraft over the Black Sea. 1965 also saw major changes in the technical organisation supporting 51 Squadron. In July 1965 the CSE was disbanded and the 51 Squadron support functions were devolved to the Signals Command Air Radio Laboratory (SCARL), responsible for airborne Sigint R&D, and the Electronic Warfare Support Unit (EWSU), responsible for installations. Both units remained at Watton.

Up until 1966 the monthly RPF programme of both the UK and US was submitted to the Foreign Secretary and Prime Minister for approval. In January 1966 the then Prime Minister, Harold Wilson, decided that, due to the repetitious nature of the majority of flights, he was content to let the Foreign Secretary and Defence Secretary authorise the routine missions. Exceptional or particularly sensitive flights would still need to be submitted to the Prime Minister for authorisation, as would routine flights where there was a difference of opinion between the Foreign Secretary and Defence Secretary.

Upgrading the Canberras

The Elint refit (SRIM 2696) applied to WJ755 during 1961-62 was intended to serve as the prototype for future Canberra installations. Plans were made to upgrade the squadron's other two Canberra B.6s (WT301, WT305), and to replace the ageing Canberra B.2 (WH698) with a similarly fitted Canberra B.6, giving the squadron four B.6s, all equipped to the same standard. In the event the Canberra Elint upgrade programme was a drawn-out affair and was not completed until the late 1960s. Part of the problem was that, in order to maintain an operational capability, the Canberras could only be withdrawn from service for upgrade one at a time. The other problem was

ELINT IN THE 1960s AND 1970s

that the Canberra Elint standard was a constantly-changing target (driven by the demands of GCHQ and other agencies), and thus by the time the last aircraft had been modified the first was out of date.

The programme got under way in August 1962 when a fourth Canberra B.6(BS), WJ768, was delivered to Watton as a replacement for Canberra B.2 WH698. The new B.6 went straight to Boulton Paul for installation of a modified nose, prior to receiving an Elint installation. While WJ768 was being fitted with its new nose, the first Canberra selected for the upgrade programme (WT301) was fed into the CSE workshops for Elint refit at the end of April 1963. By late 1963 however, with the refit half-completed, the Canberra upgrade specification was revised to a new standard, designated SRIM 2951. The major external evidence of the new change was the replacement of the aircraft's T.11 nose radome with a more rounded model. The T.11 radome had originally been selected for use by the Elint Canberras as a cheap off-the-shelf solution, overlooking the fact that it had been designed for use with the S-band AI.17 AI radar and its dielectric properties had been tuned for that application. Flight trials in WT305 following its *Breton* installation had revealed significant limitations, particularly at X-band where the radome tended to absorb signals. SRIM 2951 introduced a new wideband radome, designed to pass signals over a wide frequency range, and with a blunter shape to allow the reception of signals over a wide angle. The installation in WT301 was cleared by the beginning of December 1963 and the aircraft was returned to service in mid-month. The squadron now had two post-*Breton* Canberra B.6s (WJ755 and WT301): one with a T.11 radome, one with the new blunt nose, a *Breton* Canberra (WT305) and a fourth B.6 under structural modification (WJ768).

WJ768 returned to the CSE from Boulton Paul, complete with new nose, in September 1963 and was fed into the CSE for its Elint installation. This went well and was largely complete by mid-1964, pending electrical and radio checks. However by then changing requirements had resulted in yet another revision of the Canberra Elint fit. The installation in WJ768 was reworked to the new standard (SRIM 3088) in August, the installation was completed the same month, and WJ768 was delivered to 51 Squadron at the end of September 1964.

The evolving Canberra Elint fit meant that by late 1964 the fit in the first post-*Breton* aircraft (WJ775) was out-of-date and needed upgrading. Thus, WJ755 went to the CSE for an installation upgrade in October 1964, after WJ768 had become available for operations. By then the required Canberra Elint fit had progressed from SRIM 3088 to SRIM 3094. Work began on the installation of SRIM 3094 into WJ775 in October 1964 and the aircraft was returned to 51 Squadron at the beginning of February 1965. Canberra WT301 was then brought up to the similar standard (to a slightly revised SRIM 3095) during the first six months of 1965.

The last B.6 to be upgraded was WT305, which was fed into the CSE in February 1965,

The new rounded radome was first applied to Canberra B.6 WT301 under SRIM 2951 in 1963. The radome enhanced signal reception in the X-band and also improved D/F accuracy. Note access hatches and equipment racks in nose.
Author's collection

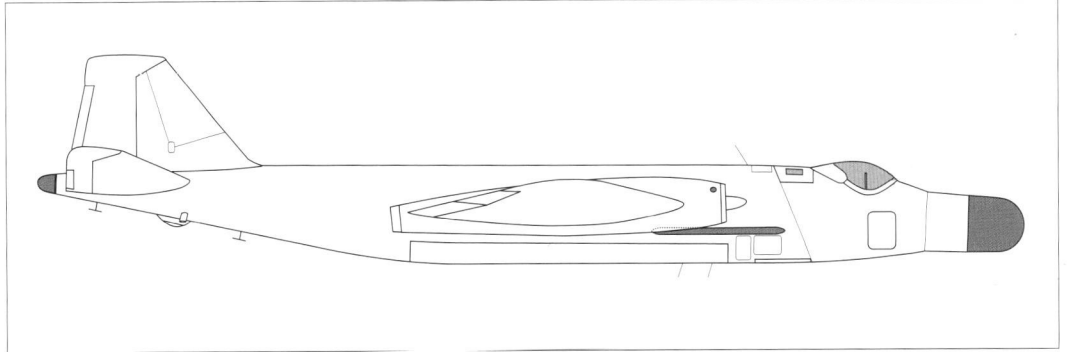

Canberra B.6(BS) with new wideband radome. Compare this shape to that of the T.11-style radome previously fitted. The new radome greatly enhanced signal reception.

119

LISTENING IN

but on arrival at the CSE, the aircraft was found to require Cat 3 repairs which delayed the start of installation until June. By then the Canberra fit had been further revised to SRIM 3144, which included the installation of a new intercom, twin PTR 175 UHF and the addition of a second *Blue Shadow* aerial on the port fuselage side. The original Canberra *Blue Shadow* installation had used a single aerial on the starboard side of the fuselage, but this only produced a picture of the ground to starboard of the aircraft, limiting the operational usefulness of the aid. Adding a second aerial (and associated switching) allowed the *Blue Shadow* to look to port or starboard as required. This gave the navigation system much greater operational flexibility, especially useful on coastal sorties, allowing the Canberra to track along a coastline in either direction. Installation of SRIM 3144 began in WT305 in June 1965 and was completed in July 1966, the aircraft resuming operations with 51 Squadron in September. Following the upgrade of WT305, a similar installation was applied to Canberra WJ768 in the period September 1966-February 1967. It is believed WT301 and WJ775 were brought up to the same standard during 1967-1968. Thus by around 1968 all four Canberras were fitted with a broadly similar Sigint and navigation suite, with twin *Blue Shadow* aerials and the new wideband nose radome.

The Canberras played an important part in the UK contribution to the joint UK-USA Sigint effort, providing the only low-altitude Elint investigation capability in the European theatre and were often employed in the investigation and analysis of Soviet Bloc low-level air defence systems.

A rare photograph of the rear cabin of a 51 Sqn Canberra B.6 to late 1960s standard, albeit with the Elint equipment stripped out. Note Blue Shadow recorder (centre) and the Special Operator's racking installed over passageway. *Author's collection*

Upgrading the Comets

Following the entry into service of the replacement Comet XK695 in late 1963, Comet XK659 was withdrawn from 51 Squadron service for a refit to a similar standard. The aircraft first went to de Havilland Chester in August 1963 for wing replacement, corrosion checks, and installation of an Ekco E160 cloud-and-collision warning radar. The special equipment racks were removed from the aircraft at Chester and returned to Watton for refurbishing and to assist in the prefabrication of parts. After some delay, XK659 returned to Watton in May 1964 for an Elint upgrade to SRIM 2943, by which time prefabrication work, including construction of the two new *Breton* racks, had been completed. The installation in XK659 is thought to have been similar to that applied to XK695 and was completed and accepted at the end of the first week in January 1965. The Ministry of Defence and Ministry of Aviation subsequently congratulated RAF Watton on meeting their target date and on their high standard of workmanship. Following ground and flight trials XK659 was returned to 51 Squadron on 19th March 1965.

The last Comet to undergo a major upgrade was XK655, which went to de Havilland in May 1965 following the return to service of XK695. This was the longest upgrade, taking nearly two and a half years to complete. Work at Chester included structural modifications, a major inspection and the installation of a LOX system, replacing the existing high-pressure bottled oxygen system, which was intended to provide sufficient gas for four sorties before recharging. The refit at de Havilland was later extended to include an upgrade of the aircraft's communications equipment via the installation of a Collins 618/T3 HF SSB and a twin PTR 175 UHF fit. The work at de Havilland's took around 18 months and XK655 was not despatched to Watton until February 1967.

During the early 1960s the UK had investigated the development of a number of Elint receivers to extend the *Breton* system down through L-band into the metric band and up above X-band into the J-band. The UK L-band receiver development programme was eventually shelved, probably on cost grounds, when authorities decided to adopt the US APR-17 receiver, as used in the RB-47H. However, by the mid-1960s the focus had switched to the ALR-8

ELINT IN THE 1960s AND 1970s

Comet 2R XK655, seen here during the early 1970s. This was the last Comet to receive a full Elint upgrade and was not completed until late 1967.
T Panopalis Collection

system, which utilised elements of the APR-9 and APR-13 receivers, and covered the frequency range 0.05- 10.75GHz (metric to X-band) in nine bands, using interchangeable tuners. The new fit in XK655 is thought to have comprised elements of *Breton* (centimetric), ALR-8 (metric and centimetric) and R.216 (VHF communications). D/F is thought to have been provided by the ALA-6 and analysis by the APA-74. The Elint refit of XK655 took eight months and the Comet finally returned to service in October 1967.

A new training Comet

The 51 Squadron training Comet (C.2 XK671) saw just over four years' service before being prematurely retired due to skin corrosion problems in November 1966. Following the loss of XK671, the 51 Squadron crews had to revert to their previous, pre-1962, training regime, travelling to Lyneham to use the simulator there and borrowing 216 Squadron aircraft for continuation training. This was clearly an unsustainable situation and the squadron made strenuous efforts to obtain a replacement Comet C.2. As luck would have it, the Comet C.2 fleet was being withdrawn from Transport Command service, resulting in the extended loan of one of the surplus aircraft (C.2 XK697) as a full-time training aircraft in March 1967. The loan appears to have been initially organised on an informal basis, and it was not until January 1969 that efforts were made to regularise the situation by formally establishing the Comet on 51 Squadron. In support of the request the squadron pointed out that a dedicated crew training and conversion aircraft was required to conserve the life of the three operational Comet 2Rs, which had to remain in service until replaced by the Nimrod in 1972 (see Chapter Eleven). If the 2Rs had to be used for continuation training then fatigue life would be consumed four times more quickly and, as a result, two of the aircraft would run out of hours in 1971. That apart, experience showed that using the 2Rs for continuation training would have an adverse impact on the serviceability of the Elint equipment and thus an impact

A fine study of Comet C.2 XK697 in Transport Command service prior to re-allotment to 51 Sqn. Even with the Elint fit installed, the elegant lines of the Comet were maintained.
de Havilland via T Panopalis

LISTENING IN

Vickers Varsity T.1 (mod) WL687 ex-97 Sqn in 1968. This aircraft was used to train the Special Operators of 51 Sqn and the D/F aerial radome can be seen under the rear fuselage.
Robin A Walker

Vickers Varsity T.1 WJ911 on 115 Sqn charge. This aircraft was used for 51 Sqn Special Operator training relieving the workload of the operational Comets 2Rs.
Peter Biggadike

on the operational task, due to increased wear and tear. The request was subsequently approved and the aircraft was retained for another three years, finally retiring in December 1972 when it ran out of fatigue life.

Supporting Detachments

As a result of 51 Squadron's regular overseas detachments a Hastings C.1 (TG530) was acquired in February 1963 for transport and support duties, replacing the training Comet (XK671) in that role. Unfortunately, the Hastings appears to have suffered from regular unserviceability and the training Comet was often used in its place. The Hastings soldiered on for a few years before retiring in November 1966. A second Hastings (TG507) was acquired as a replacement in February 1967 but had an even shorter working life.

The Hastings as a type was eventually deemed unsuitable for the year-round long-distance support and transport task, suffering from poor serviceability and operational limitations in the winter months. By late 1967 the squadron had decided to dispense with its own dedicated transport aircraft and transfer the task to Air Support Command (ASC), using scheduled sorties whenever possible. After some negotiations ASC agreed to take the commitment on and Hastings TG507 was retired in October 1968.

The transfer of the transport and support task to ASC did however raise a few concerns regarding security, since the squadron's requirements were now handled through regular movements channels. There were worries that this might give advance warning of squadron detachments, resulting in special procedures being adopted for requesting overseas support for 51 Sqn.

Training the Special Operators

Special Operator training during the 1960s was largely carried out in specialist Varsity aircraft operated, for administrative reasons, by 97 Squadron at Watton. A new SO would initially undergo a ground course, using receivers and analysers installed in racks similar to those installed in the Comets and Canberras. Ground school was followed by an airborne training phase using two Varsity T.1s (WL686, later replaced by WJ911, and WL687) fitted with six SO stations. Because of the small size of the aircraft, compared to the Comet, one Varsity was fitted for L-band and S-band search, the other for X-band search. Initial airborne training comprised cross-country sorties over the UK, with the trainee operators intercepting and plotting radars en-route. The L/S-band sorties went up the east coast, plotting the position of UK early warning radars, the SOs working in pairs and taking turns to either intercept or D/F the radars. The trainees would spend the day following each sortie analysing their data and plotting the position of radars. After completing several weeks of sorties over the UK the SOs were detached to RAF Gütersloh for a couple of weeks, for the final phase of the course, flying Varsity training sorties along the German internal border, attempting to intercept and plot radars in East Germany. Once the SOs had passed their training course (and not all did) they would be assigned to 51 Squadron and undergo further training on the operational aircraft, normally culminating in a Border sortie. When 97 Squadron disbanded in early 1967 the Varsity training task was taken over by 115 Squadron (then operating Varsities in the Calibration role) until the last specialist Varsity was retired in the early 1970s.

Scientific support

During the 1960s equipment design, modification and installation support for 51 Squadron aircraft was provided by the CSE Electronic Warfare Support Wing. The EWSW was responsible not only for the large aircraft equipment upgrade programmes but also for the provision of equipment to meet urgent operational needs under a Quick Reaction Capability (QRC) commitment. These tasks were often triggered by the identification of previously-unknown signals and might involve the construction of new equipment, or the modification of existing systems. The idea behind the QRC concept (inherited from the US) was to short-circuit normal procurement channels and produce the necessary equipment (often as a one-off) as quickly as possible. In an example of the speed of reaction possible, in one case a new Soviet data link signal was identified and its characteristics defined on Monday, a new receiver to investigate the signal was designed on Tuesday, a model of the receiver was built on Wednesday, installed in an aircraft on Thursday and flown operationally on the Friday.

In mid-1965 the CSE was disbanded, following which the EWSW was split into the Electronic Warfare Support Unit (EWSU) and the Signals Command Air Radio Laboratory (SCARL), the latter inheriting the QRC task. In January 1971 the EWSU and the Signals Air Radio Laboratory (renamed from SCARL in mid-1969) merged to form the Electronic Warfare Engineering and Training Unit (EWE&TU), moving at the same time to Wyton where it joined 51 Squadron. The unit continued to provide support to 51 Squadron, and was heavily involved in the Nimrod Elint fit (see Chapter Twelve).

Operations 1967-1970

By 1967 the squadron was operating three Comet 2Rs and four Canberra B.6s. That year saw a change in operational patterns in response to world events. Operations against Indonesia came to an end in mid-year as that country normalised relations with the West, relieving 51 Squadron of its Far East commitment. At the same time however, the Arab-Israeli war of June 1967 led to a significant increase in the effort being directed against Egypt and Syria.

By 1968 the general operational pattern was one of regular monthly sorties along the German border and over the Baltic from Wyton and Laarbruch, regular monthly operations from Akrotiri against Egypt and Syria, and three detachments a year to Sharjah for operations over Iran. In addition to the above, the squadron also flew sorties against Soviet naval units on an *ad hoc* basis, and made occasional forays into the Barents Sea. The Canberras were mainly used over the Baltic while the Comets were used on longer-range Middle East tasks, also flying Baltic sorties when required.

Co-operation with the US and NATO during the 1960s

Although the three Comets and four Canberras of 51 Squadron were a relatively modest force by comparison with the US, it was still the second-largest airborne Elint force in NATO. Some airborne Elint collection was a carried out by France and Germany, but this was limited in scope when compared to 51 Squadron's activities.

UK and US airborne Elint collection sorties over Europe were co-ordinated via joint monthly meetings, which ensured that the two countries did not duplicate each other's work and also allowed the smaller, and more specialised, UK force to be used to its best advantage. For example, the Comets and Canberras were often used to search for new signals, while the Canberras also provided the only low-altitude Elint capability in Europe. There was also some division by area, with the US and UK each having geographical areas only they covered. In joint-coverage areas, such as the Baltic, US and UK RPFs were occasionally flown in conjunction with each other to increase the Elint cover and thus improve intelligence 'take' from the operation.

The UK contribution to the joint US-UK Elint programme brought significant rewards, the principal benefit being the access it granted to large quantities of raw Elint data captured by the much larger fleet of US Elint aircraft. This was a privilege not granted to other members of NATO, who only received end-product intelligence analyses derived from the data. The UK access to the raw data was beneficial to both parties, with the US benefitting from an independent analysis of their data, and thus of the conclusions derived from it.

The tasking of 51 Squadron continued to increase during the second half of the 1960s: prior to 1965 the squadron averaged 12-13 sorties per month but by the second half of the 1960s that figure had doubled.

As might be expected, the Soviet invasion of Czechoslovakia in August 1968 had an impact on RPF operations. The plan that month was to fly nine Comet and eight Canberra sorties from Wyton into the Baltic: the first few operations were flown according to plan but the remainder were cancelled by the MoD following the invasion on 21st August. Operations resumed the following month with 30 sorties into the Baltic (22 Comet and 8 Canberra), although the Foreign Office demanded it be informed of any 'abnormal incidents' following each sortie.

The 1970s

RPF operations during 1970 continued very much as in previous years. Each month operations were flown by Comets and Canberras into the Baltic (the Comets operating from Wyton and the Canberras from either Wyton or Germany). Operations were flown over West Germany and a Comet was detached to Akrotiri for operations off the coasts of Egypt and Syria. On occasion, the Comet operating from Akrotiri flew one or two sorties off the Algerian and Libyan coasts. A monthly series of Border sorties were also flown along the East German border. The monthly programme was interrupted three times a year when a Comet was detached to Tehran for operations in the Caspian Sea area and along the Iran-Soviet border. From 1970 onwards these operations were flown almost exclusively by the Comets, leaving the short-range Canberra to concentrate on the Baltic and West Germany. In addition to the scheduled commitments, 51 Squadron also provided aircraft for maritime Sigint operations against Soviet naval forces on a demand basis.

In May 1970 the UK government under Harold Wilson announced a snap general election on 18th June. As with the previous 1964 and 1966 General Elections, there was some discussion between the Ministry of Defence and the FCO regarding intelligence operations in the run-up to, during and immediately after, the Election. The goal was to avoid any operations that might embarrass the existing government during the run-up to the election, and to avoid any operations that might result in an incoming government having to defend activities they had not endorsed. As a result, it was decided to be especially cautious in vetting operations in the Election period, and to cancel all but the lowest key operations following the Election until ministerial permissions could be renewed. Thus all RPFs were suspended from 19th June onwards until the new Ministers '…had an opportunity to take stock of the arrangements generally and to approve the continuation of flights'. In the event, June 1970 saw a change of UK government when Harold Wilson lost to Edward Heath.

1970 saw another upgrade to the Canberras when at least two of the aircraft were fitted

The Canberra B.6(BS) of 51 Squadron. Believed to have been taken on 26th June 1971, this photograph records a rare occasion when all four aircraft were serviceable. *Authors collection*

ELINT IN THE 1960s AND 1970s

with a multi-channel automatic Comint system to SRIM 3684. Canberras WT305 and WJ775 were fitted with the new system in September 1970. The so-called 'automatic voice monitoring' system comprised six ten-channel VHF receivers mounted in a crate in the rear bombbay with a second crate, forward of the VHF receivers, housing an X447 14-channel tape recorder. Control arrangements in the cabin allowed the Canberra's SO to tune, monitor and record from the six receivers.

On 1st October 1970 the Central Reconnaissance Establishment was disbanded and control of 51 Squadron moved to 1 (Bomber) Group, Strike Command. The change of control was purely administrative and the squadron remained based at Wyton.

In 1971 the UK government expelled a large number of Soviet intelligence officers, operating under the guise of diplomats, from the UK. The expulsions soured relations with the Soviet Union and led to RPFs along Soviet borders coming under increased scrutiny from the FCO, in case the Soviets were tempted to retaliate. In the normal case RPFs were cleared by the FCO on a monthly basis, but during the period of tension all sorties flying along the borders of the Soviet Union were submitted for individual approval by the FCO 48 hours in advance of take-off. By November the crisis had passed and the approvals process reverted to its previous format.

1973 saw a slight change in the pattern of operations with two Comet detachments to Masirah (February and May) for sorties over the Arabian Sea against the People's Democratic Republic of Yemen (PDRY).

During 1973 Comet aircrew were removed from operations for conversion to the Nimrod and, as a result, the squadron had to operate with only two Comet crews during the year. During the same period the Canberra flight was reduced to two aircraft (WJ775 went for Cat 3 repairs and WT305 for conversion with the *Zabra* radiometer) but despite these handicaps the squadron managed to successfully complete almost the same number of sorties as the previous year: 309 sorties out of 319 tasked, giving a success rate of 97%.

Comets and Canberras retire

Plans to replace the three Comets and four Canberras with three Nimrods had been agreed in 1968. The target date for the introduction of the Nimrod was set by the forecast fatigue life of the Comets while the Canberras, despite their low-level flying, had a much longer life-span and could (if required) continue for many more years. Predicting the actual life of the Comets was, however, an inexact science, subject to refinement as more data was accumulated on the aircraft. In 1963, when a Comet replacement was first mooted, the Comets were expected to run out of fuselage fatigue life in 1968-69. That forecast proved pessimistic and by 1968 the new dates for the three aircraft, dictated by wing fatigue and corrosion, were mid-1972, mid-1973 and mid-1976. In the event, all three Comets lasted into 1974, providing an overlap with the Nimrod.

The first Comet to go was XK659, which retired on 13 May 1974. The airframe was subsequently bought for conversion as a restaurant and flown to Manchester Airport with its new owners as passengers. The second retiree, XK655, made its last flight on 1st July 1974: a working transit from Luqa to Wyton. The last Comet (XK695) retired in January 1975, going to the Imperial War Museum at Duxford on 10th January.

Two of the Canberras, WJ768 and WT301 were retired from use in July 1974, the third, WJ775, following a few months later in October 1974. The last Canberra, WT305, was involved in flight trials of a stand-off radiometer surveillance device from 1973 onwards under a project known as *Zabra* and was not retired until November 1976.

Canberra B.6(BS) WT301 seen at Gütersloh in October 1973. The 51 Sqn Canberras flew regular sorties from a number of German airfields into the Baltic. *Spotting Group Gütersloh – Eric Westersoetebier*

Another 51 Sqn Canberra, B.6(BS) WJ768, seen at Gütersloh in October 1973. This aircraft sports a different tailcone from WT301. *Spotting Group Gütersloh – Eric Westersoetebier*

10 World-wide Operations

'The use of Andøya as a new operating base proved quite successful and the squadron is looking forward to operating from Andøya during the spring and summer months.' Officer Commanding 51 Squadron, winter 1964

'...we feel that there could be political repercussions if the Egyptians thought that our flights were following an unusual pattern and at this juncture in Middle Eastern affairs it is important that we should avoid damaging our political interests' Foreign and Commonwealth Office memo, March 1969

During the 1960s and into the first half of the 1970s the Comets and Canberras of 51 Squadron conducted operations against a wide variety of targets over an extensive geographical area, from Soviet missile installations in Novaya Zemlya to Indonesian tactical communications in the Java Sea. The following sections attempt to give an overview of the primary operational programmes by area and an indication of 51 Squaron's wide-ranging remit.

Norway and the Barents Sea

The UK had started flying Elint sorties into the Barents Sea in early 1956 and because of the long ranges involved, aircraft normally operated from the forward base of Bodø in Norway. Although Norway was a member of NATO, and carried out its own Sigint operations against the Soviet Union, it was keen to avoid antagonising its larger neighbour. The UK sorties were therefore subject to political control by the Norwegians, and monitored for adherence to the rules laid down by the Norwegian government. The UK flew less than a dozen sorties over the Barents Sea during 1956-57, operating two detachments to Bodø each year: one at the beginning of the year and one in the autumn.

The programme was continued following the introduction of the Comet, with XK663 flying its first Elint sortie over the Barents Sea in March 1958, operating in conjunction with Canberra WJ775 from Bodø. The second 1958 Bodø detachment, in November, flew four sorties (one a double Comet/Canberra RPF) from a programme originally planned to comprise nine sorties. There is some evidence that by 1958 the Norwegian government were becoming concerned regarding the conduct of foreign Elint flights and the reduction in size of the November 1958 programme may have reflected this.

The Soviet Union certainly did not welcome sorties along its northern coast and made their feelings plain to the Norwegians. In January 1959 the Soviet government lodged a formal protest with the Norwegian government regarding the activities of UK and US reconnaissance flights operating from Bodø. The Soviet complaint does seem to have had some effect on UK Elint operations from Norway as a series of five RPFs scheduled for January 1959 from Bodø were postponed until the following month 'due to forward base difficulties'. In the event only two Comet sorties were flown in February, these taking off from Lossiemouth and landing at Bodø. The unusual nature of the February operations may have been in recognition of Norwegian sensitivity regarding such flights: operations from Lossiemouth would allow the Norwegians to assert that the flights had not originated from a Norwegian base. The second series of RPFs into the Barents Sea during 1959 took place in October and comprised two single Comet sorties from Bodø. This was the first occasion on which UK sorties over the Barents were flown in daylight. Despite the switch to daylight, the Comet sorties apparently flew deep into the Barents Sea, going as far east as the coast of Novaya Zemlya.

The Norwegian concerns regarding UK and US Elint sorties operating from Norway finally came to a head in late 1959 when the UK Ambassador was informed that the number of Elint flights would have to be reduced, the flights would have to be flown at variable inter-

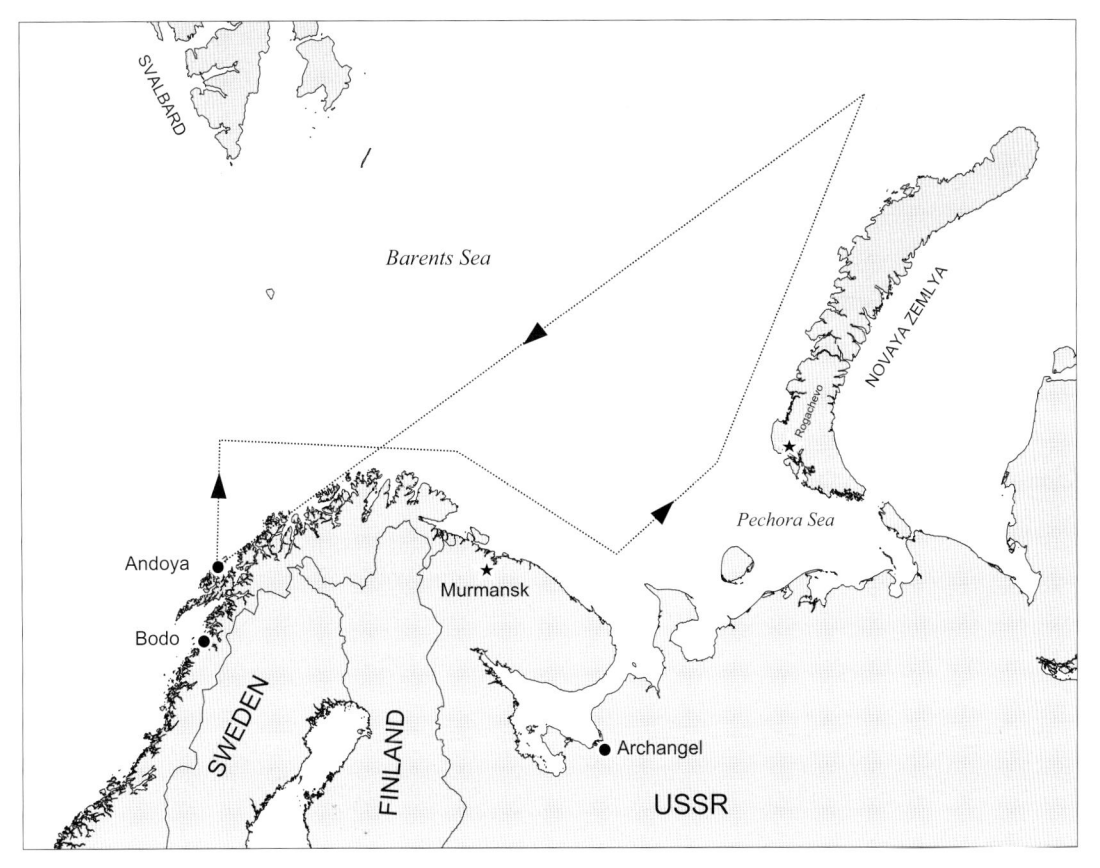

RPFs into the Barents Sea were initially flown from Bodø, later (after 1964) from Andøya. The sorties covered the heavily-militarised Kola Peninsula and the island of Novaya Zemlya.

vals and the length of detachments at Bodø kept as short as possible. Further discussions regarding future US and UK Elint programmes from Norway took place in London in February 1960 and apparently led to the Norwegians withdrawing Bodø as a forward base for UK RPFs. No further sorties were flown from Norway for over two years.

After protracted negotiations the Norwegian government eventually agreed to a resumption of RPFs from Norwegian airfields in mid-1962. The first sortie under the new arrangement (Operation *Tobias*) took place on 18th June 1962 when Comet XK655 flew a 4 hour 15 minute daylight sortie from Bodø. As might be expected, this was a fairly cautious affair, standing-off a healthy 90nm (167km) from the Soviet coast. A second operation (Operation *Encon*) took place towards the end of September when Comet XK659 flew two night sorties from Bodø, gathering intelligence on a Soviet Naval and Air Exercise. The Soviets had declared an area of the Barents and Kara Seas dangerous to aircraft and shipping and the Comet crew were instructed to remain 30nm (55km) from the danger area. Like the June operation, these were relatively shallow penetrations into the Barents Sea.

A third series of sorties were proposed in October 1962 (Operation *Agatha*). The plan was for a night sortie going much deeper into the Barents Sea than the previous two operations, penetrating as far east as the coast of Novaya Zemlya, followed by a day sortie. Both sorties would approach to within 78nm (144km) of Soviet territory. The Soviet Union had been carrying out a series of nuclear weapons tests on Novaya Zemlya from August 1962 onwards and had declared a danger area off that island, but this was scheduled to lapse on 20th October. Comet XK659 was detached to Bodø on 22nd October in preparation for the first sortie the following day. However, the operation never took place. On the morning of 22nd October President Kennedy announced the blockade of Cuba in response to the deployment of Soviet IRBMs on that island. Simultaneously, US forces went to Defcon 3 as a precaution against Soviet retaliation. This was clearly not a good time to fly what might be viewed as a provocative sortie off the Soviet coast. The operation was cancelled and the Comet returned to Watton on 25th October.

Further detachments were made to Bodø in December 1962 and January 1963 but on each occasion no RPFs were flown, probably as a result of unsuitable weather or a lack of political approval. No further sorties were flown into the Barents Sea until the autumn of 1963. Operations began with three, probably short-

LISTENING IN

Canberra B.6(BS) WJ775, seen here in the early 1960s, was fitted out to SRIM 2696 standard. Note the additional *Blue Shadow* antenna fitted on the port side to give coverage to port and starboard of the aircraft.
via Peter Green

range, Canberra day sorties from Bodø in September, followed by three long-range Comet night sorties in November, and then by a joint Comet/Canberra night operation in December. Further sorties were flown in March/April 1964, including a further joint Comet/Canberra night operation.

In the autumn of 1964 plans were made for a new series of RPFs (Operation *Plume*) into the Barents Sea. Previously all Barents Sea sorties had been flown from Bodø airfield. Perhaps in recognition of the notoriety Bodø was acquiring in some quarters, the Norwegians offered Andøya as the new forward base for Radio Proving Flights. Andøya Air Station, located on the northernmost point of the Lofoten Islands, some 130nm north of Bodø, was home to 333 Squadron, RNoAF and its Grumman HU-16 Albatross amphibians but also took civilian flights. In September 1964 Comet XK671 flew into Andøya on a liaison visit prior to the first operational detachment there, with a Comet (XK695) and a Canberra (WT301) following in October. Operation *Plume* comprised two joint Canberra/Comet night sorties (closest approach 65nm/120km) and a single Comet day sortie (closest approach 68nm/126km). All five sorties were flown as planned and the OC 51 Squadron reported: 'The use of Andøya as a new operating base proved quite successful and the squadron is looking forward to operating from Andøya during the spring and summer months'.

The next Barents Sea operation took place in April 1965. The plan for Operation *Bung* comprised two joint Comet/Canberra day sorties and one joint Comet/Canberra night sortie. This was the first multiple-aircraft daylight operation carried out over the Barents Sea by 51 Squadron. Any change in operational pattern was always carefully scrutinised by the authorising establishments: in justification, the Defence Intelligence Staff (DIS) noted that the Americans had carried out a number of similar operations during the previous year without any adverse reaction by the Soviet air defences. The Canberra (WJ768) and Comet (XK695) arrived at Andøya in marginal weather conditions and the crews then endured 26 hours of continuous snow before the weather cleared on second day, allowing the operations to be flown. A further operation was planned for July (Operation *Planetary*) but was not flown, most probably due to a lack of approval by the Norwegians.

In September 1965 Norway went to the polls and elected a new government, replacing the Labour Party, which had been in power since 1945, with a four-party coalition. The new government turned out to be even less sympathetic to foreign Radio Proving Flights than their predecessors. Back in the UK, plans were made for a new series of RPFs from Andøya in November 1965. Operation *Blountstown* would comprise three joint Canberra/Comet day sorties, approaching to within 67nm (124km) of the Soviet Coast. Planning for the operation was temporarily disrupted by the news that Ingeborg Lygren, a secretary at the Norwegian Intelligence Service in Oslo, had been arrested on suspicion of spying for the Soviet Union. As a result of the Lygren case, an assessment was carried out of the risk resulting from the possible compromise of RPF routes and timings. No unusual Soviet reactions to past sorties were noted, leading to the conclusion that, if the Soviets had indeed received advance warning of sorties, it had not appeared to make any difference to their reactions to the flights. There was thus no reason to expect a different reaction if the source of intelligence had been cut off. The conclusion was that it was safe to proceed with the operation. (Ingeborg Lygren was later cleared of all charges) However, the change in Norwegian government now made itself felt. The new Norwegian Defence Minister, Otto Tidemand, initially requested that Operation *Blountstown* be deferred until December, later declaring that Norway would no longer allow the RAF to fly covert RPFs from Norwegian airfields. The Secretary of State for Defence (Denis Healey) subsequently raised the matter with Mr Tidemand when the latter visited the UK. The protestations of the UK government appear to have had some effect and, following the refusal of a number of proposed operations, approval was eventually granted for a series of RPFs (three joint Comet/Canberra day sorties) from Andøya in May 1966.

Plans were then made for a further three joint Comet/Canberra sorties (*Literature*) in September 1966. Once again the Norwegians refused permission to use Andøya, citing

WORLD-WIDE OPERATIONS

B.6(BS) WJ775 in late 1960s configuration. Note pilot's sunshade, essential for high altitude and tropical operations as the cockpit could become unbearably warm. *RAF Wyton via Peter Green*

increased Soviet political pressure and attacks in the Norwegian press against military activities in north Norway. This time the Norwegians stood firm and further proposals for RPFs later in the year were rejected on the same grounds. As a result of the Norwegian ban, no RPFs were flown into the Barents Sea for a further period of 2½ years. Although negotiations with the Norwegians continued into 1967, there seems to have been some doubt in political circles whether it was worth the effort; the Secretary of State for Defence requesting that he: 'be assured that any useful information which is obtained from these sources is not also available from other friendly sources'. In fact, as suspected by the Secretary of State, the loss of a Norwegian forward operating base may not have been quite the disaster it initially appeared. UK operations over the Barents Sea had been at best intermittent and the UK had access to data gathered by US Elint sorties in the area. The US RB-47Hs of the 55 SRW used in-flight refuelling and could thus operate independently of Norwegian bases.

UK operations over the Barents Sea resumed two and a half years later in August 1968 following an apparent change of heart by the Norwegian government regarding the use of a forward base. A single Comet was detached to Andøya from where it carried out three sorties (Operation *Long*). There were still political problems to be overcome, however, and a request for a further operation, comprising six Comet sorties, was turned down by the Norwegian authorities in October 1968.

The last Comet RPF using a Norwegian base took place in April 1969 (Operation *Fifi*) with eight Comet sorties planned from Andøya, later reduced to four. Despite the reservations of the Norwegian government, the 51 Squadron detachment was made welcome by the RNoAF, who arranged a day out skiing during the visit. Further requests were made for permission to use Andøya in May and July but both were refused by the Norwegians. By the end of 1969 it appears the UK government had given up on attempts to use Norwegian bases and no further RAF sorties were flown into the Barents Sea.

Once again, the UK had to fall back on the US for intelligence in the Barents Sea area. In May 1967 the US had begun to replace their RB-47H/ERB-47H Elint aircraft with the RC-135C aircraft, a purpose-built Elint platform based on the C-135 Stratolifter. The aircraft carried the sophisticated ASD-1 automatic Elint system (in fact, the C-135 had been selected as an Elint platform because it was one of the few aircraft big enough to carry the ASD-1) and featured a large direction-finding aerial array either side of the front fuselage. Ten aircraft were initially completed as RC-135Bs in the mid-1960s, then being placed in storage, before being fitted with the ASD-1 system and redesignated as RC-135Cs. The advanced capabilities of the ASD-1 allowed the RC-135C to perform both the radar 'headcount' function of the RB-47H and the detailed technical analysis task of the ERB-47H. By 1970 the USAF 55 SRW were regularly operating long-range RC-135Cs in the area. These sorties were often flown from Eielson AFB, Alaska, flying over the Pole into the Barents Sea, along the Soviet coast, then south over the Norwegian and North Seas before finally landing in the UK. An alternative route started at Offutt AFB, Nebraska and flew north over Greenland. After arriving in the UK the RC-135C normally flew a number of sorties from there, finally returning back to the US via the Barents Sea.

Normally about half of the RC-135C sorties into the Barents Sea were intercepted by Soviet fighters, but in the first week of November 1970, the Soviets stepped up the pressure and intercepted every sortie. On 17th November a more worrying occurrence took place. The RC-135C flying this sortie was intercepted twice: on the second interception one of the Soviet fighters flew in close proximity with the RC-135C and fired its cannon on a heading par-

129

LISTENING IN

Taken in June 1967, this photograph shows the 55 SRW during their changeover from the RB-47H in the foreground to the RC-135B, with two of the latter being readied in the background.
T Panopalis Collection

RC-135C 64-14841 in April 1971 showing the large forward fuselage fairings housing the aerial array for the ASD-1 direction finding system.
T Panopalis Collection

allel to that of the American aircraft. Although 'demonstrative' gun firing by Soviet fighters was apparently not unknown, it was unusual for a fighter to fire from such close proximity. The Soviet Union subsequently made a formal complaint regarding the sortie, alleging that the US aircraft had infringed Soviet airspace. In response the US denied any infringement, asserting the aircraft had remained 46nm (85km) from Soviet territory and made a protest regarding the Soviet fighter's 'unwarranted discharge of weapons'.

The US informed the UK of the incident, noting that a subsequent Barents Sea sortie had passed off without incident, declaring their intention to continue with their normal RPF programme, and asking that the UK continue to authorise sorties to and from UK bases. The UK

USAF RC-135 sorties into the Barents Sea, February 1971

Date	Sortie	Type	To-From	Closest
1 Feb 71	B284	Day	Offutt-Mildenhall	50nm
2 Feb 71	C222-1	Day/Night	Eielson-Mildenhall	53nm
4 Feb 71	C232-1	Day/Night	Mildenhall-Mildenhall	68nm
8 Feb 71	C232-2	Day/Night	Mildenhall-Mildenhall	68nm
10 Feb 71	C241-1	Day/Night	Mildenhall-Eielson	53nm
10 Feb 71	A206	Night	Mildenhall-Offutt	44nm
12 Feb 71	C222-2	Day/Night	Eielson-Mildenhall	53nm
15 Feb 71	B285	Day/Night	Offutt-Mildenhall	52nm
16 Feb 71	C232-3	Day/Night	Mildenhall-Mildenhall	68nm
18 Feb 71	C232-4	Day/Night	Mildenhall-Mildenhall	68nm
22 Feb 71	B280	Day	Offutt-Mildenhall	48nm
24 Feb 71	C241-2	Day/Night	Mildenhall-Eielson	53nm

authorities agreed, noting: 'first, the flights are a necessary source of information in themselves; secondly, it would be most undesirable to allow the Russians to suppose that they can successfully exert pressure on flights taking place in international airspace'. After analysing the 17th November incident in detail, the US then decided to amend their February 1971 programme to include more Barents Sea sorties. These included a probe into the Pechora Sea (between Novaya Zemlya and the mainland) to obtain further information on the correlation between fighter and radar activity following the 17th November incident. The table opposite shows the USAF RC-135C sorties into the Barents Sea planned for February 1971 and illustrates the effort the US could bring to bear in such a remote area in a single month.

The Baltic

From the early 1960s onwards a large percentage of 51 Squadron's operations were carried out over the Baltic. Sorties were flown using both Comets and Canberras, with multiple aircraft used on some sorties. The standard route took the aircraft along the coast of East Germany, Poland and the Soviet Union. Comet sorties were nearly always flown from the UK while the shorter-range Canberras occasionally operated from a forward base in Germany, presumably when operations ventured further into the Baltic. From 1959 onwards single aircraft sorties were generally flown in daylight but multiple aircraft operations (considered more provocative) were flown in darkness. Interceptions by one or more Soviet fighters were common but normally passed off without incident.

During the late 1950s/early 1960s the Soviet Union improved its ability to deal with high-altitude bombers, forcing both the US and UK to re-examine their strategic bomber tactics. With intelligence suggesting that attack from high-altitude would be extremely costly the US began to convert their B-52s to the low-level role around 1961 and the V-Force followed in 1963-64. The low-level capability of the Soviet Air Defence system was now of great interest to both SAC and Bomber Command and thus from around 1964 onwards, the Canberras of 51 Squadron were employed on a systematic investigation of the ability of the Soviet Union to counter low-level attacks over the Baltic. The Canberra came into its own in this role since neither the Comets nor USAF RB-47Hs were suitable for operations at low level. Sorties were flown off the Soviet coast at high level, suddenly descending to low level for part of the route. Soviet air defence organisation communications were monitored by ground stations during the sorties and the resulting intercepts were subsequently used to assess the accuracy with which the Soviet radars tracked the Canberras.

In April 1965 the rules on multiple aircraft sorties were relaxed to allow operations during daylight. As part of their justification for the change the MoD noted that the US had conducted a number of daylight multiple operations over the Baltic during the previous 12 months without any adverse Soviet reaction.

In December 1964/spring of 1965 a series of Canberra and Comet RPFs (Operation *Rut*) were flown over West Germany and the Baltic in order to collect detailed technical data on a Soviet P-14 *Tall King* VHF band (150-200MHz) EW radar in East Germany. The sorties were flown along civil air routes on open flight plan, allowing the aircraft to approach within 20nm (37km) of the East German border.

Up until mid-1965 sorties over the Baltic were prohibited from approaching closer than 40nm to the Soviet coast; after May 1965 the restriction was relaxed and occasional sorties approaching to 30nm were allowed. In July 1965 the US and UK co-operated in an intensive operation to gather intelligence on a specific Soviet Air Defence Sector. Under Operation

Canberra B.6 WT301 on the approach to Watton in late 1960s configuration with rounded radome, second *Blue Shadow* aerial, tail-mounted guard receiver and tail warning receiver.
via Peter Green

LISTENING IN

Stimulation some twelve joint daylight Comet/Canberra sorties were flown over the Baltic in conjunction with a number of RB-47H and ERB-47H sorties. A second, similar, intensive joint US-UK operation over the Baltic was flown in July 1966.

The governments of Scandinavian countries were normally notified of Baltic sorties as a courtesy, even though the operations were carried out in international airspace. There were, however, the occasional inadvertent incursions into a friendly country's airspace due to navigation failures, normally leading to enquiries by the countries concerned and apologies by HM government. During 1966 there were worries that proposed Swedish defence cuts might result in the RSAF being more vigilant, in order to prove its worth, and thus more likely to publicise any accidental incursion.

Sigint sorties over the Baltic occasionally involved other aircraft and in December 1966 four Victor sorties (Operation *Brummel*) were carried out to investigate the ability of the Soviet air defence system to detect V-Force aircraft operating their H2S Mk.9A under Bomber Command 'war operations' procedures. The aim was to determine the ranges at which the Soviets could pick-up the V-bombers via passive detection and also how accurately they could track them using a combination of passive detection and radar. The USAF supplied a specialist Lockheed C-130 Aircraft Reporting Platform to watch over the Victors and provide warning of any hostile activity during the sorties.

In April 1967 two Comets were put on 30-minute standby at Laarbruch for operations to collect Soviet IFF transmissions. The Comets were scrambled for an RPF along a fixed route whenever Soviet Bloc aircraft were detected en-route over East Germany from their bases to a bombing-range on the Baltic Sea coast. A special watch was kept on the air defence systems in East Germany and Poland in order to detect an adverse reaction to the sorties, but none was forthcoming and the operation was extended into May, with Canberras replacing the Comets.

Regular operations over the Baltic by Canberras and Comets continued up until withdrawal of both types from service in 1974.

Turkey and Black Sea

From the late 1940s the UK had carried out regular RPFs from Cyprus over the Black Sea and along the Turco-Soviet border, overflying Turkey in the process. All these sorties were carried out with the knowledge and permission of the Turkish government.

The Black Sea and the Turco-Soviet border were important areas for Elint operations: not only were they likely to be overflown in time of war, but they also hosted some important Soviet defence facilities. For example, the Black Sea was home to a major Soviet Fleet and was also the base for an important radar research

Revised RPF tracks over the Black Sea and along the Turco-Soviet border. In early 1961 the Turkish government insisted that aircraft cross the coast at Silifke and keep clear of Adana.

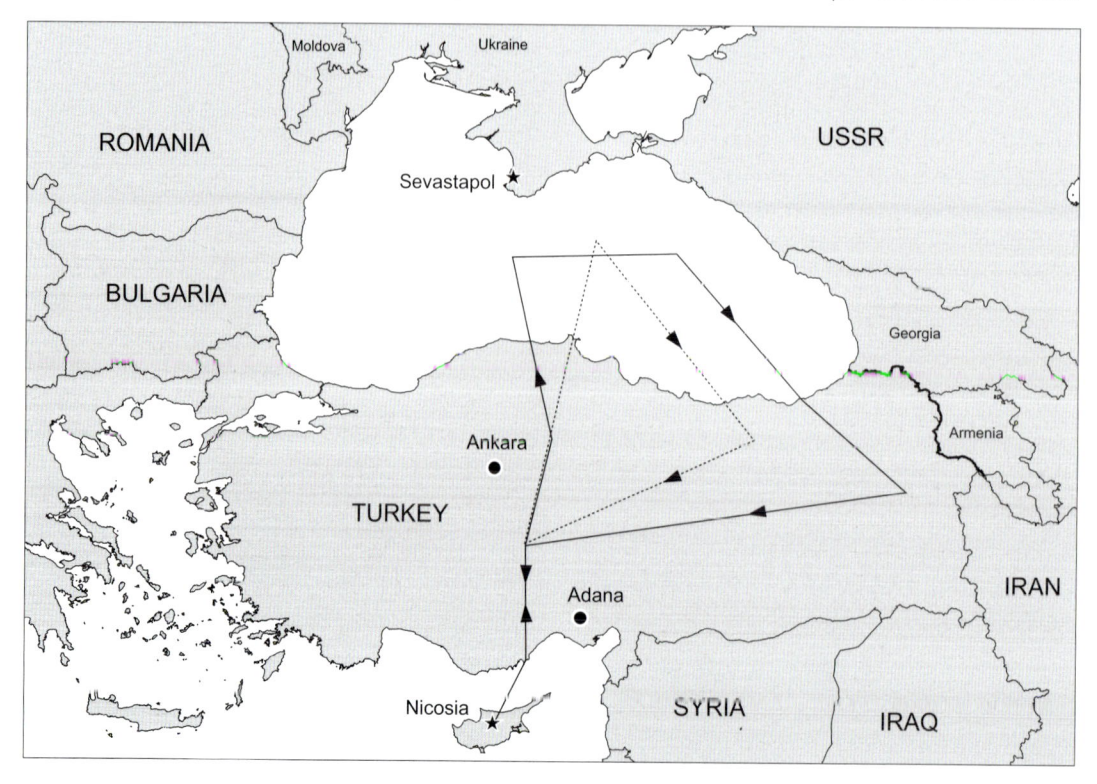

132

facility. Both the US and UK maintained ground Elint stations on the Black Sea coast to monitor these facilities.

The Comets of 51 Squadron were regularly detached to Nicosia for sorties over the Black Sea during 1958-60. The Turkish government seems initially to have taken a reasonably relaxed view of the sorties, but at the beginning of 1961 the Turks began to impose new restrictions on the conduct of the flights. In January the Turkish Defence Staff requested that the flights enter and leave Turkish airspace from/to Cyprus at a fixed point (Silifke), and that they avoid overflying the Adana area (where the US Incirlik Air Base was located). The UK agreed to these fairly innocuous conditions but resisted a third condition, which was that the flights should only leave and enter northern Turkish airspace (that is, over the Black Sea) via international air routes. Early in March the Turks proposed further restricting the flights to a maximum altitude of 40,000ft (12,192m) and limiting them to a maximum distance of 20nm (37km) from the Turkish Black Sea coast. Then, at the end of March, the Turkish government informed the UK Ambassador in Ankara that no flights could approach nearer than 55nm (100km) to the Soviet border whilst flying in Turkish airspace and that 20 days' notice should be given for all flights. Similar restrictions were imposed on American Elint sorties. The new restrictions were strongly opposed by the Air Ministry and Ministry of Defence on the grounds that that they would 'remove all value from the flights', since the Soviet air defence system would hardly react to aircraft at that distance.

In April 1961 the Chief of the Defence Staff (Lord Louis Mountbatten) and Chairman of the US Joint Chiefs of Staff (General Lemnitzer) visited Ankara and met with General Sunay, Chief of the Turkish General Staff to discuss the subject, the seniority of the participants reflecting the importance attached to the matter by the UK and US. The Turkish explanation for the restrictions was that they were worried over Soviet reactions to the previous year's U-2 and RB-47H incidents. In response Lord Mountbatten explained the importance the UK and US attached to the flights, their importance to the maintenance of the deterrent, and the intelligence benefit Turkey gained from the flights. He then explained that the flights would be useless unless they could approach to within 30-40nm (50-70km) of the Soviet coast or borders and gave assurances regarding the responsible conduct of the sorties. Although some concession was made on maximum altitude, the Turks stood firm on the main point and the 55nm (100km) restriction remained in place, applying both to sorties over the Black Sea and those along the Turco-Soviet land border.

Despite the restrictions, the Black Sea remained an important area for intelligence operations and 51 Squadron continued to fly regular sorties from Akrotiri into the Black Sea in the period 1961-65. RPF sorties entering and leaving Turkish airspace were, of course, tracked by the country's air defence system and, in November 1961, Turkish fighters intercepted a Comet returning from a Black Sea sortie due to a mix-up over clearances. Warning shots were apparently fired and the Comet was forced to land at Incirlik, where the crew was detained for half an hour before the situation was resolved and the aircraft allowed to continue on to Akrotiri. Operations were occasionally interrupted by the political unrest on Cyprus and in January and February 1964 detachments to Akrotiri were cancelled due to the 'unsettled situation'.

RPFs over the Black Sea comprised both single aircraft (normally a Comet), and multiple aircraft operations using both Comets and Canberras. From 1959 onwards the single aircraft operations were flown in daylight but joint operations, still considered more provocative, were still flown at night. Aircraft were regularly intercepted by Soviet fighters and this generally passed off peacefully, although operations were sometimes aborted if the fighter's AI radar was locked-on. During 1965, however, a Comet ran into trouble during a sortie coinciding with a Soviet naval exercise when it was intercepted by a number of the new Tupolev Tu-128 *Fiddler* long-range fighters. The *Fiddlers* boxed-in the Comet and signalled the crew to follow them and land at a Soviet airfield. The Comet pilot was forced to carry out evasive action, deployed the airbrakes and descended steeply, turning back towards Turkish airspace to escape. Although no offensive action was taken, and the Soviet fighters eventually gave up their pursuit of the Comet, these sorts of incidents raised tensions and were not popular with the Turkish authorities.

In September 1964 proposals were put forward for two special night operations including a Valiant: the first using a Comet, Canberra and a Valiant, and the second using a Valiant and two Canberras. The Valiant was almost certainly a radar-reconnaissance version operated by 543 Squadron. The purpose of the operation may have been to obtain radar mapping of the

LISTENING IN

Canberra B.6(BS) WJ768 at Luqa. Note early-type tail aerial array.
via Peter Green

This Martin RB-57F, 63-13287, was lost over the Black Sea in December 1965, following which the Turkish government banned further US and UK Elint sorties in the area. The RB-57F was a heavily-modified Martin B-57, which was a US-built Canberra.
T Panopalis Collection

Black Sea area under cover of an RPF, alternatively the presence of the Valiant may have been intended to stir up the Soviet defences. The proposals drew an objection from the Foreign Office, worried about the reaction of the Soviets, who were now 'more involved in the Cyprus question'. The concern was that a sudden deterioration in the situation in Cyprus, coupled with an RPF operation involving more aircraft than usual, and including an aircraft of unusual type, might result in an adverse reaction. As a result of these concerns the operation was reduced to two night sorties, each comprising a Comet and Valiant.

The USAF was also active in the Black Sea area, maintaining a presence at the Incirlik Air Base in southern Turkey under a joint-use agreement with the Turkish military. Incirlik was used as a base by various USN and USAF reconnaissance aircraft (including the RB-47H and RB-66C) for sorties along Soviet borders and over the Black Sea during the 1950s and 1960s.

In December 1965 a Martin RB-57F (63-13287) of the 7407th Support Squadron/7499 Support Group, normally based at Wiesbaden, Germany, was detached to Incirlik for a series of covert flights against the Soviet Union. On 14th December the aircraft, crewed by Captain L L Lackey and 1st Lieutenant R Yates, took off for a sortie over the Black Sea. Sadly the RB-57F never returned to Incirlik. The accepted explanation for the loss of the aircraft seems to be that the crew were incapacitated at altitude, probably due to an oxygen system problem: the aircraft was apparently tracked by US/Turkish radar and took over an hour to spiral down into the sea. A subsequent search of the crash area retrieved a few fragments of the aircraft but failed to find any trace of the unfortunate crew. Ten days later, on 24th December, the Soviet Deputy Foreign Minister Kuznetsov summoned the US Ambassador Kohler in Moscow and, in a prepared statement, complained about US reconnaissance flights near Soviet borders, specifically mentioning the 14th December flight. It seems likely the Soviets also complained to the Turkish government as well. As a result of the loss of the RB-57F, and the subsequent Soviet complaints, the Turkish government banned all US and UK Elint sorties over the Black Sea and along the Turco-Soviet border. UK aircraft were still allowed to fly over Turkey en-route to the Persian Gulf, but RPF operations in Turkish airspace were prohibited.

The loss of the Black Sea for airborne Elint operations was a heavy blow to GCHQ and the DIS. Although the UK still maintained a ground Sigint presence in Turkey, airborne sorties could pick up signals not seen by ground stations and sorties over the Black Sea were widely recognised as being extremely productive operations.

Canberra B.6(BS) WJ775 at Luqa in 1969 during a stop-over en-route to the Gulf.
via Peter Green

In 1967 the Air Ministry and the FCO discussed raising the subject again with the Turks, with a view to getting the ban lifted. In August, DD.Ops(Recce) at the Air Ministry wrote a letter of appreciation to the Air Attaché in Ankara: '… it is encouraging to hear that you are considering another approach to the Turks. The best of luck! As you know, it means an awful lot to us'. Unfortunately the ongoing Cyprus problem, the formation of a Turkish-Cypriot administration, and the British presence on the island, made choosing the right time to re-open talks with the Turks rather difficult. After a discussion with the US authorities, the British Embassy raised the issue of Black Sea sorties with the Turkish Ministry of Foreign Affairs in February 1968. Further discussion of the subject was then put on hold by the Foreign Office pending a renewal of the Memorandum of Understanding with the Turks covering UK and US Sigint ground stations. In the event, it appears that the Turkish government maintained their position and refused to authorise the resumption of airborne Sigint sorties over the Black Sea.

Iran and the Caspian

During the late 1940s and early 1950s the UK had flown RPF sorties over Iran into the Caspian Sea and along the Iranian-Soviet border from Habbaniya in Iraq. Although RPF sorties from Habbaniya came to an end in 1955, when the base was handed over to Iraqi control, 192 Squadron aircraft had occasionally refuelled there after that date at the end of sorties. The Iraqi coup of 1958 and the subsequent expulsion of UK forces from the country meant the RAF could no longer overfly Iraq en-route to Iran and needed to find a new base for operations in the Caspian area.

From January 1962 onwards 51 Squadron began to fly RPFs from RAF Sharjah, Trucial States, in the Persian Gulf. Operating from Sharjah the Comets and Canberras flew across the Gulf and over Iran, then out into the Caspian Sea or along the eastern Soviet-Iranian border. Iran was then member of the Central Eastern Treaty Organisation (CENTO), and thus an ally of the UK, and the RPFs over Iran were flown with the knowledge and permission of the Iranian government. Like other western-oriented countries abutting the USSR, Iran had to balance its alliances with NATO countries against the maintenance of a working relationship with its powerful neighbour. The fact that UK Elint aircraft only overflew Iran and did not need to land there undoubtedly helped negotiations with the Iranians.

Sharjah, located over 700nm (1,280km) from the Caspian coast and the Iranian-Soviet border, was not an ideal base for these operations. The long distances involved severely reduced the length of time a 51 Squadron aircraft could spend in the operational area. In general a 5½ hour Comet sortie from Sharjah was only 'on-station' along the Soviet border for 1½ hours. However, the UK had no alternative to Sharjah and had to make the best of it. The only other British base in the region was RAF Muharraq in Bahrain, which although slightly closer to the area of operations, was not normally used for flights over Iran for diplomatic reasons, due to Iranian territorial claims on Bahrain. Conditions at Sharjah could be challenging for the ground crew, with high temperatures and humidity limiting the time that could be spent working on the aircraft. The same conditions made life uncomfortable for the aircrew at the beginning of each sortie, even when ground coolers were used to try to keep the aircraft cabin temperature within reasonable limits.

The first 51 Squadron detachment to Sharjah in 1962 staged through El Adem, flying south to 'Nasser's Corner' and then via Khormaksar in Aden. However, the later, more nor-

LISTENING IN

Between 1962 and 1969 sorties were flown from Sharjah along the Soviet-Iranian border. Operations were also conducted against Iraq on the transit from Akrotiri. From 1969 onwards sorties were flown from Tehran.

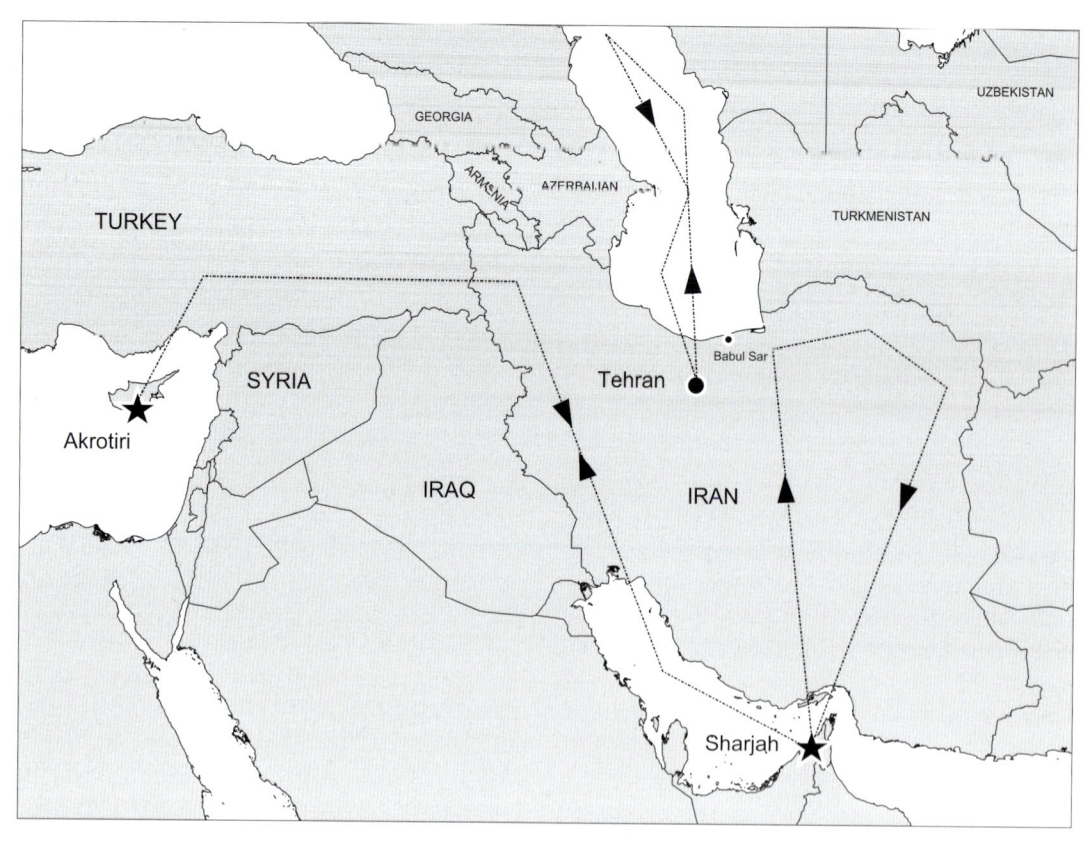

Comet R2 XK695 on the approach to Luqa, showing the aircraft's relatively clean lines.
Victor Pisani – Malta

mal route, was from Akrotiri across Turkey and Iran. Flying this route the Comet or Canberra could carry out an RPF along the Soviet Trans-Caucasus border (Georgia, Armenia and Azerbaijan) before turning south for the Gulf. Operations against Iraq were often flown on the same sorties. The routes varied slightly from one flight to another with aircraft keeping 55nm (100km) or more from the Soviet border. A similar route, and similar operations, were flown on the return trip to Akrotiri. RPFs from Sharjah were often integrated with operations in other localities. Thus a detachment at Akrotiri, operating over the Black Sea, might deploy a Comet to Sharjah for a few days. Likewise, Comets en-route to and from Changi for operations against Indonesia, would carry out RPFs from Sharjah.

In November 1963 an opportunity arose to mount an RPF operation from Iran itself. During 1962/63 Iran had been supplied with two modern Decca *Hydra* air defence radars, funded by the UK under the CENTO Mutual Assistance Programme. The first *Hydra* was installed at Tabriz in north western Iran in mid-1962 and the second at Babul-Sar, near the Caspian Sea, in the summer of 1963. In the winter of 1963 arrangements were made to calibrate the

WORLD-WIDE OPERATIONS

Canberra B.6(BS) WT305 on detachment in the Persian Gulf sometime in 1967. Note the extent of the sun shield for the cockpit and, of course, the pilot, a necessity in the Gulf.
via Peter Green

radars using two 51 Squadron Canberras: Canberras were used to calibrate UK air defence radars and were thus ideal as calibration targets for the Iranian radars. The plan was to carry out RPFs during the calibration sorties, the latter providing a cast-iron cover story for the operation. Thus Canberra B.2 WH698 and B.6 WJ775 were deployed to Teheran/Mehrabad (home of the Iranian AF 1st Tactical Fighter Wing and its North American F-86F Sabres) in the second week of November. The joint calibration/RPF operation, comprising 16 sorties, was carried out as Exercise *Liver* during mid-November, both Canberras returning to the UK on 21st November. The detachment at Teheran was supported by Comet XK671, which transported stores and ground crew.

The Shah of Iran was not only aware of, but seems to have been positively in favour of, RPFs over his country against the USSR. In March 1965 the Shah visited 51 Squadron at Wyton during his state visit to the UK. The Shah was escorted by CAS (Sir Charles Elworthy), the AOC, Bomber Command (Air Chief Marshal Sir Wallace Kyle, KCB, CBE, DSO, DFC) and the AOC CRE (Air Commodore B P Young CBE). The Shah was shown over a Canberra and Comet and His Imperial Majesty apparently 'expressed great interest in the work of the Squadron and appeared to be impressed with the standards of efficiency maintained'.

In December 1965 the Turkish government halted all US and UK RPFs over Turkey and the Black Sea following the loss of a US RB-57F. This had the effect of increasing the importance of operations over Iran, since these were now the only way to investigate defences along the southern borders of the USSR.

In early 1966 the UK was offered a further opportunity to fly RPFs from Teheran, once again as part of a joint radar calibration/RPF operation as Iran needed to work up a third new air defence radar and needed airborne radar targets. The new radar was conveniently situated close to the Soviet border and the opportunity was taken to carry out RPF operations in the area, using the calibration task as a cover story. Operation *Nadir* took place in February 1966 and the OC 51 Squadron remarked: 'A great deal is expected of the detachment's co-operation with the Iranian Air Force, especially after the great interest taken in the squadron's operations by the Shah of Iran during his visit to Wyton in March 1965'. Canberra WJ775 was employed on *Nadir* and operated from Mehrabad airbase, Teheran. The first two flights were unsuccessful due to *Blue Shadow* navigation radar failures, but, after rectification, seven successful sorties were flown. The personnel detached at Mehrabad returned to the UK 'suitably bronzed'. Operations over Iran against the Soviet Union and Iraq continued to be flown from Sharjah in the period 1966-68 at approximately four-monthly intervals.

In January 1969 the MoD(Air) asked the FCO to make enquiries regarding the occasional use of an Iranian base for RPFs 'in the Caspian Sea area'. The FCO raised the matter with the Iranian government in April 1969 via the British Embassy in Teheran. The reply was that the Shah agreed to the flights 'without hesitation'. Although the Shah was apparently unconcerned about the political consequences of hosting Elint sorties from an Iranian air base he was anxious that secrecy be maintained, and thus requested that knowledge of the flights limited to the bare minimum of Iranian personnel. Plans were subsequently made for a series of three RPF programmes per year from an Iranian base, replacing the three programmes previously flown from Sharjah.

The first detachment to operate from Iran following the agreement took place in August 1969. This was actually another dual calibration/RPF operation, using the calibration of an Iranian air defence radar (supplied by Marconi) at Mal, close to the Turkmenistan border, as an opportunity to fly RPFs against the Soviet

LISTENING IN

Union. Two Canberras (WJ775 and WT301) were detached at the beginning of the month and spent three weeks operating from Teheran/Mehrabad, completing 15 sorties. A tented base was established on the airfield and the crew accommodated in a hotel in Teheran, but unfortunately a large number of the detachment contracted a stomach bug, detracting somewhat from their enjoyment of the exotic location.

The first use of an Iranian forward base for a regular RPF proper (in place of Sharjah) was in November 1969 when Comet XK695 was detached to Teheran/Mehrabad for Operation *Thorn*, a series of ten day sorties, apparently approaching to within 27nm (50km) of the Soviet Border. The detachment went well, despite a number of problems with spares and servicing. Following *Thorn* all further operations in the Caspian areas took place from Teheran rather than Sharjah. The programme entailed around three detachments each year and continued up until 1974 when the Comet was replaced by the Nimrod.

Iraq

In 1961 the State of Kuwait gained independence from the UK and shortly afterwards neighbouring Iraq reasserted its long-held claim on the territory, declaring that Kuwait was historically a province of Iraq. Iraqi threats against Kuwait, coupled with Iraqi troop movements on the Kuwaiti border, resulted in the UK deploying military forces in the area under Operation *Vantage*. The threatened Iraqi offensive never materialised and *Vantage* was wound down during July. Despite the apparent success of *Vantage* in discouraging Iraqi aggression, the UK continued to keep an eye on the country with regular fortnightly border reconnaissance sorties by Canberra PR.7s and PR.9s operating

Chart for the last five of a series of 10 operations flown against Iraq during February 1968. The sorties were flown by a single Comet at 35,000ft.

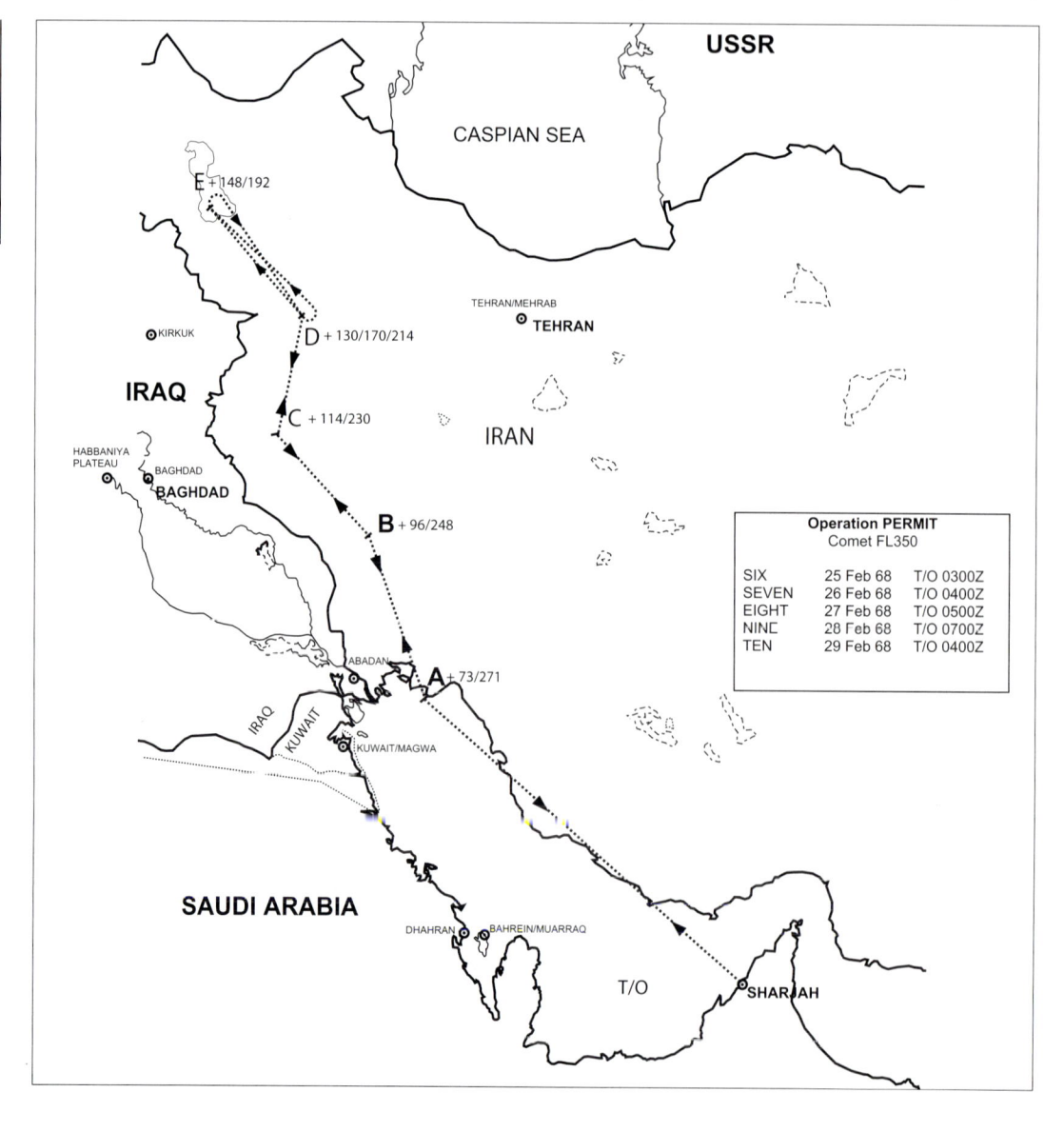

from Bahrain on detachment from the UK and Cyprus.

Airborne Sigint operations against Iraq appear to have begun in April 1962 when Canberra B.2 WH698 was deployed to RAF Muharraq, Bahrain for Operation *Genus*. The Canberra flew a series of six daylight sorties (*Genus* 1-6) from there along the Iraqi border, presumably compiling an Iraqi electronic order of battle. The operation was continued in June when WH698 returned to Bahrain for a further seven sorties (*Genus* 7-13). Other RPFs against Iraq were flown during the regular Comet RPF sorties over Turkey and Iran between Akrotiri and Sharjah. The proposal for the May 1962 RPF programme included two RPF sorties (Operation *Saint*), passing within 38nm (70km) of the Iraqi border en-route between Akrotiri and Sharjah. The proposed route was subsequently modified at the request of the Foreign Office 'so that it will not be obvious to the Iranians that we are undertaking operations against Iraqi targets'. The worry seems to have been that the Iranians (who had their own disputes with the Iraqis) might ask the UK to supply intelligence on their neighbour.

The UK continued to monitor Iraq during the 1960s, principally via sorties operating over Iran from Sharjah, and on transits between Akrotiri and Sharjah, covering both the Turkey-Iraq and Iran-Iraq borders. During February 1968 a series of ten Comet sorties (Operation *Permit*) were flown from Sharjah along the Iran-Iraq border, collecting order of battle information on Iraqi radars and VHF communications intelligence.

In the spring of 1969 arrangements were made with the Iranian government to relocate 51 Squadron's principal operating base in the Gulf from Sharjah to Iran itself. At around the same time a long-standing dispute between Iran and Iraq over the Shatt al-Arab waterway escalated, with Iraq threatening war over the Iranian abrogation of a long-standing treaty. The Iranian authorities were by now well aware that British RPFs over Iran also targeted Iraq and in mid-April 1969 the Iranian Air Force asked the Air Attaché at the UK embassy in Teheran for 'all possible information in our possession regarding Iraqi radar coverage'. This request caused concern in the Ministry of Defence, with the Secretary of State apparently asking for an assurance that that moving operations from Sharjah to an Iranian base 'will not result in further embarrassing attempts to involve us in local quarrels'. The Secretary of State was subsequently reminded that UK Elint operations along the southern border of the Soviet Union were already completely dependent on Iranian goodwill, and thus the relocation of operations to an Iranian base would hardly increase the obligation. Information on Iraqi radar coverage was supplied to the Iranians.

The Yemen

During the 1960s the United Kingdom maintained a military presence in the British Protectorate of Aden and uneasy relations with the state of Yemen to the north. In September 1962 the feudal government of Yemen was overthrown in an Egyptian-backed military coup, sparking a civil war between the royalists (backed by Saudi Arabia) and the republicans (backed by Egypt and the Soviet Union) that dragged on until 1968. The September coup, and subsequent reports of Egyptian troop movements and arms supplies, was a source of concern to the British government and Canberra photo-reconnaissance (PR) sorties were flown along the Yemeni border, and over the Red Sea, looking for Egyptian arms shipments.

In November 1962 Canberra B.6 WT305 and Comet XK659 were detached to RAF Khormaksar, Aden for Elint operations against Yemen under Operation *Flame*. The intent was presumably to establish an order of battle and determine the build-up of Egyptian arms. The Comet was probably employed in the Comint role, attempting to determine the presence of Egyptian troops by voice intercepts. Following *Flame* no further RPF sorties appear to have been flown against Yemen for almost a decade.

In 1967 the UK left Aden, following which the Western and Eastern Aden Protectorates amalgamated to form the People's Republic of South Yemen (PRSY). The new regime developed close ties with the Soviet Union and China and somewhat strained relations with its immediate neighbours. In particular the PRSY supported the Dhofar rebellion in Oman and was also involved in disputes with Saudi Arabia. From 1970 onwards the UK provided military support to the Sultanate of Oman under its new, reformist, Sultan in its struggle against the Dhofar rebels. Since the People's Democratic Republic of Yemen (PDRY), as the PRSY was renamed in 1970, provided arms and training for the rebels it became a UK intelligence target. Regular Canberra photographic reconnaissance sorties were conducted along and over the border looking for training camps and troop movements and early in 1973 plans were made for a programme of RPF operations against the

LISTENING IN

During the 1960s RPFs against Yemen were flown from Khormaksar. By the 1970s operations had shifted to Masirah. The Arabian Peninsula was of particular interest to 51 Sqn as the political landscape was changing quickly in the area.

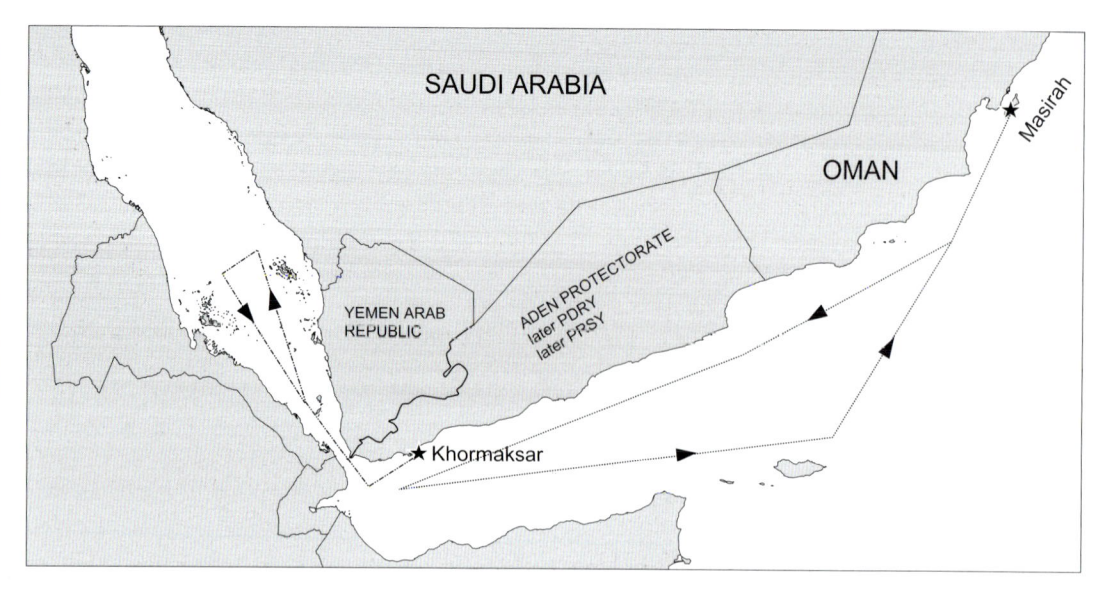

country. The first operation took place in February 1973 when Comet XK695 was detached to RAF Masirah. Three day sorties were flown over the Arabian Sea under Operation *Egma*, standing off 22nm from the PDRY coast. A similar operation (*Angle*) was flown three months later in May 1973, again using a single Comet operating from Masirah.

Indonesia

At the start of the 1960s the island of Borneo was split into four main parts. The larger, southern part, (about two thirds of the island) comprised the Kalimantan provinces of Indonesia. The north of the island was divided into the British Colonies of Sarawak and North Borneo, and the British Protectorate of Brunei.

In 1961 the UK proposed the incorporation of North Borneo, Sarawak and the British protectorate of Brunei into a federation with Malaya and Singapore, the resulting entity to be known as 'Malaysia'. In January 1963 the Indonesian government, under President Sukarno, declared its opposition to the union and announced a policy of 'Confrontation' with the proposed federation. The Indonesian government had used a similar policy of Confrontation to wrest control of West New Guinea from the Dutch the previous year.

The British government took the Indonesian declaration of opposition to the formation of Malaysia seriously and at the end of April 1963, Comet 2R XK659 of 51 Squadron was detached to Singapore for a series of RPF sorties against Indonesia under Operation *Tarnish*. The

The first two RPFs against Indonesia (*Tarnish* and *Olympic*) operated from Singapore (Tengah) and Australia (Darwin). *Olympic* also involved sorties from the Cocos Islands, covering the southern coast of Java.

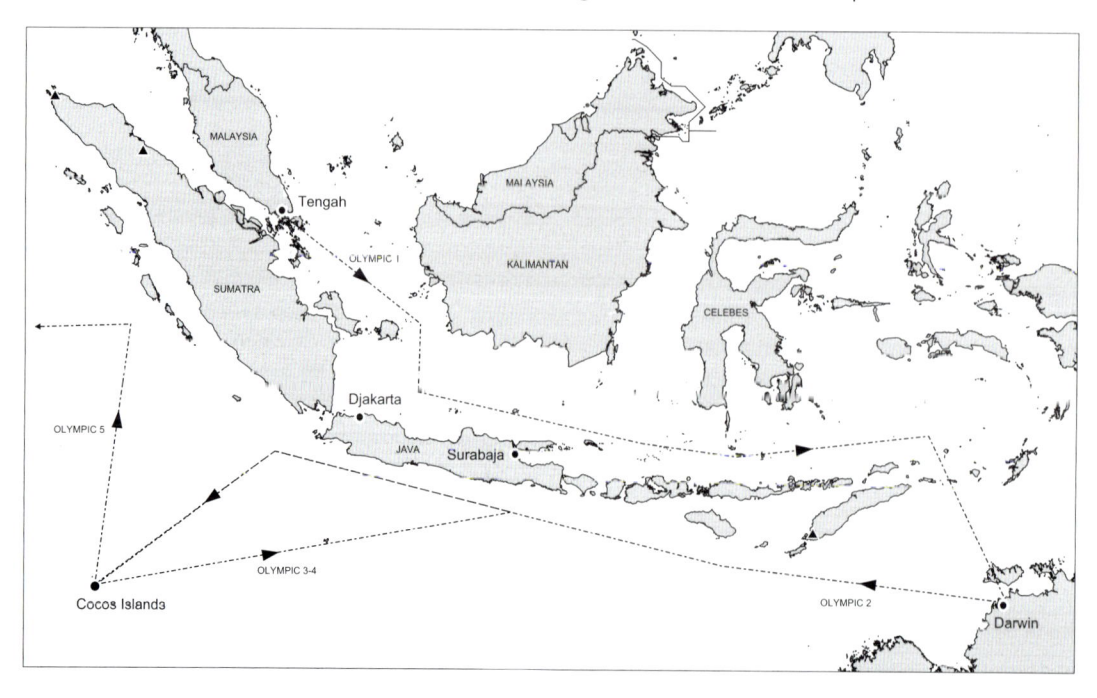

140

aim of *Tarnish* was most likely to compile an Indonesian Order Of Battle, as a precaution against future hostilities. The Comet flew 11 'very successful' sorties in the period 26th April to 20th May, operating from Singapore and Darwin, Australia and flying the length of the Indonesian archipelago.

Despite Indonesian opposition the Federation of Malaysia duly came into being in September 1963, although without Brunei. A second RPF operation using a 51 Squadron Comet was mounted, at short notice, against Indonesia in December 1963 as Operation *Olympic*. The aircraft flew out to Tengah, Singapore via Libya, Khormaksar and Gan. The first sortie took the Comet along the north coasts of Sumatra and Java, landing at Darwin, Australia, while the second sortie went along the south coasts of Java and Sumatra landing on the Cocos Islands in the Indian Ocean. Three further sorties were flown from the Cocos Islands, again along the south coast of Indonesia and the Comet returned to the UK on 22nd December.

Amongst the targets of operations *Tarnish* and *Olympic* were the dispositions of the Indonesian Air Force (AURI) and the Indonesian air defence system. The AURI was in fact one of the more potent air forces in the region, equipped with a variety of Soviet aircraft, including MiG-19S *Farmer-Cs*, MiG-17F/PF *Fresco-Cs*, MiG-21F *Fishbed-Cs*, Ilyushin Il-28 *Beagles* and Tupolev Tu-16KS *Badger-Bs*. The C&R element of the Indonesian air defence system comprised around 15 Polish *Nysa* radars and two Soviet *Big Mesh* radars associated with the S-75 *Dvina*/SA-2 *Guideline* SAM defences being constructed around Djakarta. However, the air defence radar system was in the process of being upgraded by Decca (a UK company) under a £6M programme known as 'Parrot', the contract for which had been signed in 1958. By late 1963 the radars had been delivered but not yet installed and commissioned. Following the declaration of Confrontation by Indonesia in 1963 the UK government took steps to halt work by Decca on the commissioning of the system. Since no statutory powers existed to terminate the contract the government was obliged to ask, rather than order, the company to cease work on the project. Following negotiations, Decca complied with the request and terminated its contract with Indonesia in December 1963 in return for a fairly hefty compensation package.

In early 1964, shortly after the integration of Borneo and Sarawak into Malaysia, Indonesian forces began to carry out small-scale attacks across the Kalimantan border into Sarawak and Sabah (North Borneo). In response the British government built up military forces in the area and air defence patrols were initiated to discourage incursions by the Indonesian Air Force into Malaysian air space.

The UK government also made arrangements to improve its intelligence on Indonesia and as part of this programme formed a ground-based mobile Sigint unit (54 Signals Unit) at Seletar, Singapore in March 1964. Staff included linguists trained in the Bahasa Indonesia language and detachments were mounted at Labuan and throughout the Malaysian peninsula, mainly targeting Indonesian communications traffic.

Airborne Sigint operations against Indonesia resumed in November 1964 when Comet XK695 was detached to Singapore via Akrotiri, Sharjah and Gan (Operation *Glidden*). One target of the detachment was probably the disposition of the 18 MiG-21Fs (*Fishbed*-C) delivered to Indonesia during the previous year. The first RPF in the series was actually carried out on open flight plan on the Gan-Changi leg as the Comet flew round the north of Sumatra and into the Strait of Malacca. From Singapore the Comet carried out a daylight sortie over the Java Sea, standing off between 90 and 60nm (167 to 111km) from the Indonesian coast. The Comet then relocated to Labuan, from where it flew a sortie east into the Celebes Sea, south through the Makassar Strait, then north-west over the Java Sea back to Changi. This was followed by a series of five sorties over the Java Sea from Changi, one of which was carried out at night. The detachment flew 80 hours in 25 days, returning to the UK at the end of the month. A couple of linguists from 54 Signals Unit were temporarily attached to the squadron during the detachment and flew in the Comet during some of the sorties, assisting with the Comint task.

The May 1965 RPF programme (Operation *Amami*) followed a similar format to the previous year, with a single Comet deployed to Changi. RPFs were flown from Changi over the Java Sea, with a single sortie from Labuan going around the top of Borneo, over the Celebes Sea and down the Makassar Strait. All the sorties were flown over international waters, approaching to within 90nm (187km) of the Djakarta Defended Area (Java) by day and 42nm (78km) by night, approaching to within 55nm (102km) of the Surabaja Defended Area by day and 45nm (83km) by night, with the

LISTENING IN

closest approach to the Borneo border being 30nm (56km). The expectation was that the Comet would be picked up by Indonesian radar but the likelihood of an interception was thought to be low, particularly at night since the AURI lacked an all-weather capability, and routes were planned to minimise the success of any intercept attempt. In the event there was no Indonesian air defence reaction to the initial *Amami* sorties resulting in a proposal that sortie *Amami 9* should be allowed to permit closer approaches to the Javanese defended areas of Djakarta and Surabaja: 'By changing the route slightly we will improve the intelligence gathering capability of the operation'. The *Amami* detachment flew 98 hours and 17 operational sorties in 30 days and the detachment was apparently very much enjoyed by all who took part in it.

The fifth programme of operations against Indonesia took place in January 1966, when Comet XK659 flew out to Changi via Agra, India, carrying out a series of operations against China en-route. From Changi the Comet flew a series of seven RPFs (Operation *Mirimar*) against Indonesia. As with previous detachments, the majority of sorties were flown from Changi and a single sortie was flown from Labuan. As before, the risk of the operation was assessed as fairly low: analysis of previous operations suggested that, although the Comets had theoretically been within Indonesian radar cover for the majority of each sortie, the Indonesian air defence system had taken little notice and, with one possible exception, had apparently not detected any of the covert RPFs.

In March 1966 political instability within Indonesia, and an attempted communist coup, led to a change of regime and the rise of General Suharto. Although Suharto initially proclaimed that Confrontation would continue, he later indicated that the door to peace talks was open and in April positive indications emerged that Indonesia was looking for a settlement to the dispute. Finally, a meeting was arranged between the Indonesian Foreign Minister and the Malaysian Deputy Premier in Bangkok on 30th May.

A sixth series of RPF operations against Indonesia were planned for June 1966 (Operation *Crawford*) and the ground crew to support the detachment were ferried out beforehand in three Transport Command aircraft. Comet XK659 left Wyton for Changi on 30th May. However, the change in the Indonesian position, evidenced by the planned Bangkok talks, led to an objection to the proposed RPF programme by the Foreign Office. In a letter to the Secretary of State for Defence, George Thomson (Minister of State, Foreign Office) protested: 'It does not seem to me that we would at present be justified in mounting any operations against Indonesia which are not immediately necessary and which might prejudice the outcome of the present delicately poised negotiations to end "confrontation"'. He noted that Operation *Crawford* was principally intended to collect intelligence on the Indonesian air defence system for the Commander-in-Chief, Far East. Whilst undoubtedly useful, the information was not urgently needed. On the other hand, detection of the flights by the Indonesians might have an adverse effect on the planned talks. The Minister thus declared: 'I am afraid I cannot therefore agree to the proposed June programme, at least until we see how the Bangkok talks between Malik [Adam Malik, Indonesian Minister of Foreign Affairs] and Razak [Tun Abdul Razak, Malyasian Deputy PM] this weekend go'. George Thomson went on to say that if the talks showed continued military action was likely, then he would reconsider his decision.

As a result of the Foreign Office objection XK659 and crew were held at Sharjah en-route while the political situation was assessed. The Bangkok talks went well and it soon became clear that Indonesia wanted a rapid end to Confrontation. After a few days delay, Operation *Crawford* was formally cancelled and the Comet left Sharjah for the UK on 7th June. The ground crew detachment, having spent a pleasant week at Changi, were flown home on various charter flights.

Confrontation eventually came to an end in mid-year and a peace-treaty was signed in August 1966. Despite the cessation of hostili-

51 Squadron carried out a small number of RPFs against China. One series was flown from Agra at the invitation of the Indian government. The remainder were flown on transit sorties to and from Hong Kong.

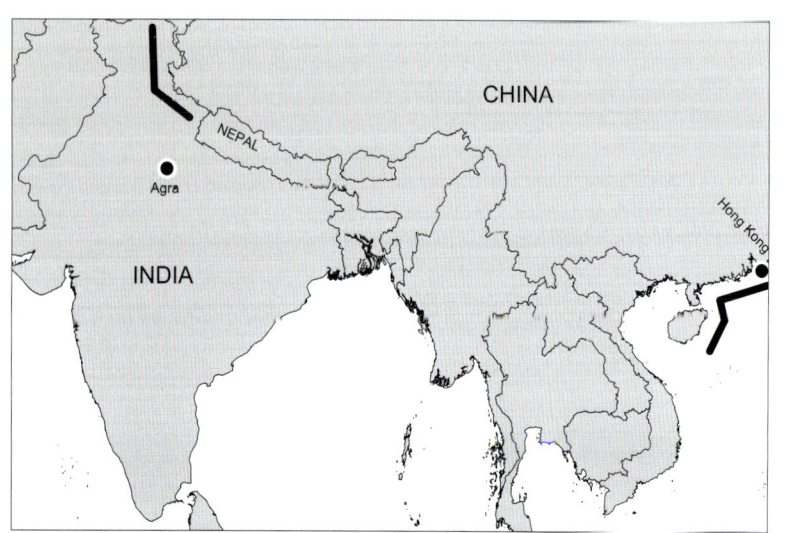

ties the UK continued to monitor Indonesia as a potentially hostile state and a new RPF operation (*Shelbourne*) was proposed for November 1966. The Secretary of State for Defence was cautious, commenting that: 'The consequences of an incident with the Indonesians during the flights over the Java Sea could of course be serious now that [Confrontation] has ended'. However, he agreed to authorise the programme on the understanding if any attempt at interception was made, then the Comet would abort its sorties and that further flights in the programme would be cancelled. A Comet was deployed to Changi on 5th November from where six RPFs were flown, some sorties being lost to unserviceability.

Another, final, operation against Indonesia was mounted in May 1967, revealing, *inter alia*, that one of the Decca 'Parrot' radars, supplied in 1963 was now operational in East Java. A second operation was planned for November 1967 but this was postponed by the Defence Minister for political reasons. The fact was that by mid-1967 Indonesia was restoring its relations with the West and the Foreign Office were keen not to disrupt the process. However, the MoD still considered the country an intelligence target: as the MoD(Air) explained to Ministers, Indonesia remained the strongest military state in the region next to China, and it was thus prudent to keep an eye on it. Another consideration was that the UK supplied intelligence on Indonesia to the US, bolstering the UK's contribution to the US-UK Sigint agreement. Despite the military intelligence arguments, the political view seems to have prevailed and no further RPFs were flown against Indonesia.

Peeking at the Chinese

In January 1966 an operation was conducted against China at the invitation of the Indian Air Force. Comet XK659, supported by XK671, was detached to Agra, Uttar Pradesh in northern India. From there it flew four RPFs (Operation *Polymenus*) along the Indian/Chinese border, flying on open flight plan some 30nm from Kashmir cease-fire line. On completion of the series the Comet proceeded to Changi for operations against Indonesia.

During the deployment of Comet XK695 to the Far East for operations against Indonesia in November 1966 a couple of RPFs were also flown against China. On 10th November the Comet carried out Operation *Shelbourne Delta* en-route from Labuan to Hong Kong and a second operation (*Shelbourne Echo*) was flown on 14th November on the return flight from Hong Kong to Changi. A similar series of RPFs were flown during the May 1967 Comet detachment to Changi (Operations *Lyric X-Ray* and *Zulu*) when a Comet flew from Changi to Hong Kong and return, collecting information on radars in China en-route. The sorties were flown on open flight plan along airways at 35,000ft (10,668m) and approached to within 72nm (132km) of the Chinese coast.

Egypt and Syria

During the mid-1960s the UK kept an eye on Syria and Egypt, flying occasional Comet RPFs from Cyprus to update order-of-battle data on the two countries and to look for the deployment of new Soviet weapons systems. The operations normally comprised a small number of daylight Comet sorties every three or four months. Each flight went along the coasts of both countries, standing off at least 30nm (55km) from shore. Occasional operations were also flown along the Turkey-Syria border on transits from Akrotiri to Sharjah in the Gulf.

In May 1967 relations between Egypt, Syria and Israel began to deteriorate and in mid-month Egypt moved tank and infantry divisions into the Sinai, announcing the closure of the Straits of Tiran to Israeli shipping a few days later. During the same month 51 Squadron detached a Comet to the Far East for operations against Indonesia and the UK intelligence staff took advantage of the Comet deployment to schedule an Elint sortie against Egypt in order to monitor the military build-up. The sortie was a working transit, flown during the return of the Comet to the UK, on the Akrotiri-Luqa leg of the trip. In view of the tensions in the area the sortie was flown at high altitude on open flight plan, with a nearest approach of 37nm (68km) to the coast of Egypt.

At the end of May plans were made to fly two covert Comet sorties towards the end of June against Egypt and Syria from Akrotiri, approaching to within 44nm (81km) of the Egyptian coast and 30nm (55km) of the Syrian coast. This drew an immediate objection from the Foreign Secretary: 'In view of the situation in the Middle East I cannot agree at present to the two covert flights over the Mediterranean Sea against Syrian and Egyptian targets, though I note they are in any case not planned until fairly late in the month; I will review them in a fortnight's time in the hope that the situation may by then have improved'. The Ministry of Defence agreed to review the plans in mid-

LISTENING IN

USS *Liberty* receiving aid from a US Navy SH-3 Sea King. *Liberty* had been attacked and severely damaged by Israeli forces while monitoring electronic activity during the Six Day war. In all, 34 crew were killed and 171 wounded in the attack.
US Navy

June. Unfortunately other events intervened.

On 5th June, in a surprise attack, Israel launched large scale air strikes against Egypt and Syria, destroying the majority of both countries' air forces on the ground. Israeli ground troops then pushed into the Sinai, the West Bank and the Golan Heights. Israeli forces took the Sinai Peninsula by 8th June and on 10th June they overran the Golan Heights.

The dangers of Elint operations in a war zone were graphically illustrated on 8th June when Israeli naval and air forces attacked and seriously damaged the USS *Liberty*, a US Elint ship, positioned in international waters approximately 20nm (37km) off the northern coast of Sinai in an attack that killed 34 crew members and wounded another 171. The incident is still mired in controversy, with Israel claiming the attack was a case of mistaken identity, an assertion strongly disputed by surviving crew and some members of the US intelligence community. One theory is that Israeli forces attacked the *Liberty* in an attempt to destroy communications intercepts of controversial IDF operations. US Elint intercepts of the incident remain classified.

The war, subsequently dubbed the 'Six Day War', ended with a ceasefire on 11th June but despite the cessation of hostilities the situation remained extremely tense. Even flights on open flight plan along the Egyptian/Israeli/Syrian coasts were considered risky and the Foreign Office asked that subsequent Comet transit flights Luqa-Akrotiri en-route to Sharjah during July should take a more direct route than usual, to avoid going anywhere near the Egyptian coast.

Within two weeks of the cease fire the Soviet Union began to resupply Egypt and Syria with arms to replace those lost during the war, with priority given to replacing aircraft and air defence systems. By October the situation had cooled sufficiently for the Ministry of Defence to propose an Elint operation against Egypt and Syria from Cyprus. The operation would comprise three Comet sorties (*Intact 1-3*), approaching to within 32nm (59km) of the Syrian coast, and 45nm (83km) of the Egyptian coast. As justification for the sorties the Ministry noted: 'Since the Arab/Israeli War there has been a serious lack of intelligence information on the United Arab Republic (UAR) Air Defences which it is understood are being built up and improved with Russian assistance. For this reason we are anxious to resume RPFs against Egypt, in particular, and Syria at the earliest opportunity'. The Foreign Office were adamantly against the idea, pointing out that that Anglo-Egyptian relations were at a sensitive phase and the proposed flights risked damaging an improvement in relations.

The Ministry of Defence deferred to the Foreign Office but gave notice that the proposal would be re-submitted as soon as the political situation had improved. Despite initial accusations of UK complicity in the Israeli attack, the aftermath of the Six-Day War saw a distinct improvement in UK-Egyptian relations. Diplomatic relations between the two countries, which had been broken off in 1965 over a perceived failure to act against Rhodesia, were restored in October/November and the long-standing ban on UK overflights of the country was also relaxed.

In December, as promised, the Ministry put forward plans for a series of six daylight Comet sorties (*Harley 1-6*) from Cyprus during January 1968. The proposal noted that: 'Notwithstanding the recent resumption of diplomatic relations with Egypt, there is an urgent requirement

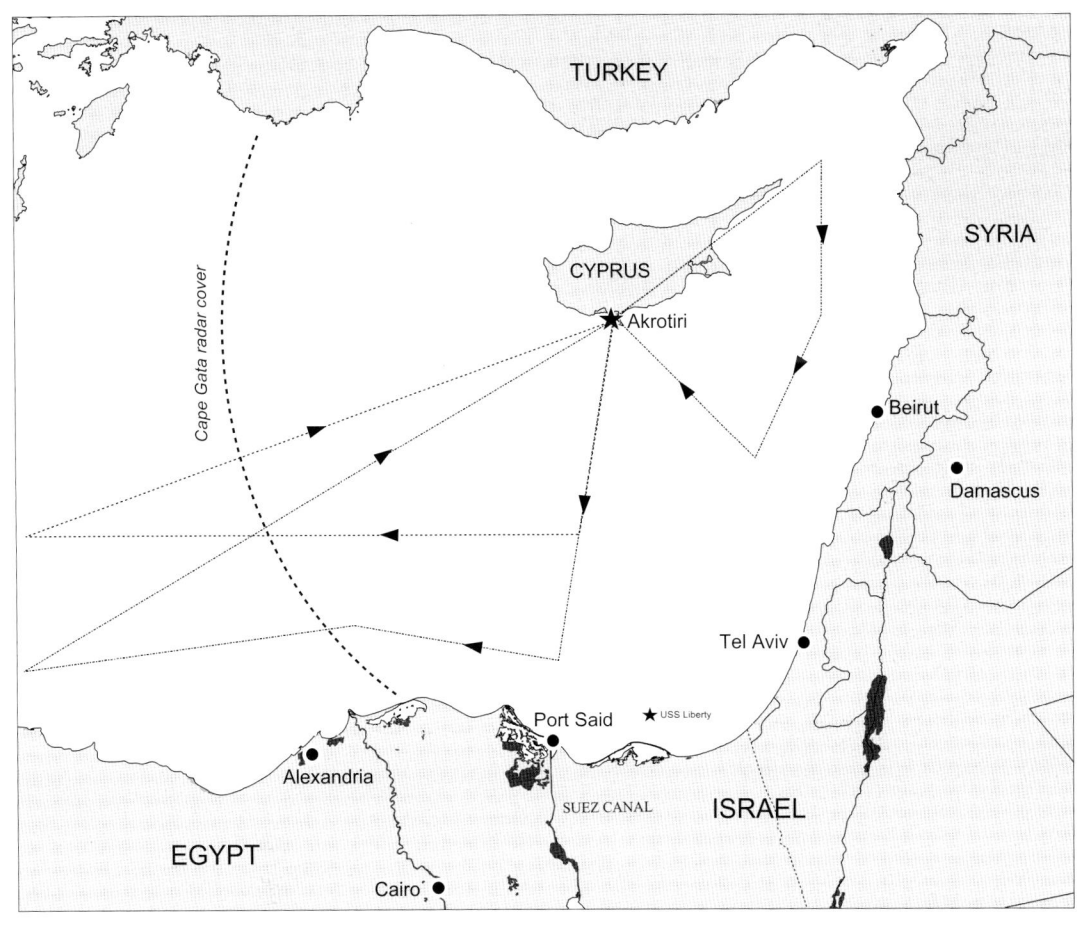

Regular operations were flown against Egypt and Syria from 1967 onwards. Initial operations approached to within 35nm of the Egyptian coast. However, following the War of Attrition the Comets maintained a distance of 80-90nm.

for up-to-date intelligence on the UAR. The UAR, and in particular, Egypt have received a large quantity of Soviet military equipment since the war and have had to make major redeployments. The immediate requirement is to establish the current UAR order of battle, including search for new types of equipment, with a view to assessing the Arab capability to renew hostilities'. The proposal also noted that the DIS had assessed the risk of an attack on the Comet flights as low. Both Egypt and Syria had the capability to detect and track high-altitude aircraft but the likelihood of them attacking an aircraft operating at high altitude over international waters was thought to be small. This was supported by the fact that there had been no reports of Egyptian reaction to recent US high-level covert Elint sorties. The proposal was approved by the Foreign Office and Comet XK659 was detached to Akrotiri in mid-January. All six sorties were flown at 35,000ft (10,668m) and approached to within 45nm (83km) of the Egyptian coast and 35nm (65km) of Syria. The Israeli government was notified before each sortie, doubtless to avoid any misunderstandings regarding the target of the flights.

A further two sorties against the UAR took place in February 1968 during the outbound and return legs of a Comet deployment to Sharjah. These were non-covert 'working transit' sorties, flown on open flight plan between Cyprus and Malta, and stood off a healthy 55nm (102km) from the Egyptian coast. Unfortunately one of these sorties accidentally infringed the Beirut FIR en-route to Akrotiri. The subsequent investigation revealed that, although the flight was monitored by 280 SU radar, they were not given the full aircraft track in advance and thus had been unable to issue a warning regarding the divergence. It was subsequently agreed that the radar controllers would be provided with sufficient route information to allow track deviations to be spotted.

The end of Six Day war in June 1967 had left Israel occupying the Sinai Peninsula and Israeli troops dug in on the Bar Lev Line along the Suez Canal. The end of the war had not however settled the conflict: Egypt was determined to regain the Sinai from Israel and hostilities continued, albeit at a low level, with intermittent exchanges of artillery fire and sporadic raids across the Suez Canal. As a result the political situation in the Middle East remained tense. A further series of RPFs against Egypt were proposed in November 1968, comprising six Comet covert sorties from Cyprus but the For-

LISTENING IN

Boeing RC-135Cs of the 55 SRW flew long-range sorties from the US and UK, the latter covering the Baltic, Europe and the Mediterranean. Note the size of the fairings for the ASD-1 Elint system. *T Panopalis Collection*

eign Office vetoed the operation on the grounds that: '... the Middle East situation is particularly delicate at present and the UN General Assembly is in session discussing the subject'.

Curiously, despite the concerns over RPFs off the Egyptian coast, the UK government did not object to US Elint flights against Egypt taking off or landing in the UK. From 1967 onwards USAF Boeing RC-135Cs were deployed to the UK on a monthly basis for operations over Europe and by 1968 these aircraft were also covering the Mediterranean (Algeria and Egypt) and the Adriatic (Albania, Bulgaria). During 1968-69 the RC-135Cs flew a regular schedule of two Mediterranean sorties one month and four the next. Two standard Mediterranean profiles were flown, the aircraft taking off from Offutt AFB, flying across the Atlantic, covering Egypt and the Adriatic, then landing at RAF Brize Norton. The second profile took off from Brize Norton and flew a similar route in reverse, landing back at Offutt.

A further series of UK sorties were proposed against Egypt and Syria in February/March 1969 under Operation *Wick*. The plan was for four Comet sorties from Cyprus against Egypt and Syria, two sorties on open flight plan, approaching to within 20nm (37km) of the Syrian coast and 48nm (89km) of the Lebanese coast and two covert sorties standing off 45nm (83km) from Egypt. The expectation was that the Comet would probably be detected by radar and there was a possibility it might be intercepted by Egyptian fighters patrolling off the Egyptian coast. As safeguard, the Comet would be kept under positive radar surveillance from Cape Gata on Cyprus until it reached 30°East, and the crew would be warned of the approach of unidentified aircraft. An Elint ground station on Cyprus would also listen out on Egyptian frequencies.

The proposal for flights against Egypt did not go down well with the Foreign & Colonial Office (FCO), who were anxious not to compromise the UK's ability to contribute to any dialogue resulting from Soviet proposals for a settlement in the Middle East. In their opinion the covert RPFs represented a risk and they wanted the programme substantially curtailed. After an explanation from the DIS regarding the importance of the flights, and the low risk of interception, the Foreign Office agreed to the two covert sorties, providing they were rescheduled on open flight plan. The FCO still, however, harboured doubts regarding the remaining two sorties, which closed to within 20nm (37km) of the coast, noting: '... we feel that there could be political repercussions if the Egyptians thought that our flights were following an unusual pattern and at this juncture in Middle Eastern affairs it is important that we should avoid damaging our political interests'. The major objection seems to have been that the sorties included a racetrack pattern off the Egyptian coast which the Foreign Office apparently found provocative.

The FCO objections to Operation *Wick* seemed to confirm to the Air Ministry that the Foreign Office: '... did not fully understand the implications of the various stipulations and embargoes on which they insisted from time to time'. In order to clear the air a meeting was convened in late February, attended by representatives from the DIS, Air Ministry and FCO. The DIS and Air Ministry attendees stressed the importance of obtaining intelligence on Egypt and Syria, and the advantages of flying covert RPFs. They explained that a covert RPF was not only more effective (since the element of surprise was retained) but it was also likely to be less politically sensitive since the nationality of the aircraft was concealed. Of course, this would not apply if the aircraft was intercepted, but the US had been flying an average of six covert high-altitude sorties a month off Egypt and Syria for a period of two years without a single interception. Finally, it was explained that covert flights were perfectly legal as military aircraft were not bound to file flight plans for sorties over international waters. These arguments seem to have done the trick: the FCO agreed in principle to the resumption of regular RPFs against Egypt and Syria, thus making possible a 'significant improvement in our Order of Battle intelligence on both countries'. The *Wick* sorties were flown by Comet XK659 at the end of February/beginning of March.

In March 1969 the Egyptians escalated military operations against Israel, declaring a 'War of Attrition' in an attempt to force the Israelis to the negotiating table. Artillery and air strikes were carried out against the Bar Lev Line along the Suez Canal and against Israeli positions in the Sinai.

Sorties against Egypt and Syria were flown in April and May. The sorties alternated operational areas, flying off the Egyptian coast one day, the Syrian coast the next. While 51 Squadron Comets loitered off the Egyptian coast the Egyptian Air Force occasionally repaid the compliment, flying sorties off Cyprus. Some of these flights infringed the Sovereign Base Area (SBA) airspace, prompting a complaint to the Egyptian government. A further set of four Comet sorties were flown against Egypt in July (Operation *Ewe*) and the squadron was warned that: 'In view of our recent informal protest to the UAR about flights off the SBA, Akrotiri, we attach particular importance to the flights off the UAR not accidentally flying closer to the UAR than the 25nm authorised'.

In July Israel struck back against the Egyptian 'War of Attrition', mounting large-scale air strikes against Egyptian positions along the Suez Canal. By the autumn of 1969 Israel had largely destroyed the Egyptian SA-2 *Guideline* air defence system and had achieved complete air superiority over the Canal. Israeli air superiority was given an additional fillip in September 1969 with arrival of McDonnell Douglas F-4E Phantoms from the US, which facilitated long-range bombing raids into Egypt.

During the second half of 1969 51 Squadron flew regular Comet sorties against Egypt and Syria from Akrotiri every other month. In November, in a break from routine, two Canberras (WJ775 and WJ768) were detached to Akrotiri and eight sorties flown (Operation *Dachs*). The Canberras flew their early morning sorties on the same days but with take-offs staggered by a couple of hours. Flying under similar rules to the Comets, the Canberras approached to within 25nm (46km) of the Syrian and Egyptian coasts.

In January 1970 Israel upped the pressure on the Egyptian government in an attempt to force a conclusion to the War of Attrition, using its Phantoms to carry out long-range air strikes against urban centres in the Nile Valley and the Nile Delta. The Egyptians had no defence against these raids and, in desperation, Nasser flew to Moscow at the end of January to appeal for Soviet military assistance. The Soviet leadership agreed to provide the Egyptians with what was effectively a complete PVO air defence division, including three S-125 *Pechora*/SA-3 *Goa* surface-to-air missile brigades, four P-15 *Flat Face A* early warning radars and two MiG-21MF *Fishbed-J* regiments. The SAM battalions would be manned by Soviet troops and the MiG-21MFs flown by Soviet pilots.

The Soviets began to arrive in Egypt at the end of February, initially deploying SA-3 *Goa* units around Alexandria, Cairo, the Nile Delta, and the Aswan Dam and the first SA-3 site was

One of the air defence systems being deployed in the Eastern Mediterranean was the S75 *Dvina* (SA-2 *Guideline)* SAM and associated radars. This Egyptian example is about to be unloaded from its transporter onto the launcher in the centre. *USAF*

declared operational in mid-March 1970. Once missile cover had been established, Soviet MiG-21MF *Fishbed-J* units flew in to airfields in the Alexandria and Cairo areas and were operational by mid-April, taking over the defence of Egypt south of Cairo. The arrival of the Soviets, and the international condemnation following the bombing of an Egyptian school in April, resulted in the Israelis halting their deep-penetration raids. The focus now moved back to the Suez Canal area.

The arrival of a Soviet PVO Air Defence Division in Egypt was naturally of immense interest to both UK and US intelligence agencies: surveillance was stepped up and a Comet was detached to Akrotiri in May 1970 (*Mogul*) and June (*Adage*). As before, these sorties approached to within 25nm (46km) of the Egyptian and Syrian coasts with at least one sortie flown on open flight plan.

The Israelis initially tried to avoid the new Soviet-manned Egyptian defences, with some strike missions turning back rather than risk interception by the Soviet-flown MiGs. However, a serious clash between the Soviet and Israeli forces was inevitable and this duly occurred at the end of June when two Israeli F-4E Phantoms were ambushed and shot down by SA-3 batteries. Two Canberras (WJ775 and WT301) deployed to Akrotiri for a series of four sorties (Operation *Careen*) against Egypt at the beginning of July. However, the loss of the two Israeli Phantoms a few days earlier looked like the start of a new, and dangerous, phase of the conflict and, probably for that reason, *Careen* was cancelled by the UK government a day or so later. The crews spent a pleasant weekend in Cyprus before returning to the UK the following week.

Hostilities continued through July with various clashes between the Soviet/Egyptian and Israeli air forces, but by the end of the month both sides were beginning to run out of steam. At the same time, international pressure was growing for an end to the conflict. After much diplomatic pressure a ceasefire was finally signed on 4th August 1970. The intelligence on Soviet movements and capabilities obtained by 51 Squadron during May/June seems to have been greatly appreciated by the Defence Intelligence Staff and references were made to the 'excellent results' obtained by the squadron. The C-in-C, Strike Command noted: 'I have watched the progress of the recent series with particular interest, not only because the tasks have been of current intelligence importance but also because they represent flexible reaction to a new situation and new target deployments'. He went on to observe that the operations were effectively tactical in nature and thus provided some experience in the squadron's future role (see Chapter Twelve).

Both the DIS and GCHQ were extremely anxious to resume Elint sorties against Egypt following the ceasefire and proposed a new operation (*Capella*) for August. Their justification for the operation noted that 'The Capella series of flights are of the utmost importance in determining the current deployment of Soviet equipment in the UAR. Both the DIS and GCHQ consider that there is a need to maintain continuity of information and strongly support the programme'. Although the shooting war had finished, both sides remained at readiness and the danger of a flare-up remained. After some negotiation, permission was obtained for *Capella*, but strict conditions were imposed: the sorties would be flown in the Nicosia Flight Information Region (FIR) on semi-open flight plan under positive radar control from Cape Gata, keeping a minimum of 90nm (166km) from the Egyptian coast and 40nm (74km) from the Syrian coast. Operations were duly resumed in August 1970 and sixteen Comet sorties were flown during *Capella*. Due to the large number of sorties the crew and aircraft were rotated half-way through the detachment.

Sorties against Egypt and Syria became a monthly commitment for 51 Squadron from August 1970 onwards for the next few years. Operation *Capella* continued through the last quarter of 1970, with approximately ten Comet sorties each month. In January 1971 the programme was renamed *Salom* (the programme routinely changed its name at six-monthly intervals) and reduced slightly in intensity with between five and ten sorties flown each month. At the same time the limit on the closest approach to Egypt was relaxed slightly and the Comet was allowed to close to within 83nm (153km) of the Egyptian coast.

The Middle East remained tense and the UK government wanted to keep a close watch on any military developments and so in March 1971 authority was granted for a contingency operation in the event of a sudden flare-up. Codenamed *Brim*, the operation allowed for up to five Canberra sorties in the event of a crisis, thus providing short-notice cover outside of the regular monthly Comet detachments. The agreement was that the intelligence authorities would provide the Foreign Office with four days advance warning in the event of a deployment from UK.

The monthly RPF programme against Egypt and Syria continued more-or-less unchanged through 1971, 1972 and into 1973, with an average of around six Comet sorties per month. Operations were interrupted later that year by the Yom Kippur War of October 1973.

Comet XK655 was positioned at Akrotiri at the start of October 1973, on a regular monthly detachment, and two sorties were flown. The surprise attack on Israel on 6th October by Egypt and Syria led to the immediate cancellation of the remainder of the operation. This also impacted operations in November, with the FCO Minister noting that: 'The present ceasefire arrangements in the Middle East are still fragile. They could break down at any time. ... At present I consider it would be prudent for us to continue to keep RAF intelligence-gathering aircraft well away from belligerent states. But in view of the very valuable information collected by these flights I would not wish to ask for a complete ban'. As a result, sorties were confined to an area close to Cyprus, north of 34°N and west of 34°E, keeping the aircraft 100nm (185km) from the Syrian coast and 150nm (277km) from the Egyptian coast. The programme returned to normal in December and the monthly series of sorties continued up until June 1974 when the Comet was withdrawn from service.

Algeria, Libya, Tunisia and Spain

During the early 1960s the UK scheduled occasional Elint sorties against Algeria, which had re-equipped with Soviet arms following its war of independence with France. Sorties were normally slotted into operations from Cyprus, sometimes as working transits en-route to and from Akrotiri. A number of RPFs were mounted against Algeria in 1966 following the delivery of Soviet air defence radars and SA-2 *Guideline* missiles.

Surveillance was also carried out against Spain, which was in dispute with the UK over the ownership of Gibraltar. During February 1967 a Comet was detached to El Adem in Libya from where it flew a number of sorties looking for Soviet shipping, also carrying out RPFs against Spain, Tunisia and Algeria. In May 1967 operations were conducted against both Algeria and Spain during a transit sortie (*Lyric Golf*) en-route to Cyprus. Ministers were assured that the sortie, approaching to within 15nm (28km) of the Spanish coast: 'does not infringe the new Spanish restricted flying area around Gibraltar'. When operations against Egypt were stepped up in 1967 the opportunity was taken during the monthly Comet detachments to Cyprus to carry out one or two sorties into the Western Mediterranean each month, flying along the coast of Algeria and Tunisia.

From late 1969 onwards, RPF operations in the Western Mediterranean also took in Libya. During the 1960s the UK had enjoyed good relations with the Libyan monarchy and had maintained an RAF base at El Adem, 18 miles (24km) south of Tobruk, in the Libyan Desert. The Libyan monarchy came to an end in September 1969 following a military coup which eventually brought Colonel Muammar al-Qaddafi to power and six months later the British withdrew from Libya, followed shortly after by the Americans. Libya's alignment with Egypt, and subsequent links with the Soviet Union, made it an intelligence target, and thus subject to RPF operations. Occasional RPFs were flown off the Libyan coast during the regular monthly Comet detachment to Cyprus, with the Comet operating from Akrotiri against Egypt and Syria, then relocating to Luqa at the end of the detachment for a sortie against Libya. RPFs were also carried out against Libya on open flight plan during the Comet's transit back to the UK, flying 26nm (48km) off the Libyan coast. A request was made during early 1973 for permission to fly closer to the Libyan coast during the covert operations to ease the navigation problem, but this was rebuffed by the Secretary of State for Defence. The Israelis had shot down a Libyan Boeing 727 airliner in February and it was thought likely the Libyan AF might be patrolling more aggressively. The Secretary of State's refusal was remarkably prescient: on 21st March Libyan AF Mirage 5s attacked a US C-130 Hercules Sigint aircraft flying in international airspace off the Libyan coast. Although shots were fired, the C-130 managed to escape. The incident resulted in a ruling, based on 'intelligence advice', that RAF aircraft should avoid the Tripoli FIR and not fly south of 34°20' N off Libya, keeping the aircraft 90nm (166km) from the coast. This restriction remained in force until mid-1974 when the Comets were replaced by Nimrods.

Maritime Elint operations

During the 1950s the relatively small Soviet Navy had mostly been confined to home waters, with only the occasional transfer of units between Fleets. From the mid-1950s how-

LISTENING IN

ever the Soviet Union began to build up its naval strength, also developing new ships. By the early 1960s the Soviet Navy had begun to transform itself into a 'blue water' navy.

The first evidence of the change in Soviet naval strategy was an exercise held in the Norwegian Sea in the summer of 1961, the first major post-war Soviet naval exercise held in international waters. The exercise took place during a transfer of naval units from the Baltic Fleet to the Northern Fleet and vice versa. Both flotillas met en-route in the Norwegian Sea where a short naval exercise was carried out before the units continued on to their respective destinations. The following year's exercise, carried out in July 1962, not only involved a transit of ships between the Baltic and Northern Fleets, but also included a transfer of units from the Black Sea Fleet to northern waters. On 5th July the Admiralty noted: 'There are strong indications that the Soviet annual summer Fleet exercise, in the North East Atlantic and the Norwegian Seas, is imminent. This annual exercise provides our sole opportunity for gathering intelligence on Soviet naval tactics and of observing the capabilities of their ships in the open sea'. As predicted, all three Soviet flotillas joined up in the Norwegian Sea for a large scale exercise. Comet XK655 carried out two *Marcasite* sorties against the Soviet naval units during July, operating from Kinloss. Both sorties were flown under strict rules which allowed a single close approach for visual identification, after which the aircraft was required to stand off 30nm (55km).

The Soviet Union mounted a similar naval exercise during the summer of 1963, closely watched by the UK, US, Canada and Norway. As part of the UK contribution to the joint surveillance operation (which involved ships, maritime aircraft and submarines) a number of Comet and Canberra Elint sorties (Operation *Limbo*) were flown against Soviet ships from Leuchars during August 1963

By 1965 the Soviet Navy was mounting a number of naval exercises in international waters each year and in March authorisation was sought for a small number of Comet RPFs (Operation *Fitzroy*) in connection with Soviet Northern Fleet Spring Exercise in the Norwegian Sea. Permission was granted, but operations were embargoed during the visit of the Soviet Foreign minister to the UK. Although a Comet was detached to Andøya, Norway in April and the crew spent Easter at 30 minutes readiness 'in the snow', no sorties were flown. A larger, multi-national, surveillance operation was carried out in July during the Soviet Northern Fleet's Summer Exercise and two Comet sorties were flown from Lossiemouth under *Fitzroy*. The US detached two Douglas EA-3B and two Lockheed RC-121M aircraft to Lossiemouth

The *Moskva* helicopter cruiser was formidably armed with air defence weapons and thus sported an array of radars and other systems, making it an important Elint target.

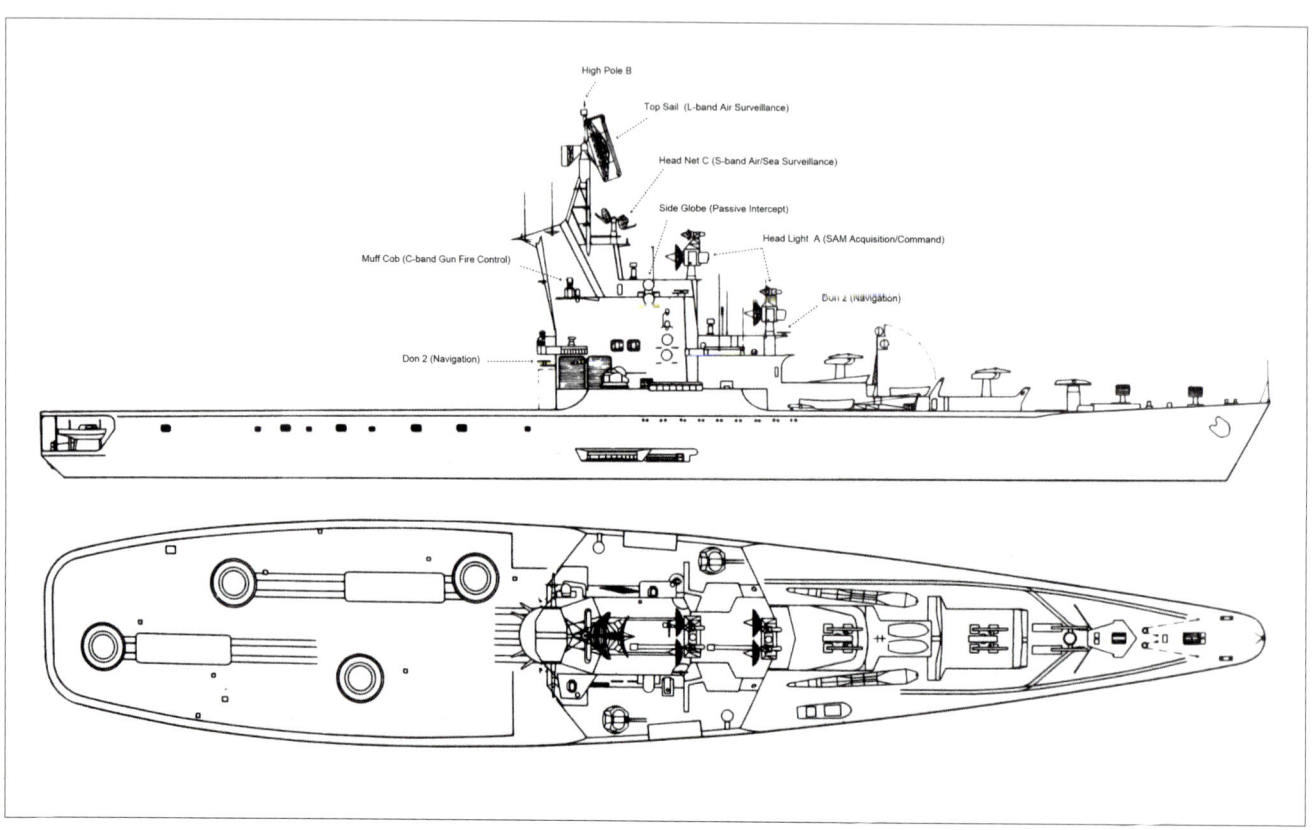

during the exercise period, flying Elint sorties against the Soviet fleet under the same rules as the Comets.

Similar operations were carried out during 1966: in March/April 1966 Comet and Canberra sorties (Operation *Limbo*) were flown against Soviet naval units in the Atlantic, both from Wyton and from a forward base at St Mawgan; further sorties (Operation *Fervency*) were carried out later in the year over the Norwegian Sea during the Soviet Northern Fleet's Summer Exercise. The aircraft remained above 10,000ft (3,048m) in the vicinity of the Soviet Fleet, shadowing from a distance of 30nm (56km). Maritime patrol aircraft tracking the Soviet ships were withdrawn to a distance of 100nm (185km) during the operations.

The tempo of maritime Elint operations changed significantly in 1967. As in previous years, a number of sorties were flown against the Soviet Northern Fleet (Operation *Kendal*) in the Norwegian Sea, the Comets operating from Andøya. The big change was the number of operations conducted in the Mediterranean. The Soviet navy had first exercised in the Mediterranean in 1964 on a small scale, but in May 1967 the Arab-Israeli crisis resulted in a large-scale deployment of the Soviet Black Sea Fleet to the Mediterranean. The build-up began in May 1967 and at its peak the Soviet force comprised eight submarines and 13 surface vessels, some three times the size of previous deployments.

Authority for a six-month programme against the Soviet fleet in the Mediterranean (Operation *Pastry*) was granted in early 1967 and the first Comet deployment to Cyprus took place in March. A number of fixed routes were prepared, some on open flight plan: routes 1 and 2 were off Tunisia, while routes 3, 4 and 5 were over the Aegean Sea. Approximately six Comet sorties were flown each month in the period May-October 1967, with the aircraft operating from Akrotiri (Cyprus), Luqa (Malta) or El Adem (Libya) depending on the route being used. Sorties were flown on an opportunity basis, based on intelligence from maritime surveillance and other sources, often taking off and landing at different bases. Although the *Pastry* operation was subject to blanket authorisation, each sortie required final approval by the DIS, GCHQ and the Foreign Office 24 hours before take-off. As ever, caution was exercised in planning the sorties. Two of the proposed routes were on covert flight plan, approaching close to the Tunisian coast, but in June the Foreign Minister ruled that these could not be used, given the Middle East situation and the 'current Algerian attitude'.

Elements of the Soviet fleet remained in the Mediterranean through the winter of 1967 and into 1968, resulting in an extension of the *Pastry* operation. Comet sorties were mounted on demand, based on intelligence requirements and the movements of Soviet naval units. In November 1968 a high-value target appeared when the *Moskva* helicopter carrier first deployed to the Mediterranean from the Black Sea. Launched in 1965 and commissioned in 1967, the *Moskva* carried 14 Kamov Ka-25 *Hormone* helicopters, and was designed primarily as an ASW platform. Further operations were flown against the *Moskva* in 1969 and in January 1970, but by then the call on the Comets for maritime Sigint operations against Soviet naval units had lessened, probably due to the entry into service in the maritime reconnaissance role of the Nimrod MR.1, which had its own limited Sigint suite. However, the Comet was still used for particularly interesting targets and a number of Comet sorties were mounted against the *Moskva's* sister ship, the *Leningrad*, in early 1971 when it left the Black Sea for the Mediterranean.

The *Moskva* in the Mediterranean, showing some of her complement of Ka-25 *Hormone* helicopters and some of the antennae for its weapons systems, most of which were mounted for'ard. (US Navy)

11 New Directions

'The US are not at present working on this project as far as can be discovered, hence for once we are ahead, and this may later prove a useful "quid pro quo".'
Ministry of Aviation, February 1963

When the Comet 2R entered operational service in late 1958 its initial equipment fit comprised the same US-made, manually-operated, Elint receivers and analysers installed in its Washington predecessor. Even the new UK-developed *Breton* system, developed to replace the US-built Elint equipment, was a largely manual system. The Special Operator had to identify a target signal either by audio or visual means, then tune a superheterodyne receiver and operate an analyser to obtain the signal's frequency and other properties.

The problem was that by the end of the 1950s the signals environment was becoming denser and more complex. In simple terms there were more radars and more variation in radar type. The main Elint tasks were to both locate, classify and 'head-count' radars, in order to compile an electronic order of battle; and to detect and investigate new signals. Unfortunately the large number of radars picked up by the Comet's Elint suite made it increasingly difficult for an operator to perform either task, and the situation was only expected to get worse as time progressed.

The CSE were aware of the problem and by the late 1950s were examining ways to overcome it. One obvious line of attack was to automate the Elint process. As described in earlier chapters, a requirement for an automated Elint system had first been issued in the late 1940s but had proven impossible to meet with the equipment then available (see Chapter Two). Even though the requirement for an automated system had remained, little work had been done in the area due a lack of technical staff, and the high-cost of development.

The United States had more money and men to throw at the problem and significant work was done there to try to automate the Elint collection process. The first steps were taken as early as 1948 when the Federal Telephone Laboratories started work on a wide-band automated centimetric intercept and direction finding (D/F) system providing 360° azimuth

Canberra WT305 in action over the Baltic. The turret behind the cockpit housed a radiometer and video camera that allowed the operator to track a target aircraft and acquire information on its infrared signature. Adrian Mann

cover. An initial operational assessment of the system was carried out in 1950 and a developed version subsequently entered service as the AN/APD-4 in 1955/56. The APD-4 comprised 36 horn aerials, arranged to give 360° cover, a number of wideband crystal DV receivers covering a range of frequencies, and a signal analysis and a recording unit. The system output, displayed on a CRT and recorded on film, provided information on the time of intercept, direction-of-arrival (to within 10° or so), a rough indication of frequency (±500MHz), and an indication of signal PRF, scan rate and polarisation. The system was entirely automatic in operation and required no manual intervention after switch-on. However, APD-4 was a strictly analogue system and subsequent post-flight analysis required an analyst to examine and transcribe each film to produce a record of the sortie.

The APD-4 system was fitted to the new RB-47Hs of Strategic Air Command and the RB-66Cs of Tactical Air Command circa 1955/56. In both aircraft the APD-4 was supplemented by a suite of manually-operated receivers and analysers. The idea was that the APD-4 would automatically record the signals environment, providing both radar head-count (order of battle) information and also identifying new signals requiring a subsequent manual investigation. The system would run unattended while the Special Operators (SOs) would use the more sensitive manual systems to investigate in detail radars identified by APD-4 on previous sorties. In practise the APD-4 was not a success: analysis of the films was a laborious and time-consuming business and the system was easily swamped by a number of radars operating in the same band, making analysis of the record extremely difficult.

However, even before the APD-4 had entered service, the US had embarked on the development of an even more advanced automatic Elint system, incorporating a suite of 200 DV receivers and three scanning superheterodyne receivers. Known as the AN/ALD-4, and intended for the RB-58 reconnaissance version of the B-58 Hustler, the new system overcame the principal failing of the APD-4 by using digital techniques, recording intercepted signals as digital data records on magnetic tape, thus allowing automated post-flight analysis by the associated GSQ-17 'Finder' ground-readout system. The combination of the ALD-4 and GSQ-17 produced a completely automatic, integrated, signal collection and analysis system. Although the RB-58 was eventually cancelled, the ALD-4 showed such promise that development was continued and a number of RB-47H aircraft were modified circa 1959 under the *Silver King* programme to carry a 25ft (7.6m) by 5ft (1.5m) ALD-4 pod under the fuselage. After some delay the system entered service in 1961 and apparently proved very successful, doubling the collection capability of the RB-47H and recording many signals not previously intercepted. In parallel with the ALD-4 the US developed the even more ambitious DLD-1 Elint system. Although the DLD-1 never entered service it was subsequently developed into the ASD-1 receiver system for the RC-135C.

CSE Sideways Looking Intercept System (SLIS)

When the CSE came to consider the automatic Elint problem in 1959 it was clear that developing a system along the lines of the ALD-4 or DLD-1 was a non-starter: the resources simply did not exist in the UK to develop anything so complex. In any case, the ALD-4 had not completely solved the problem of collection in a high-density signals environment. Although it was not as easily swamped as the APD-4, it still had problems when many radars operated on the same band.

The solution proposed by the CSE development team was to limit the number of signals received at any one time, thus preventing overloading of the receivers and analysers. An Elint aircraft typically flew parallel to a border or coastline, looking for signals in one direction only, but the omni-directional nature of many existing aerial systems meant that aircraft actually picked up signals from all directions, including signals from friendly radars. The CSE proposed using a highly-directional sideways-looking narrow-beam aerial, thus limiting the number of signals received at any one time to those emitted from transmitters in a narrow swathe beam-on to the aircraft. This arrangement would also provide a basic D/F capability. The proposed system would automatically record signals in analogue form on one track of a magnetic tape recorder while a second track recorded time and aircraft heading in digital form. The tapes would be analysed post-flight by a Ground Read-Out system, using signal PRF and scan-rate to correlate signals and identify individual radars, and using heading and time data to pinpoint the location of the radar.

The CSE proposal was submitted to the LSIC Working Party on Collection and Processing in

LISTENING IN

the second half of 1959 and was accepted in principle. The key selling points of the CSE proposal appears to have been a possible solution to the problem of working in high signal densities and an improvement in D/F, rather than the automatic nature of the system. Following approval work began at the CSE to build an experimental Sideways Looking Intercept System (SLIS), initially working on S-band, for feasibility trials.

At the end of 1960 the Air Staff issued AST.3614, calling for: '… an examination of the problems involved in providing airborne search receiving systems employing narrow beam sideways-looking aerials in order to reduce the signal density presented to the receiving system at any one time whilst providing a high order of system sensitivity and direction finding accuracy'. Target date for the initial study was 31st March 1961.

The experimental SLIS was developed in the period 1959-61 and installed in Varsity WL690 of the CSE Development Squadron in August/September 1961. Flight trials of the system were carried out during late 1961/early 1962. However, by 1962 the analogue CSE SLIS appears to have been eclipsed by a more advanced digital SLIS proposed by the RAE Radio Department.

RAE SLIS

The RAE team (Pavey and Hampton) first published their proposals for an automatic SLIS to meet AST.3614 in an RAE Technical Note (RAD 798) in May 1961. The TN noted that different techniques would be required depending on the frequency band covered as no single technique would work across the entire spectrum. The proposal thus concentrated on the development of a system covering the S-band in the first instance, since this band was crowded with a lot of high-power pulse radars, with an increasing use of frequency dispersion techniques. The paper also pointed out that it was not practical to design a system providing both accurate D/F and high-accuracy measurements of signal characteristics. The proposal thus concentrated on providing a radar 'head-count' system, providing accurate plotting but with sufficient analysis of received signals to determine the basic radar type only.

The system proposed by the RAE was a wideband crystal detector-video (DV) receiver, fed from a Luneberg lens aerial, using a twin-beam system providing instantaneous bearing of a signal by amplitude comparison (that is, by measuring the signal strength at each beam). The range of an emitter would be determined by measuring the rate of change of bearing for a train of pulses. The system would also incorporate an instantaneous frequency measurement unit to provide frequency information on received pulses, a pulse-width measurement system, and a pulse polarisation detector. A pulse correlation unit would use direction-of-arrival to determine that a received group of pulses (designated a 'scan') originated from the same radar, thus deriving a PRF measurement. A recording system would digitise 'scan' characteristics (bearing, PRF, pulse width, frequency, strength and polarisation) and store them on a multi-track magnetic tape (along with timing information, commentary and analogue signal data) for later analysis (by digital computer) on the ground. The RAE team also suggested that an airborne processing unit could analyse received pulse-trains in real-time and alert a SO to unusual signals.

The RAE proposal was presented to the LSIC Technical Committee in late 1961. The Committee had previously decided, in 1960, that the UK should not follow the US down the fully-automatic Elint route, commenting: 'In view of the limited UK resources and the very large US effort now concentrated in this field, it is concluded that with the exception of special purpose equipment in quiet bands, it is neither practicable nor desirable for the UK to undertake the development of fully automatic methods for the collection of strategic Elint material at the present time'. However, the RAE proposal appeared to offer a cheaper, and more practical, approach to automatic Elint than the complex DLD-1 type of systems being developed in the US. After some discussion, the Technical Committee agreed to approve funding of the project on a research basis. AST.3614 was subsequently rewritten to call for a system matching the RAE proposal and reissued in early 1962.

As part of the RAE investigation an experimental X-band receiver incorporating elements of the proposed SLIS had been constructed at the RAE during 1961. Following ground trials, the system was installed in Varsity WL679 of the RAE Radio Department and test flown in the spring of 1962. A £50,000 contract was subsequently issued to Plessey in April 1962 for an initial study of the SLIS system. The main objectives were to establish the feasibility of obtaining an instantaneous bearing accuracy of 0.5° over the range of signals likely to be encountered, and the to investigate the feasibility of

NEW DIRECTIONS

DH Comet 4 XV814 was used by the Royal Aircraft Establishment for various system trials related to the Nimrod. It is seen here with Nimrod fin and Kalki aerial below the forward fuselage. Mick Freer/Touchdown Aviation

sorting and correlating data received by the system in a ground-based digital computer. The inclusion of the SLIS study in the EW R&D budget, at a cost of £40,000, was reviewed in February 1963, when the Ministry of Aviation noted that: 'The US are not at present working on this project as far as can be discovered, hence for once we are ahead, and this may later prove a useful "quid pro quo"'.

By early 1964 the Plessey study had shown that it should be possible to produce a system handling 1,000 pulses per second, providing accurate emitter location and a reasonable indication of emitter parameters, even for radars incorporating frequency agility. The original suggestion had been to carry out some sorting of pulses in the airborne equipment, but the Plessey study suggested it would be more practical to record all possible pulse data and perform the sorting on the ground.

The success of the initial study resulted in the LSIC authorising further work on the SLIS project during 1964. At the same time, OR3614 was reissued as ASR.817. The reworked requirement called for a sideways-looking signal intercept system for use in the Comet 2R replacement, initially working on S-band but eventually covering the frequency range 0.4 to 11.5GHz. The system was required to accurately measure and digitally record signal frequency, bearing and other parameters for analysis on the ground. A facility providing in-flight warning to the SO of new signals, based on combinations of frequencies, PRFs and pulse-widths, was also required.

The plan was to develop a prototype model for flight trials in a Canberra, carried in a wingtip pod, during the second half of 1967, leading in turn to the delivery of three service models on S-band in the period 1969-71. The system would use a 7ft (2.1m) linear aerial array for maximum D/F accuracy, coupled to an accurate navigation system (Doppler, Decca and twin Gyros) to provide an accurate ground position. Projected total cost was £1.255M and contracts were subsequently placed with Plessey, Hawker Siddeley Aviation (HSA) and EMI Electronics.

Development of the system seems to have been delayed, and by the end of the 1960s the system had still not flown. By then plans to use a Canberra for initial flight trials had been abandoned, possibly due to the increasing size of the system, in favour of an RAE Comet 4C. An aircraft (XV814) became available in 1971 and the developed SLIS system was installed, coupled to a linear aerial under the forward fuselage, during 1971-72. To counteract the aerodynamic effects of the SLIS aerial, the aircraft was also fitted with an enlarged dorsal fin. Flight trials of the system (now codenamed *Kalki*) began during the second half of 1972 and apparently proved the D/F accuracy of the system during flight trials extending into 1973. Unfortunately problems were experienced correlating *Kalki* data with aircraft position and heading, preventing the location of emitters with the required level of accuracy. By 1974 service interest in *Kalki* had lapsed and the programme appears to have come to an end shortly afterwards.

IR Elint

The UK Elint programme was primarily concerned with gathering intelligence from the radio frequency portion of the spectrum but on occasion the intelligence gathering extended to shorter wavelengths.

In the late 1960s Hawker Siddeley Dynamics (HSD) began development of a highly-manoeuvrable short-range IR-guided air-to-air weapon under the designation *Taildog*. At that time the

LISTENING IN

Canberra XH132 engaged on Operation *Harpoon*, acquisition of infra-red signature information of Soviet bombers. This photo was taken from a Soviet aircraft, presumably the target of the sortie. *Author's collection*

XH132 at the conclusion of Operation *Harpoon*. The *Harpoon* sensors were forward facing and this could only gather data from directly astern of the target. *via Ron Henry*

quoted lock-on ranges of IR-guided air-to-air weapons were based on theoretical estimates. In order to prove these figures, and to provide information to assist in the design of *Taildog*, the Defence Intelligence Staff (DIS) requested that an operation be mounted over northern waters to collect IR data from Soviet *Badger*, *Bear* and *Bison* aircraft.

A contract was subsequently issued to HSD covering the modification and support of Canberra SC.9 XH132 on this operation. Initial work-up flights were carried out against UK targets during 1970, the Canberra being flown by a Ministry of Aviation pilot with an HSD observer. The operational phase of the project (Operation *Harpoon*) began in early 1971 and involved the short-notice deployment of XH132 to Leuchars from where sorties were made against Soviet aircraft under the operational control of HQ Strike Command. The interceptions took place over international waters under the standard rules governing intelligence-gathering sorties, but the minimum distance was relaxed to allow the Canberra to close to within 800 yards (731m) astern of the targets to obtain the necessary IR recordings.

Twenty sorties had been planned for the period Jan-Jun 1971 but, due to a lack of targets and other factors, only one successful sortie was flown during this period. As a result the operation was extended until the end of 1971.

Information on the IR signatures of Soviet aircraft was may also have been collected on an opportunity basis by the Comets of 51 Squadron. Representatives from the EWSU visited RAE Farnborough in May 1970 to obtain information on a Comet IR installation used at the Establishment, and in October a meeting was convened at Wyton to discuss the installation of a passive IR sensor in the Comet 2R.

Project *Zabra*

In the early 1970s joint US-UK plans were made to fit a Canberra B.6 of 51 Squadron with a US-made IR radiometer for a programme of measurements against Soviet aircraft under Project *Zabra*. A mock-up of the proposed installation was applied to Canberra B.2 WG789 at RRE Pershore in mid-August 1972. The *Zabra* system comprised a steerable nitrogen-cooled IR radiometer head, bore-sighted with a TV camera, along with display, control and recording units. A joystick controller allowed an operator to steer the sensor head in azimuth and elevation to acquire a target, using a TV display for tracking purposes. Both video and IR data could be recorded to magnetic tape for subsequent analysis.

The problem was how to install all this equipment in the limited space available in the Canberra. The solution was to mount the steerable sensor/optics assembly on the navigator's hatch underneath a 'top hat' fairing and shoehorn the control and display units into the rear cabin. Since the sensor head obstructed the operator's ejection path a rocket system was

NEW DIRECTIONS

Canberra B.6(BS) WT305 shortly after *Zabra* installation. The white turret on the navigator's hatch houses a US-made IR radiometer and associated TV camera. Note hinges at rear of hatch and rocket fairing at front, added to ensure clean separation of hatch in an emergency. *Author's collection*

provided to separate the hatch and sensor from the aircraft should ejection be necessary.

By March 1973 plans had been made to install the *Zabra* sensor (now designated ARI 23263/1) in Canberra B.6(BS) WT305 of 51 Squadron. After being stripped of most of its existing Elint equipment the aircraft was delivered to RRE Pershore in mid-1973 where the system was installed. Work involved the installation of a nitrogen pack in the bomb-bay to provide coolant for the sensor head, uprated power supplies, and the sensor fit in the rear cabin. Probably for operational reasons the cabin layout applied to WT305 differed from the mock-up in WG789. The sensor controls, including a joystick for steering the head, were located in a console on the starboard side of the rear cabin, with the TV monitor/display unit hung from the cabin cross-tube in front of the starboard ejection seat.

The installation was completed by mid-December 1974, following which the aircraft went to the A&AEE for acceptance trials. Initial handling trials, without the radiometer fitted, were carried out during the second and third weeks of January. These went reasonably well although movement of the rear hatch during one sortie gave the test pilot a few nervous moments. Once the handling trials were completed the radiometer was fitted and navigation and radio tests flown with two 51 Squadron aircrew as navigator and system operator. The acceptance programme was completed at the end of March 1975 and the aircraft was then delivered to 51 Squadron on 2nd April 1975.

Flying on *Zabra* began in mid-February 1976 when WT305 was detached to Laarbruch for a series of sorties under Operation *Lock*. The first flight, a 3 hour and 40 minute daylight sortie, was flown on 18th February and five further sorties were flown during the month before the Canberra returned to Wyton. The *Zabra* Canberra was almost certainly used to measure and record the IR signatures of Soviet aircraft exercising in the Baltic, with some sources suggesting the primary targets were the Sukhoi Su-15 *Flagon* and Mikoyan-Gurevich MiG-25 *Foxbat* fighters. The programme continued through 1976, being renamed *Burglary* in July. From August onwards the Canberra used NAS Nordholz, on the German Baltic coast, as a forward base, possibly to extend its time in the operational area. The final operational flight took place on 13th October 1976 and the Canberra returned to Wyton the following day.

Following the conclusion of *Zabra* operations WT305 was withdrawn from service and by November had been re-allotted as the Wyton gate guard, thus marking the end of the Canberra in 51 Squadron service.

The *Zabra* radiometer turret on WT305. The sensor was slaved to a video system that allowed the operator to acquire and track the target from a variety of angles, not just astern. *Author's collection*

157

12 Replacing the Comet

'We are unlikely to prevent the Russians from knowing what 51 Squadron are up to, and they will be aware of the existence of the HS801(R). However, there is no point in making things any easier for them, or for other potential enemies with less sophisticated intelligence organisations by providing the new aircraft with a special name' Air Vice-Marshal Leslie D Mavor ACAS(Policy), August 1968:

Plans had existed for a Comet replacement as far back as 1955, shortly after the Comet 2R had been ordered, and several years before it entered service. At that time the Comet was viewed as an interim Elint platform and the intention was to replace it with a larger aircraft sometime in the early 1960s.

The original candidate for the long-term Elint aircraft was the Vickers V.1000 transport to Spec. C.132/OR.315, a large, long-range, jet providing 14 Special Operator positions. Estimates suggested the total cost of procuring three Elint V.1000s would be £4.05M, comprising £3.75M for the basic aircraft and another £0.9M for the Elint fit. The slightly less capable turboprop Bristol Britannia 300LR transport (another contender for OR.315) was also considered as a cheaper fall-back option, with three aircraft, complete with Elint fit, coming in at £3.3M.

In the event the V.1000 project was cancelled in August 1955 and interest in the turboprop-powered Britannia lapsed not long after, probably because by then it had become clear that the only suitable replacement for the Comet would be another, larger, jet transport aircraft. Only a transport aircraft could provide the necessary space to accommodate the Elint fit, and only a jet aircraft could provide the necessary altitude and speed performance.

The question of a future Elint aircraft seems to have languished for a few years before being reopened in 1959 after the loss of Comet 2R XK663 in a hangar fire. A Working Party, convened to determine if the burnt-out aircraft should be replaced, also considered successors to the Comet. By that date a new contender for the role had emerged in the form of the Vickers VC10 airliner. Expected to enter production in the early 1960s, the latter was a large, long-range, high-performance jet with ample room for a comprehensive Elint fit. The Working Party duly concluded that the ideal future Elint platform was the VC10, with the Comet acceptable as an interim type.

Although the VC10 met all the requirements for an Elint platform, it was going to be expensive: initial estimates put the cost of the aircraft at £1.5M each, with the total cost of procuring and converting three aircraft likely to exceed £5.5M. The LSIC Technical Committee considered the subject again in November 1961 but by then, there were major doubts regarding the ever-increasing cost of the UK signals intel-

The Bristol Britannia long-range transport was considered as a Comet 2R replacement in the mid-1950s. By 1959 however, interest had shifted to jet transports such as the DH Comet 4, Vickers VC10 and Boeing 707. *Author's collection*

REPLACING THE COMET

A hypothetical VC10R of 51 Squadron cruises high above the Barents Sea, monitoring a Soviet Northern Fleet exercise. The VC10 was proposed by BAe as an Elint platform to release the three Nimrod R.1 airframes for conversion to AEW.3 airborne early warning platforms.
Adrian Mann

ligence programme. As a result, a decision on the future Elint aircraft was put on hold pending the outcome of a comprehensive investigation into the funding of Sigint.

The Hampshire Report, 1962

By 1961-62 the cost of the UK Signals Intelligence programme (which included airborne Elint operations) had reached around £20M per annum: a figure approaching the annual running costs of the Foreign Office. The escalating cost of Sigint led the government to commission a review under Professor Stuart Hampshire. Professor Hampshire's report, delivered in May 1963, confirmed the importance of Sigint and, *inter alia*, the continued need for an airborne Elint capability, and thus the need for a replacement for the Comet 2R. The Hampshire Report's endorsement of airborne Sigint seems to have resulted in the establishment of yet another Working Party to consider the successor to the Comet 2R. The report was delivered in the second half of 1963, by which time the fuselage fatigue life of the Comet was expected to run out during 1968-1969.

The Working Party concluded that the ideal Comet 2R replacement was a large jet transport with a range of 5,000nm (9,260km), a ceiling of 40,000ft (12,192m) and a payload capacity of 30,580lb (13,870kg). Unfortunately by 1963, probably as a result of the Hampshire Report, the budget for the future Elint aircraft had been restricted to a frugal £1.5M over the three financial years 1967-70 and £1.5M did not buy much in the way of a modern transport aircraft, even in the mid-1960s. An obvious solution was to try to acquire an aircraft on the cheap from Transport Command, but unfortunately projections showed that no suitable surplus aircraft would be available in the required time period. Thus the choice was effectively restricted to second-hand aircraft bought on the civil market. Three jet airliners were considered: the previously-recommended VC10, the Boeing 707 and the Comet 4.

The preferred aircraft remained the VC10. This could carry the maximum required payload over a distance of 4,000nm (7,408km), met the maximum space requirements and had a layout eminently suitable for an Elint fit. It was also designed to operate from hot/high airfields, making it ideal for operations around the world. Furthermore, since the VC10 was scheduled to enter service with Transport Command (five had been ordered in September 1961), the RAF would already have the necessary support organisation to maintain the aircraft. Second choice was the Boeing 707 which could carry the maximum required payload over the maximum range of 5,000nm (9,260km) and also met the space requirements. The two disadvantages of the Boeing were the under-wing podded engine layout, which, it was thought, might interfere with signal reception and D/F, plus the fact that the aircraft would be unique in RAF service, and thus be expensive and difficult to support. Last in preference came the Comet 4. This could only carry the minimum acceptable payload over a limited 2,400nm (4,444km) range. Cabin space was sufficient for an Elint fit but could not accommodate an additional communications intelligence fit. On the plus side, the aircraft was already in RAF service and thus would be easy and economical to maintain.

The above ranking of the candidate aircraft only considered their suitability against the operational requirements. When cost was fac-

159

LISTENING IN

The DH Comet 4C was the initial choice for the Comet 2R replacement. Although range and internal space were marginal, the aircraft had the major advantage of being in service with Transport Command. Author's collection

tored into the decision the ranking was reversed. The Comet 4 was expected to be available on the second-hand market for around £375,000 each and the Elint fit would add an additional £200,000 per aircraft. Thus three Comet 4s would come in at an estimated total cost of £1.75M, which was just about within budget. Next came the Boeing 707, expected to cost £660,000 per airframe, £300,000 each for Elint conversion, totalling £2.09M for three aircraft. Finally came the VC10, which was not expected to be available second-hand, and would (on revised estimates) cost a hefty £2.6M each. After conversion the total cost for three VC10 would be an eye-watering £8.4M. The Working Party concluded that the only affordable aircraft for was the Comet 4, but that the operational advantages of both the VC10 and Boeing 707 were such that they should not be discarded from the selection process until it was absolutely clear that they were unaffordable.

Choosing a replacement, November 1963

The replacement for the Comet 2R was further considered by the Air Ministry DD.Ops(Recce) in a report issued in November 1963. By then the VC10 had been eliminated from the list of candidate aircraft, almost certainly on cost grounds. The expected price of both a second-hand Boeing 707 and Comet 4 had also been revised: the Comet estimate was down from £375,000 to £175,000, and the Boeing down from £660,000 to somewhere between £300,000 and £500,000. The revised figures suggested that the Boeing 707 would just be affordable circa 1967. Not surprisingly, given the Boeing's obvious advantages, DD.Ops(Recce) recommended obtaining a firm estimate for three second-hand Boeing 707s and commissioning a design study into the modifications necessary to convert the aircraft into Elint platforms.

The Vickers VC10 was the Air Ministry's preferred choice for the Comet 2R replacement. It had excellent range, plenty of space and was due to enter RAF service in the transport role. Unfortunately it was too expensive. Author's collection

160

The Air Ministry's proposal for the Comet 2R replacement was submitted to the LSIC in the first half of 1964. Unfortunately, cost seems to have trumped operational capability, and by 1965 the Boeing 707 had been rejected in favour of the cheaper Comet 4. As previously noted, the Comet 4 was not only the cheapest candidate on capital costs alone; it was already in RAF service (in Transport Command) and the necessary technical support organisations were in place, thus making an Elint version reasonably economical to operate and maintain. By comparison, the more expensive Boeing 707 would require its own dedicated spares and maintenance organisation. Furthermore (and despite previous assertions to the contrary) it appears that there was a possibility of obtaining three Comet 4C from Transport Command, which would further reduce costs and, in theory, would keep the provision of three Elint-modified aircraft well within the £1.5M budget.

In September 1965 the Air Force Department, Ministry of Defence (the successor to the Air Ministry) contracted HSA to carry out a feasibility study into the conversion of three Comet 4Cs for the Elint role and agreement was reached on the release of three Comet 4Cs from Transport Command for the role. Unfortunately the HSA report, delivered at the end of May 1966, estimated the total cost of converting three aircraft at £4.58M, a sum greatly exceeding the £1.5M originally envisaged by the Air Ministry. Dismayed at the high cost of the Elint Comet 4C, the Air Force Department decided to completely re-examine the case for the Comet 2R replacement. This led the LSIB to invite the MoD and GCHQ to examine the future requirements for airborne signals intelligence, its scale and its cost. A committee was duly formed in June 1966 to carry out the investigation, headed by the chairman of the LSIB and comprising MoD and GHCQ representatives.

MoD/GCHQ Study, June-October 1966

The MoD/GCHQ committee observed that, although the UK obtained the majority of its Sigint from ground stations, airborne collection was still important for a number of reasons. One major justification was its importance as a contribution to the joint US-UK Elint programme. The US appreciated the UK efforts in airborne Elint collection (particularly the focus on low-level Soviet Air Defence in Europe) and this helped to guarantee continued UK access to large quantities of US Elint and other intelligence data. It was thus possible to argue that the cost of the UK airborne Elint programme was repaid many times over by access to US intelligence. Airborne collection also had a number of practical advantages over ground collection. An aircraft could cover targets along the entire periphery of the Soviet Union, including targets inaccessible to ground stations; an aircraft was the only way to reliably intercept signals from short-range emitters for example, at frequencies above J-band and in the infra-red where intercept ranges were limited by atmospheric attenuation; an aircraft could map the polar diagrams of target radars in three dimensions; and an aircraft provided a rapid-response facility, to quickly target areas of interest. Considering the above arguments, the committee concluded that an on-going UK contribution to the UK/US Elint programme was necessary to ensure continued access to US Elint data, and thus that an independent UK airborne signals intelligence effort should be continued for the foreseeable future.

The committee then looked at possible replacements for the Comet 2R. The use of former Transport Command Comet 4Cs was ruled out for a number of reasons. The primary objection to the Comet 4C seems to have been its poor value for money, rather than absolute cost. An Elint Comet 4C would have a limited service life, a less-than-ideal range and support only a limited Elint fit. There was also a concern that taking three Comet 4Cs from Transport Command would create a replacement problem for that force. Although the committee was mainly concerned with the Comet replacement, the question of the Canberra was also examined. The Canberra provided a very useful low-altitude Elint capability, the requirement for which was likely to continue in the future. Another argument against the Comet 4C was that it could not operate at low altitude, and so could not replace both the Comet 2R and the Canberra.

Having ruled out the Comet 4C, the committee then looked at alternative replacements for the Comet 2R. These were clearly going to be more expensive than the Comet 4C, but would hopefully provide a much more cost-effective Elint platform. On the face of it this would seem to have been an ideal time to re-examine the Boeing 707 and VC10 but by then an even more promising candidate had emerged in the shape of the HS801 Nimrod.

The Nimrod was a maritime reconnaissance development of the Comet, designed to meet

LISTENING IN

The HS801 maritime reconnaissance aircraft was an ideal candidate for the Comet replacement. In addition to a long-range, it also possessed a very good low-level performance, and was thus able to replace the Canberra. *Author's collection*

MR.254/OR.381 for an interim maritime patrol aircraft and due to enter service with Coastal Command in 1969. Investigation revealed the aircraft had a longer range and better load-carrying capability than the Comet 4C, had plenty of room to accommodate Elint equipment and operators, was a new build and thus would have a minimum life of ten years, and was also capable of operations at low-altitude (unlike the Boeing 707 and VC10) and thus might also be able take on the Canberra's low-level role.

The MoD/GCHQ committee completed their study in October 1966, subsequently recommending that (a) an independent UK airborne signals capability should be maintained, (b) that the Comet 4C was not a suitable replacement for the Comet 2R and (c) that the MoD should examine 'with all speed' the possible use of the Nimrod as a Comet replacement.

The recommendations of the committee were endorsed by the Chiefs of Staff in November 1966 and the Air Force Department (AFD), Ministry of Defence initiated an investigation into the Nimrod as a follow-on Elint platform during the first half of 1967.

AFD/HSA Study, March 1967

As part of their investigation, the AFD commissioned HSA to carry out a Design Study for an Elint version of the HS801 Nimrod, designated the HS801(R). The study found that the Nimrod was well suited as an Elint platform, mostly due to the fact that the long-range maritime patrol and Elint tasks had overlapping requirements. The Nimrod had a range of 4,600nm (8,519km) with a payload of 14,000lb (6,350kg), good speed and altitude performance, a large cabin able to accommodate the required numbers of Special Operators (SOs) and their equipment, a large weapons bay providing space for aerials, and a sophisticated navigation suite, providing high accuracy over long distances. These qualities made the Nimrod an excellent choice as a Comet 2R replacement. In addition, the Nimrod also had good low-altitude performance, giving it the ability to carry out the Canberra task as well. The potential of the Nimrod to replace both the Comet and Canberra in 51 Squadron service was a large plus in its favour as it would allow 51 Squadron to operate a single aircraft type, producing savings in manpower and support equipment.

The HSA study then examined how the Nimrod could be adapted to the Elint role. The conclusion was that the basic airframe, systems and engines could be retained: thus relatively modest modifications would be required, allowing R&D costs for the new version to be kept to around £2M. The main changes would be a new cabin layout, with an upgraded air conditioning system, accommodating the SOs and their equipment, a remodelled weapons bay accommodating a new aerial suite, and the

162

installation of various other aerials in the tail, on the fin and on the wings.

The AFD considered that three HS801(R)s would be required to replace the three Comet 2R and four Canberras of 51 Squadron. However, the HS801(R), although a very capable aircraft, was also very expensive and there was some argument regarding the numbers required (echoing similar doubts regarding the earlier Comet procurement). As a result, cost estimates were produced for both the two and three aircraft cases with HSA estimating the total costs of the HS801(R) programme would be £8.55M for two aircraft, £11.4M for three, these prices including the R&D cost. Adding on engine spares and support equipment brought the capital cost of the programme in the period 1967-78 up to £9.68M (two aircraft) or £13.02M (three aircraft). Expenditure on Elint equipment for the aircraft was expected to come in at an additional £1-1.38M while running costs were expected to be £0.65M per year for two, £0.9M for three. The HSA estimates assumed an Intention to Proceed (ITP) was issued before November 1967, as that would allow the HS801(R) to immediately follow the last maritime Nimrod on the production line, also providing economies of scale in materials ordering.

The AFD investigation then looked at the key dates in the HS801(R) programme. The life of the Comet 2Rs was dictated by wing fatigue life and corrosion and, if no serious corrosion occurred, the aircraft would run out of life in mid-1972, mid-1973 and mid-1976 respectively. Thus the first HS801(R) would be required by mid-1972. Installation of Elint equipment was expected to take around 18 months, so the first HS801(R) really needed to be available by late 1970 in order to meet a mid-1972 in-service date. This could be achieved by adding an additional three aircraft to the existing Nimrod production order. As mentioned above, HSA timescales meant that an ITP would have to be issued by no later than November 1967, ensuring that the firm could order materials and equipment for the three HS801(R) as part of its main production 'buy' and thus keep costs down.

Air Staff Requirement 389

In June 1967 the requirement for the HS801(R) Elint version of the Nimrod was formalised via the issue of the first draft of Air Staff Requirement 389. At that time the aircraft's primary role was strategic Elint and thus the main performance requirements were dictated by the requirement for global operations:

- Total Elint payload (equipment, racking and operators) of 19,000 lb (8,618kg)
- Accommodation for up to 23 Special Operators
- Normal range of 4,100nm (7,593km)
- Extended (reduced safety factors) range of 4,440nm (8,222km)
- High-Low-High capability, radius 1,500nm (2,778km) with 2-hour loiter at low altitude.
- Transit speed of at least 400kts (740km/h) and a loiter speed of 170-220kts (314-407km/h).
- Ability to operate from runway length of 7,300ft (2,225m), 9,000ft (2,743m) in hot conditions.

The requirement envisaged the aircraft spending 75% of its life flying 8-11 hour RPFs at 30,000-40,000ft (9,144-12,192m), 20% of its life flying 8-11 hours hi-lo-hi RPFs (including a 3-4 hour low level loiter) at altitudes between 1-15,000ft (304-4,572m) and 30-40,000ft (9,144-12,192m) and 5% of its life on 5 hour pilot training sorties at altitudes between 0-20,000ft (0-6,096m).

The original ASR.389 proposal involved the provision of a new under-fuselage pannier shape in order to provide an optimal 'view' for the aircraft's aerial suite. HSA noted that since the new pannier would change the shape of the Nimrod lower fuselage it would need aerodynamic testing, requiring both wind-tunnel and flight trials. These and other costs associated with a new pannier shape would add about £0.5M to the programme R&D costs. It would clearly be a lot cheaper to use a pannier with the same shape as that used in the Nimrod MR.1. After some investigation the use of the standard pannier profile was accepted as technically feasible and adopted for ASR.389.

When ASR.389 was issued in mid-1967 the number of Nimrod conversions required was still undecided. The conclusions of the AFD study suggested that three Nimrods could replace both the three Comet 2Rs and the four Canberras of 51 Squadron. However, the Nimrods were very expensive aircraft, and in order to ensure cost-effectiveness the Chief of the Defence Staff, Field Marshal Sir Richard Hull, asked the Chief Advisor (Studies), Dr Alan Cottrell, on behalf of the LSIB, to form a working party to examine whether two aircraft would suffice. The original intention had been for the Working Party to report by the end of the year, but the urgent requirement to get an ITP issued by the end of November 1967 resulted in the report being delivered in October.

LISTENING IN

Airborne Tactical Elint Requirement

Before looking at the conclusions of the CA(S) Working Party study it is necessary to first examine the emerging requirement for airborne Tactical Elint. During the 1950s and early 1960s the UK's Elint aircraft had been employed largely in a Strategic Elint role. The information obtained via Elint sorties (technical intelligence and order-of-battle information) was not considered highly time-sensitive and was expected to remain valid over a period of weeks or even months. This was predicated on the relatively slow introduction of new radar and missile systems and the relatively static nature of defences. These assumptions applied equally to Soviet Bloc defences and to those of other countries against which the UK might have to fight.

The US experience in Vietnam during the mid-1960s came as a rude wake-up call. Although fighting in a limited-war scenario against a third-world nation, the US came up against a sophisticated air defence system using relatively modern Soviet equipment. Furthermore, the North Vietnamese air defences, including missile systems, were highly mobile, regularly moving between previously-prepared positions. The combination of visually and radar-laid AA guns (effective at low altitudes) and SA-2 *Guideline* SAM systems (effective at medium and high altitudes) initially proved lethal to US fighter-bomber aircraft. Attempts to route aircraft around defences were hampered by their mobile nature: a route skirting SA-2 defences one day might cross the middle of an SA-2 defended zone the next, an attempt to fly at low-level through a known SA-2 zone might encounter a previously un-plotted radar-laid AA gun battery. In response to these problems the US was forced to develop a Tactical Elint capability, able to deliver up-to-the-minute information on the disposition of the North Vietnamese air defence system.

The US experience in Vietnam was watched closely by the UK and resulted in adaptations of ECM and Elint policy to counter the new threat. It became clear that future strike aircraft would require a comprehensive suite of radar warning and active ECM systems, and that a requirement existed for an airborne Tactical Elint capability to provide an up-to-date picture of enemy defences. The British Army was planning to field its own truck-borne Tactical Elint system (to GST 3021) but this would only have a limited range, and was primarily designed to detect battlefield radars. Only an airborne capability could provide the Tactical Elint required to support strike aircraft.

The initial proposal for a UK Tactical Elint capability was based on the use of a podded Passive ECM Tactical Aid (PETA) carried by strike/reconnaissance aircraft. The idea was that a stand-off reconnaissance sortie by a fast jet carrying the PETA Elint/ESM pod could capture information on enemy radar frequencies, PRFs and bearings, automatically identifying and plotting the location of threat radars. This information could be used to route strike packages around the defences, determine the nec-

Had the Air Ministry adopted the Trent Andover, this scene could have played out. An Andover R.3 of 51 Squadron operating with a pair of Buccaneer S.2B in the tactical Elint role over West Germany in the early 1980s. The Andover would have stood off while the Buccaneers made approaches to the Inner German Border to trigger a response from Warsaw Pact air defence and fire-control radars.
Adrian Mann

The Armstrong Whitworth Argosy was briefly considered as a Tactical Elint platform, no doubt because it was available for free as the Lockheed Hercules C-130K entered service. It was rejected in favour of the Nimrod.
T Panopalis Collection

essary ECM fit and settings to counter the defences, and provide targeting information for an attack on the defences themselves. The RAF was planning to introduce its first anti-radar missile (the AS.37 Martel) into service in the early 1970s and the effective use of this weapon would require accurate information on enemy radar frequencies and locations.

The joint Naval/Air Staff requirement for the PETA pod was issued as NAST 852 circa 1966. Pod size constraints and the limitations of 1960s technology meant that the system would be restricted to identifying only three radar types on any one sortie: however there were typically only a few AA and SAM radar types in any one area, so this was considered an acceptable limitation. Unfortunately the usefulness of the system was also dependent on a reasonably high level of D/F accuracy to pinpoint the location of enemy radars. The requirement to locate a target radar to within 5nm (9km) meant a bearing accuracy of 2° would be required and the pod would also have to be tied into the aircraft's navigation system to derive positional information. A review of the requirements during late 1966 suggested that the pod would be very complex and could not meet the necessary D/F accuracy goal, since that would require an aerial size bigger than anything a podded system could practically carry and so NAST 852 was abandoned around January 1967.

The abandonment of the pod concept led to consideration of a full-blown Tactical Elint aircraft. The use of a manned aircraft, with a comprehensive Elint fit, offered a number of advantages over a podded system. It could, in time of tension, maintain an up-to-date intelligence picture over the full frequency spectrum, accurately fix mobile and static emitters for targeting and route planning purposes, monitor emitter frequencies to determine ECM pod settings, fix the position and frequencies of emitters for attack by anti-radar missiles, use Comint to determine enemy tactics, provide real-time warning of new threats and assess the effectiveness of ECM on enemy defences.

Two different aircraft types were studied as possible Tactical Elint platforms. The first was a conversion of the Armstrong Whitworth AW.660 Argosy transport aircraft. There was already a plan to convert a number of Argosy C.1 transports to the radar calibration role around 1970 and the aircraft was large enough to take the fixed-fittings for both the calibration and Tactical Elint roles, allowing a quick change of role when required. Optionally, a small number of aircraft could be fully-equipped for both roles. Although the Argosy had a relatively limited altitude range and speed, its ceiling of 25,000ft (7,620m) was just sufficient to give it an acceptable Elint pick-up range. It would of course have to stand off a considerable distance from defences, and would need to be escorted in some areas. The alternative to the Argosy was to use the HS801(R), already planned for the Strategic role, in the Tactical role. The latter proposal had an obvious cost advantage, combining Strategic and Tactical Elint in the same aircraft.

Initial studies on Tactical Elint had assumed its main application would be in limited war scenarios overseas, against countries such as Iraq, Egypt and Indonesia. In theory Tactical Elint had limited application in the European theatre against the Soviet Union given NATO's nuclear tripwire policy, which called for massive nuclear retaliation in the event of a war. The only application of Tactical Elint in Europe would be in the period of tension before the outbreak of hostilities, but this was expected to yield limited results due to Soviet signals security measures. By 1967 however the NATO tripwire policy was in the process of being replaced by a more flexible strategy, which allowed for an initial conventional response to Soviet Bloc aggression. The new policy was promulgated in September 1967 as MC 14/3. The emergence of a flexible response in Europe meant that Tactical Elint assumed a new relevance in that theatre since NATO air forces might be called upon to carry out a number of conventional strikes against Soviet Bloc forces in the early stages of a conflict. These conventional strikes would need to be planned using up-to-date Electronic Order of Battle intelligence in order to enhance their chances of success, and preserve the strike force for later, possibly nuclear, operations. The Soviets would find it more difficult to preserve signals security during the, possibly lengthy, conventional phase of a war, providing an opportunity for Elint to contribute to operations.

Chief Advisor (Studies) Working Party, October 1967

The Chief Advisor's working party (the Elint Assessment Working Group) had been constituted to examine whether two or three HS801(R) were required. In order to expedite the report the Chief Advisor (Dr Cottrell) asked ACAS(R) of the Air Force Dept(MoD) to produce a paper outlining the airborne Sigint case and recommending a future course of action. Thus the working party effectively reviewed (once again) the continued case for airborne Elint collection and the means to meet it.

The Working Group broke Elint down into four classes of operations: Strategic Elint on a long time-scale, Strategic Elint on a short time-scale (weeks/days), Standoff Elint on a very short time-scale (hours) and Tactical Elint in the battle zone (hours). Their conclusion was that the benefits of operating three HS801(R) purely in the Strategic Elint role did not warrant the expenditure of nearly £14M from the Sigint budget. However, the Working Group then looked at the usefulness of the aircraft in the Tactical Elint role, providing intelligence directly to ground and air forces. The conclusion here was that airborne Tactical Elint was an essential support component to a modern air force, and could make a significant difference to the effectiveness of the RAF's limited front-line forces. This was especially true in Europe, where large numbers of mobile and frequency-agile Soviet radars and associated missile systems were deployed. The report effectively redefined the primary role of the HS801(R) as that of Tactical Elint, noting that: 'it would be right to regard these aircraft as an essential component of the strike and reconnaissance forces, which would, in addition, be able to make a contribution to our collection of Strategic Elint in peacetime'.

The Working Group then considered various alternatives to the HS801(R) in the Elint role. The first option was to use the proposed new OR.387 AEW platform (which was still under discussion) as a dual-role aircraft, carrying out both the Elint and AEW tasks. This was quickly dismissed as impractical due to the difficulties of equipping and operating an aircraft in both roles. It was hard enough to fit the required Elint suite into a

The Trent Andover, originally proposed for the AEW role, was later considered for Tactical Elint. The Elint antennae would have replaced the AEW radar equipment in the fore and aft radomes. The AEW Andover was an attempt to save money that was fundamentally flawed from the start as it was not, as advised, merely a jet-powered Andover.

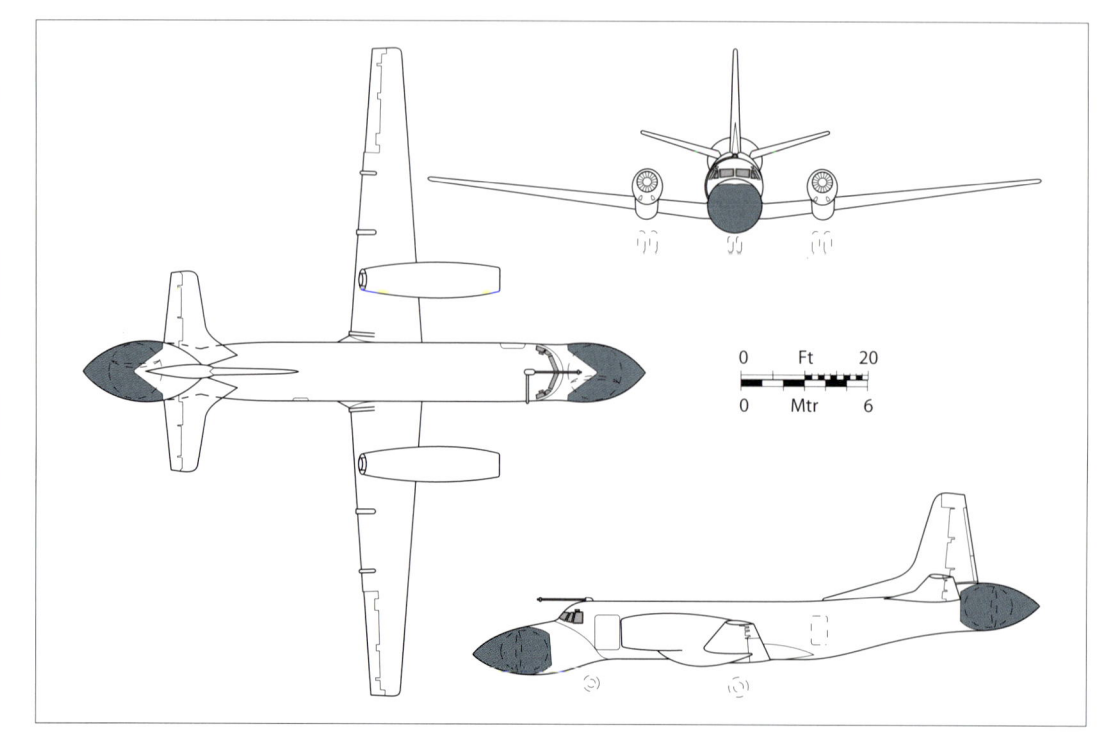

dedicated aircraft, never mind into one already carrying a large AEW radar. The Working Group then looked at using the aircraft types under consideration for the AEW role as dedicated Elint platforms. The first type considered was the much-modified, Trent-engined Andover. The major problem with this proposal (also highlighted during the AEW evaluation) was the aircraft's lack of range: it would need to be air-to-air refuelled three times on maximum-range Elint sorties. Apart from the logistical complications, the use of AAR in sensitive areas was considered unwise. The Andover was also too slow, was likely to be only 50% as effective as the HS801(R), and would not be that much cheaper. Another type under consideration for the AEW role was the BAC 1-11, but this suffered from the same range problems as the Andover and, again, would not be much cheaper. An alternative to a dedicated Elint platform was to use Elint pods fitted to Strike and Reconnaissance aircraft, but this solution, technology problems apart, would only produce basic intelligence, might also require AAR, would probably require an additional buy of strike aircraft, and development of the Elint pod alone would probably cost between £2M and £5M.

Having dismissed all the foregoing options as unworkable, the Working Group then produced five options based around the Nimrod. The preferred option (Option A) was to procure three Nimrod(R) at a cost of £13.9M. This would provide a full Sigint capability (Elint and Comint) at the lowest possible cost. Option B reduced the number of Nimrod(R)s to two aircraft. Although this would save about 25% on the three aircraft option (once running costs and the like had been taken into account), it would halve the operational value of the force, and was thus of doubtful cost-effectiveness.

The remaining three options (Options C, D and E) attempted to reduce costs by giving the Long Range Maritime Reconnaissance (LRMR) Nimrods already on order for Coastal Command a dual role, thus avoiding the purchase of three dedicated Elint aircraft. Option C proposed giving three Nimrods from the MR force a primary Sigint role using Nimrod(R) Elint equipment, also retaining a secondary LRMR role. The problem with this proposal was that there would only be room to fit the aircraft with an extremely limited LRMR mission equipment fit and so the aircraft would therefore be of limited usefulness in their secondary role. Option D was the converse of Option D, the aircraft having a primary LRMR role and a secondary Sigint role, using a subset of the Nimrod(R) Elint fit.

The three aircraft would carry about 25% of the full Elint equipment fit, but would only have 10% of the capability. The aircraft would also lose some of its LRMR capability.

The final option (Option E) was to essentially beef-up the planned limited Elint kit in all 38 LRMR Nimrods. The LRMR Nimrod was already scheduled to carry the French ARAR/ARAX 10B as an interim ESM suite, giving it the capability to intercept, analyse and D/F signals from maritime radars. The ARAR 10B was an intercept and D/F receiver covering 2.3-11.1GHz band and ARAX 10B was the associated pulse analyser. ARAR/ARAX would thus give the Nimrod a limited Elint capability in the S, C and X-bands. Option E proposed extending this capability via the French RAK-2 system which covered the J, K and Q-bands, theoretically giving the LRMR Nimrod a reasonable Elint capability covering S to Q-band. Option E was costed at £2M to equip all 38 aircraft and appeared, in some quarters at least, to offer an economical solution to the Comet 2R replacement problem. However, the Working Group pointed out that Option E would provide only two SO positions per aircraft and would actually give a rather poor, low-grade, Elint capability compared to that obtainable using the Nimrod(R). The fact was that ARAR/ARAX provided only basic D/F accuracy, RAK-2 only provided D/F in the J-band, ARAR/ARAX/RAK-2 would be swamped in a high-density signals environment and the aircraft would lack an L-band, Comint and multi-frequency radar intercept capability. There was a plan to replace ARAR/ARAX with an improved ESM system to AST.833 in 1973, but even this system would be sub-functional compared to the proposed Nimrod(R) fit, and would struggle in high signal densities. Furthermore, it seemed unlikely that Coastal Command would be willing to divert scarce LRMR resources from their primary maritime task in time of tension or war.

After considering all the above the Working Party endorsed Option (A) for the procurement of three Nimrod(R) in a report (LSIB/EAWG/20/67) delivered to the LSIB on 11th October 1967.

Paying for the Nimrod

The future Sigint budget was discussed at a meeting of the PSIS Committee (responsible for Sigint funding) towards the end of July 1967. Although the attendees were all in favour of the Nimrod(R), there was considerable opposition to their cost being met from the relatively small Sigint budget. It was evident that if the Nimrod(R) programme proceeded, then the cost would have to be met directly from the Air Force Vote (that is, from Air Force funding). In reality this was something of an administrative point, as the cost of Elint aircraft and equipment had always been met from the Air Force Vote anyway via a 'hidden' contribution to the Sigint budget. In fact the cost of the Nimrod(R) had already been factored into the Air Force Vote long-term costings. However, the proposed change to direct funding from the Air Vote would mean that the Nimrod(R) and its equipment would have to be justified as an Air Staff requirement, rather than a more general intelligence requirement.

As it turned out, the proposed change in funding arrangements mirrored the change of aircraft role, from Strategic to Tactical proposed by Dr Cottrell. Following the Cottrell report, agreement was reached that, given the Tactical role of the aircraft, none of the capital cost of the aircraft, and only part of the running costs, would be attributed to GCHQ, and as a result the RAF would have to justify the case for the Nimrod(R) itself.

Elint Vulcan

During August 1967 there was a brief informal investigation at the Air Ministry into the use of the Avro Vulcan B.2 as a 'last-ditch' Comet replacement, should the Nimrod proposal fall through. The conclusion was that the Vulcan had a number of shortcomings in the Elint role and could only offer, at best, a limited capability. The major problems were the limited accommodation in the cabin and the Elint equipment installation problem. The Vulcan would only be able to accommodate 2-3 SOs, resulting in a much-reduced intercept capability compared to the Comet which could only be rectified by (an expensive) redesign of Elint equipment to increase automation. Fitting Elint equipment in the Vulcan would also be challenge, the only practical arrangement being a bomb-bay Elint equipment pack, fed from an under-fuselage aerial array covered by a new bomb-bay radome. Since most Elint equipment was not designed for operation in an unpressurised environment some sort of capsule would also be required. There were also other concerns: the Vulcan range in the Elint role was estimated at 3,000nm (5,556km), against the 4,000nm (7,408km) of the Nimrod, the low-level fatigue life would be limited and the large delta wing would probably impact on D/F accuracy. In the event, the approval of the Nimrod as the future Elint platform made further work on the Vulcan proposal unnecessary.

Endorsement of the Nimrod(R), 1967

During the 1960s the authorisation of a large defence programme required a number of discreet administrative steps. A requirement was first issued, the proposal to meet the requirement was then presented to the Operational Requirements Committee (ORC), the programme then went to the Weapon Development Committee (WDC), and thence to the Secretary of State and the Treasury.

The Chief Advisor (Studies) Working Party report had established the need for an airborne Elint capability and the case for three Nimrod(R) aircraft. The next step was to present the proposal to the ORC, and the Nimrod(R) was duly discussed at a special restricted meeting of the committee on 16th November 1967. The requirement for the aircraft, set out in an AFD paper, noted that the aircraft's primary role would be the gathering Tactical Elint immediately before and during hostilities; the secondary, peacetime role, would be Strategic Elint. All members of the Committee except one agreed that the requirement should be endorsed. The dissident was ACAS(P), Air Vice-Marshal Leslie Mavor, who took the view that the Chief Advisor(Studies) investigation had only established the 'qualitative' case for the aircraft, that is, it had established that the aircraft were useful; what it had not done was investigate the 'quantitative' case, that is, it had not tried to establish the cost-benefit of the Nimrod(R) in terms of strike aircraft saved from destruction due to the provision of tactical Elint. In ACAS(P)'s view this needed to be established before committing £14M to the programme. The refusal of ACAS(P) to endorse the Nimrod(R) meant that the programme had to be referred to the Chiefs of Staff for a decision before proceeding to the WDC.

The Chief of Air Staff, Sir John Grandy, discussed the Nimrod(R) with the CDS at a meeting on 22nd November, prior to submitting the proposal to the Chiefs of Staff. Amongst the attendees at the meeting was Sir William Cook, the Chief Advisor (Projects), who voiced doubts as to whether the Nimrod(R) was the most cost-effective way of meeting the requirement. In particular, he questioned the Air Force Department's opposition to air-to-air refuelling as a means to extend range and suggested that Trent Andover using air refuelling might be a cheaper solution. Although the Trent-engined Andover option had been discussed before and rejected, Sir William's doubts had be assuaged and thus CAS agreed to produce a paper for

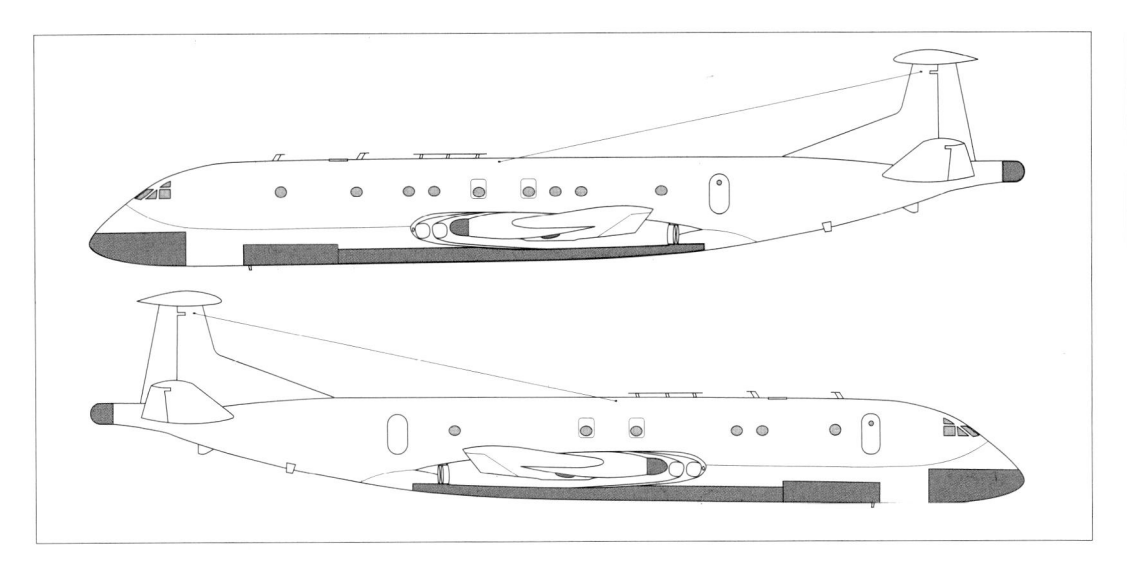

The Nimrod R.1 as originally configured by Hawker Siddeley as the HS801R. The weapons bay was converted into an equipment bay, with large dielectric panels covering an array of D/F aerials. The nose housed an ASV-21 radar as used on the Nimrod MR.1.

CA(Projects) examining the various options for meeting the Elint requirement in advance of the Chiefs of Staff meeting the following week. The Chief of Air Staff's paper, revisiting the various alternatives to the Nimrod(R), including the Trent Andover, was duly presented to the Chiefs of Staffs in the last week of September. Although it seems to have simply restated the arguments presented in previous Working Party and AFD papers it did the trick, and the proposal for three Nimrod(R) aircraft was duly approved by the Chiefs of Staff, clearing the way for a submission to the WDC.

The Nimrod(R) programme was submitted to the WDC on 8th December 1967, by which time costs had increased slightly due to delays in ordering the aircraft. The joint Ministry of Technology (MinTech)/MoD(Air Force Department) paper proposed the procurement of three Nimrod(R) aircraft at a cost of £14.05M. The proposal was to initially fit the aircraft with the Comet Elint suite, supplemented by new aerials, aerial turning gear, recorders and passive warning equipment. The initial cost of the new equipment would be £0.73M, with a further £1.25M required up to 1978. The proposal was endorsed by the WDC in January 1968.

Defence Cuts, 1968

Following the endorsement of the WDC the proposal was then prepared for submission to the Secretary of State for his approval. Unfortunately this was not an ideal time to propose major expenditure on a military project. In November 1967 a large UK trade deficit had led the Prime Minister, Harold Wilson, to devalue the pound and institute spending and credit controls. In January 1968, in a further measure to contain the financial crisis, the government announced large defence cuts and the withdrawal of UK forces from the Far East and Middle East. The cuts led to a large-scale reorganisation of the RAF front-line and fresh doubts regarding the requirements for the three Nimrod(R)s. One view was: 'I think that before putting forward the submission, we need to take a general view of whether the three HS801(R)s will still remain a firm feature of the RAF programme during the restructuring of our future force plans on which we are about to embark'. Ideally the AFD wanted to wait until the plans for the future shape of the RAF had become clearer before proceeding with the submission to the Secretary of State and Treasury. However, the problem was that this would entail a delay of several months and the aircraft needed to be ordered fairly quickly in order to maintain continuity of Nimrod production and thus get the three aircraft at a reasonable price. ACAS(Pol) suggested that the submission should proceed, pointing out that the case for Nimrod(R) in Europe was strong and that the three aircraft would enhance the UK contribution to NATO by providing a much-needed Tactical Elint capability. In his view the only unknown was whether the three Nimrods would be authorised as an additional purchase, or come out of the already-committed maritime Nimrod order for 38 aircraft (that is, replacing three Nimrod MR.1). In the latter case, the contract for the additional three aircraft could probably be cancelled with only a small penalty.

At the end of February 1968 F.6b(Air) of the Ministry of Defence wrote to the Ministry of Technology noting that no firm decision could be made on the Nimrod(R) until the future structure of the RAF had been settled. He therefore agreed that a six month contract to a value of £250,000 should be placed with HSA for design work and production planning for three Nimrod(R), hopefully keeping the project on schedule and the unit production costs within estimates.

By February 1968 all concerned with the Nimrod(R) programme were convinced that the case had been comprehensively made for the procurement of the three aircraft, with the only question remaining being whether the aircraft would have to come out of the existing maritime Nimrod order, or whether the three aircraft would be additional to the maritime Nimrod order. It therefore came as something as a shock to discover that the Secretary of State was not completely convinced of the need for the aircraft, and was expecting the on-going Future Combat Aircraft Study (examining future strike aircraft) to clarify the usefulness of the aircraft in the European theatre. He also requested information on the Tactical Sigint capability of other NATO countries and the amount of Sigint data shared within NATO. As a result of the doubts the Air Force Dept. (MoD), in conjunction with the Army, Navy and Signals Command, prepared a paper restating the Air Staff concept for the tactical use of the Nimrod(R).

The Future Combat Aircraft Study looked at the strike and reconnaissance aircraft requirements of NATO up to 1980 and by mid-1968 had concluded that strike operations in a European environment were feasible, but likely to suffer high losses. However, it was clear that the situation would be significantly worse without the effective use of tactical routing to avoid defended areas, and the use of ECM to degrade defence effectiveness. The successful prosecution of both of these tactics was reliant on the availability of up-to-date intelligence on the type and disposition of enemy air defences that is, on Tactical Elint.

The only NATO country with a Tactical Elint capability was the United States: West Germany, Norway, France and Turkey were known to have a small number of Strategic Elint aircraft but did not possess any tactical capability. The processed results of airborne Elint intelligence gathered by the UK was shared with a number of other NATO partners (Norway, Denmark, Turkey and West Germany) under bilateral agreements, in return for which the UK obtained intelligence from those countries' ground Elint collection stations. Although Elint information was not generally shared with NATO, plans were being drawn up in conjunction with the US to allow the general sharing of Elint in the build-up to war.

The Nimrod(R) concept of operations was described in an MoD(Air) paper (AF/W.1623/68/DD.Ops(Recce)) dated May 1968. This described the primary role of the aircraft as Tactical Sigint in the European theatre, with a secondary peacetime role collecting Strategic Sigint. The Nimrods would operate in both the Comint and Elint roles during the conventional phase of a war. In peacetime the UK's primary source of Comint on Soviet forces in Europe was 26 Signals Unit, stationed in Berlin and in the run up to, and during, war its primary goal would be to maintain a Soviet Air Order of Battle by monitoring the VHF communications of the Soviet 24th Tactical Air Army. However, 26 SU might have to relocate in a hurry in the immediate prelude to war, thus breaking the flow of intelligence. The Nimrod(R) could plug the gap by providing a continuous Comint monitoring capability during periods of tension. In the Elint role the Nimrod would monitor the disposition and operating parameters of radars associated with Soviet air defences, providing support to NATO's dual-role strike aircraft. The main targets would be the *Spoon Rest* metric early warning and *Fan Song* S-band fire control radars associated with the SA-2 *Guideline,* and the *Flat Face* L-band target acquisition and *Low Blow* X-band fire control radars associated with the SA-3 *Goa.* The aircraft would also fly in support of the Army, identifying and locating enemy battlefield radars for targeting purposes. During operations the Nimrod would fly sorties of 6-8 hours duration, operating 50nm (93km) from the forward edge of battle in order to keep it out of range of SAMs. It was envisaged that the aircraft would carry out operations on the Northern Flank (probably from Kinloss), in the Central Region (from Wyton), and on the Southern Flank (from Luqa or Akrotiri).

In August 1968 a further submission was made to the Secretary of State, reaffirming the requirement for the Nimrod(R), also pointing out that the Air Staff considered the three aircraft would be an essential feature of any front line, regardless of the contraction of the RAF and the withdrawal from overseas bases East of Suez. Although GCHQ had expressed some doubt regarding the cost-benefit of the aircraft in the Strategic Elint role, the main case for the Nimrod(R) rested on their use in the Tactical Role and the importance of Tactical Elint in the European theatre had been confirmed by the Future Combat Aircraft study. The uncertainties regarding the size of the total Nimrod procurement had also been settled with the requirement for 38 aircraft in the maritime role having been confirmed by the force structure planning exercise. The submission also pointed out that the £250,000 design contract issued to HSA

Radars associated with Soviet SAM and AA weapon Systems of the 1960s and 1970s

NATO Codename	Frequency Band	Type/Function	Range (km)	Associated Air Defence Weapon
Back Net	E	Early Warning	300	SA-5 *Gammon*
Back Trap	E	Early Warning	400	SA-5 *Gammon*
Bar Lock	E/F	Early Warning	200	SA-5 *Gammon*
Big Back	L	Early Warning	600	SA-5 *Gammon*
Fan Song	E/F/G	Fire Control	60-145	SA-2 *Guideline*
Flat Face	C (UHF)	Target Acquisition	250	SA-3 *Goa*
Gun Dish	J	Fire Control	20	ZSU-23-4 gun
Knife Rest	A (VHF)	Early Warning	70-75	SA-2 *Guideline*
Long Track	E	Target Acquisition	150+	SA-4 *Ganef*, SA-6 *Gainful*
Low Blow	I	Fire Control	40-85	SA-3 *Goa*
Pat Hand	H	Fire Control	128	SA-4 *Ganef*
Side Net	E	Height Finder	28	SA-2 *Guideline*, SA-3 *Goa*, SA-5 *Gammon*
Spoon Rest	A (VHF)	Target Acquisition	N/A	SA-2 *Guideline*
Square Pair	H	Fire Control	255	SA-5 *Gammon*
Squat Eye	C	Early Warning	130	SA-3 *Goa*, SA-5 *Gammon*
Straight Flush	I	Fire Control	60-90	SA-6 *Gainful*
Tall King	A	Early Warning	600	SA-5 *Gammon*
Thin Skin	H	Height Finder	240	SA-4 *Ganef*, SA-6 *Gainful*

Note: The variety of radars illustrates the need for a sophisticated Tactical Elint capability.

had bought some time by allowing the main order to be delayed without an adverse effect on the aircraft unit price. However, that contract would expire in August, and failure to authorise procurement of materials and components for the aircraft at that point would jeopardise both the aircraft price and delivery date. The submission urged the Secretary of State to support the project and authorise an approach to the Treasury.

The Secretary of State approved the ASR at a meeting on 7th August and grudging Treasury approval for funding for the three aircraft was obtained in early September 1968. This allotted £10.46M from the Air Votes and a further £1.95M from the Ministry of Technology (Min Tech) vote. The Min Tech then began negotiations with HSA for the production of the three aircraft to Specification 266 D&P. Unfortunately, the delay in placing the order meant that the delivery date of the first aircraft went back from November 1970 to March 1971.

The Sigint version of the Nimrod was initially referred to in official documents and correspondence as both the HS801(R) and the Nimrod(R). Following approval of the purchase of three aircraft for the RAF in September 1968, thoughts turned to the aircraft's service designation. The initial suggestion was to designate the aircraft the 'Nimrod R Mk 1'. However, this was opposed in some quarters due to possible confusion with the 'Nimrod MR Mk 1', and it was suggested that, in view of the considerable changes between it and the maritime Nimrod, the aircraft should be given a new name. This this was quickly ruled out, with ACAS(Policy) noting: 'We are unlikely to prevent the Russians from knowing what 51 Squadron are up to, and they will be aware of the existence of the HS801(R). However, there is no point in making things any easier for them, or for other potential enemies with less sophisticated intelligence organisations by providing the new aircraft with a special name'. ACAS(Policy) also noted that a new name would inconveniently publicise the presence of a Sigint aircraft when filing flight plans and the like. Alternative designations of 'TR Mk 1' and 'SR Mk 1' were also suggested but rejected and the designation 'Nimrod R Mk 1' was officially approved in June 1969.

The Nimrod Elint fit

Planning the Elint fit for the Nimrod(R) took place during 1967-68. The original Nimrod(R) submission in late 1967 contained only a broad outline of the requirement for tactical operations, making it difficult to draw up specifications for the Elint fit. The paper describing the full operational concept for the aircraft

(AF/W1623/68/DD.Ops(Recce)) was issued in June 1968 and the requirements for Elint equipment to match the operational concept were specified in October 1968 in AF/W1622/68. The final requirements included a capability in both Elint and Comint.

The initial plan for the Nimrod Elint suite was to fit the aircraft with a combination of equipment then installed in the Comet, equipment still under development for the Comet, and completely new Nimrod-specific equipment.

The receiver element of the proposed Elint fit comprised eight manual search and D/F positions using a combination of ARI 18050 (Upper L, S, C and X-band) and ALR-8 (Metric, L, S, C and X-band) superheterodyne receivers transferred from the Comets. The actual equipment installed at each of the eight stations could be selected to suit the particular search task. The ARI 18050/ALR-8 receivers would be supplied with signals via a new Ferranti aerial suite, specifically designed for the Nimrod, comprising seven wideband 'spinner' D/F aerials in the equipment bay, a smaller aerial in the tailcone and an aerial in the nose of each wing slipper tank.

The heart of the Nimrod Elint fit was the new SCARL-developed Integrated Display and Measurement System (IDMS). An IDMS unit was provided at each special operator's station to analyse signals intercepted by the receiver suite, providing facilities for the measurement, digital display, storage and management of key signal parameters (frequency, PRF, pulse-width, scan rate, direction of arrival, time of arrival). Signal parameters recorded by the IDMS could be stored digitally on magnetic tape or output on paper tape, facilitating subsequent processing by digital computer. The output could also be shared with other operators and supervisors. IDMS provided a semi-automatic Elint capability: it was still up to the SO to determine signals of interest, but once a signal had been selected, the IDMS relieved the operator of the drudgery of signal analysis and logging.

The main receiver suite was supplemented by three guard receivers (X.449, X.450 and X.451) and a new Instantaneous Frequency Indicator (IFI) unit, providing warning of signals in normally quiet portions of the S, C and X-band. Higher frequency (millimetre band) cover would be provided by a French RAK-2 system, providing search facilities in J, K and Q-bands and D/F in J-band.

By the late 1960s a number of new radars incorporating advanced techniques (such as frequency hopping, FMCW and pulse compression) were entering service. The proposed Nimrod search and analysis suite provided a capability against frequency-agile radars by allowing signals from four search positions to be correlated. A further capability against so-called 'complex' signals (that is, signals with an intrapulse modulation) would be provided by the new Complex Signal Recognition Unit (CSRU) which provided advanced analysis facilities. Video and still cameras, plus high-bandwidth video recorders, would be provided to record signals analysed by the CSRU.

The biggest single item of expenditure in the Nimrod Elint suite was the new aerial array. The intention was to fit an array of Ferranti hydraulically-powered rotating aerials in the under-fuselage equipment pannier, providing both intercept and D/F facilities. The array would comprise a total of seven spinner aerials in two groups, four forward and three aft, with each aerial group driven via its own hydraulic power pack. The first six 33in (84cm) diameter aerials had a look angle of 60° either side of the aircraft while the smaller 24in (61cm) diameter rear aerial had a field of view of 200° to the rear of the aircraft. For performance reasons preamplifier, tuner and RF switch units would be located in the pannier, adjacent to each aerial. Provision would also be made for the installation of a 10ft (3m) long SLIS (see Chapter 11) aerial in the forward part of the pannier, replacing the forward spinner aerial.

The pannier aerials would be supplemented by a 15in (38cm) spinner aerial in the aircraft tail cone (driven from the rear pannier hydraulic pack), an electrically-driven 20in (51cm) spinner aerial in the nose of each wing tank, four window aerials (two each side), a spiral aerial in the fin tip radome and a wire aerial between fin and fuselage top.

Determining overall responsibility for the expensive aerial suite proved tricky. Previous Elint equipment development for the Comet and Canberras had been undertaken by the Signals Command Air Radio Laboratory (SCARL), using the Ministry of Technology to place contracts with outside firms where appropriate. However, in the case of the Nimrod aerials, the Ministry of Technology would be theoretically responsible since they were the funding agency. The problem was, the MinTech had no expertise in the field, and they had no authority over SCARL, which was controlled by Signals Command. In the end a compromise was agreed under which MinTech would be responsible for development policy, SCARL would be the R&D authority and the contractor would be the design authority.

Initial estimates suggested the cost to fit the aircraft would be around £0.73M, of which £0.5M would be for the completely new equipment. This comprised £397,000 for the aerial array (£200,000 R&D and £197,000 production), £28,000 for production of a Magnetic Tape Recorder and £75,000 for production of the Passive Warner. The IDMS and IFI system were already under development and had been partly paid for, but required a further spend of £230,000 to complete and bring into production. The above equipment covered the projected Nimrod fit up until 1972. Other equipment might be fitted in the future (for instance, SLIS), but this was considered outside the scope of the initial planning and costing exercise.

Preparing for the Exploitation of Tactical Elint

The emergence of Tactical Elint as the primary role of the Nimrod(R) highlighted shortcomings in the existing arrangements for the ground exploitation of Elint data. The intelligence take from the Comets and Canberras was distributed to the Defence Intelligence Staff (DIS) and GCHQ for analysis, with GCHQ handling the subsequent distribution of the finished product to interested agencies. These arrangements were expected to continue when the Nimrod replaced the Comets and Canberra in the Strategic Elint role. Although arrangements existed to provide special assessments of topical events (via *ad hoc* Current Intelligence Groups) there were no mechanisms in place to process and distribute time-critical Tactical Elint to commanders in the field.

In April 1968 the MoD (Air Force Dept) EW Committee established Working Party No.4 to investigate the exploitation of Tactical Elint data from the Nimrod R.1, the facilities required to analyse such data, and how such facilities could be integrated with existing tactical intelligence organisations. The Working Party consulted widely during their investigation, soliciting input from the MoD(Army), the MoD(Navy), HQ Army Strategic Command, HQ Air Support Command, HQ Coastal Command and the Joint Warfare Establishment. The Working Party soon established that the HS801(R) could not be considered in isolation, but rather had to be seen as one piece in the bigger tactical intelligence picture. The eventual recommendation of the Working Party was that the Tactical Elint data gathered by the Nimrod R.1 should be subject to first/second phase technical analysis at a new Tactical Sigint cell based at Wyton (air-transportable for deployment to a forward operating base, if required), then fed into a Central Intelligence Exploitation Centre (CINTEX) under the control of the DIS. The CINTEX would act as a central clearing house for all tactical intelligence data, also taking inputs from GCHQ, 54 Signals Unit (in the Far East) and 26 Signals Unit (in Europe), various RAF Commands, Naval intelligence, the USAFE European Electronic Intelligence Centre, and other sources, including visual reconnaissance. The Nimrod data would be input into a computer database, collated and cross-referenced with other data sources (including an Elint database) and then disseminated to the field. The output of the CINTEX would be fed to a tri-service Tactical Intelligence Centre (TIC) located at a Field Commander's HQ, based either in the UK or overseas. In addition to the above flow, the Nimrod(R) Elint data might also be fed directly to the TIC, depending on the location of operations.

The Nimrod(R) might be called upon to fly long sorties (up to 8 hours) in the Tactical role in order to capture a continuous picture of the enemy's Electronic Order of Battle (EOB). Thus, in order to meet the requirement for timely dissemination of Tactical intelligence, the aircraft would have to relay intelligence data to a ground station whilst still in flight. Some intelligence could be exploited in the aircraft, and a summary could then be relayed via a voice link. For example, in-flight analysis of Elint data could confirm the EOB of specific targets in a given area and be reported verbally. Similarly, Comint could be exploited in flight (for instance to determine movement of a military unit) and a summary reported to a ground station. However, the real solution to the data download requirement was to fit the aircraft with a secure datalink facility. This would allow the raw output of the aircraft's IDMS recording and analysis system to be transmitted to a ground station and then fed into the CINTEX computer. In January 1969 the Working Party recommended that SARL examine existing datalink systems for suitability, also investigating the development of a custom system as a last resort.

Communication Intercept Role

By April 1969 studies into the Nimrod(R) tactical role suggested that the aircraft's capability could be enhanced by the provision of an extended communication intercept (Comint) fit. Signals staff suggested that the aircraft could accommodate 17 Special Operators Voice (SOV) and four Supervisors along with a

LISTENING IN

Comint receiver suite. Estimates put the additional cost of the Comint fit at £0.552M. Although it was considered desirable to introduce the equipment before the aircraft entered service, there were concerns that the additional cost might lead to the Nimrod(R) project being referred back to the WDC. As a result the procurement of the Comint fit was deferred whilst the operational requirement was further investigated. There were also concerns regarding the supply of SOV personnel. The plan seems to have been to increase the 51 Squadron SOV establishment and also make arrangements to borrow suitably-trained personnel from ground Comint units in an emergency.

In January 1971, ASR.871 was issued for a tactical communications intercept suite for the Nimrod R.1. This was envisaged as an alternative to the standard strategic intercept suite and called for the installation of 48 VHF/UHF receivers in four banks of 12, with one bank at each of four operator positions. In addition, four of the aircraft's existing receivers were to be given a dual-role VHF/UHF capability by a change of RF head and frequency read-out unit. Each of the 52 receivers would be associated with its own twin-channel audio cassette tape recorder (recording a timestamp on the second channel) and facilities would also be provided for playback in the air. A new aerial distribution system would be required to distribute the required frequencies to the receivers: 20-1GHz to eight receivers, 40-80MHz, 100-150MHz and 200-400MHz to the remaining 44. The suite would also feature a new internal communications control system which would allow a supervisor to allocate VHF/UHF channels to operators for manual logging, to a tape recorder for recording, or both.

Designing the Installation

In June 1969, co-incident with the change of role of RAF Watton, the EWSW was reformed as an independent unit and renamed the Electronic Warfare Support Unit (EWSU). The unit's main function was to provide engineering, installation and training support to 51 Squadron and its aircraft although it also carried out signals equipment installations in other aircraft.

The EWSU was allocated the new Nimrod Elint installation task (SRIM 3623) in mid-1969. This was a massive task, dwarfing the previous Comet 2R installation in complexity and scale. By June 1969 the delivery schedule called for the first Nimrod R.1 to arrive at the EWSU in February 1971, with the second and third aircraft following in November 1971 and September 1972 respectively. The EWSU thus had around 20 months to plan the installation and prefabricate the necessary parts. Work started on preliminary installation drawings for the Nimrod in June 1969 and a meeting was held the same month attended by MoD(Air), HQ 90 Group, SARL and EWSU to cost the Elint suite. There still seems to have been a few unknowns and so drawing work on the installation was halted in July due to 'uncertainty in equipment installation configurations'. This was apparently resolved in August when an equipment development programme was received from 90 Group. By November 1969 the EWSU had issued a schedule for the SRIM 3623 installation design and manufacturing tasks, accompanied by detailed schedules describing the timing of, and dependencies between, the various sub-tasks.

The EWSU worked closely with HSA, who were carrying out the preliminary modification of the aircraft, to ensure that the aircraft would be

The proposed Nimrod R.1 layout, circa 1971. Two distinct layouts were envisaged: the principal Elint fit with 16 Special Operators and an alternative Comint fit incorporating 23 Special Operators). The station names give some idea of the division of tasks on operations.

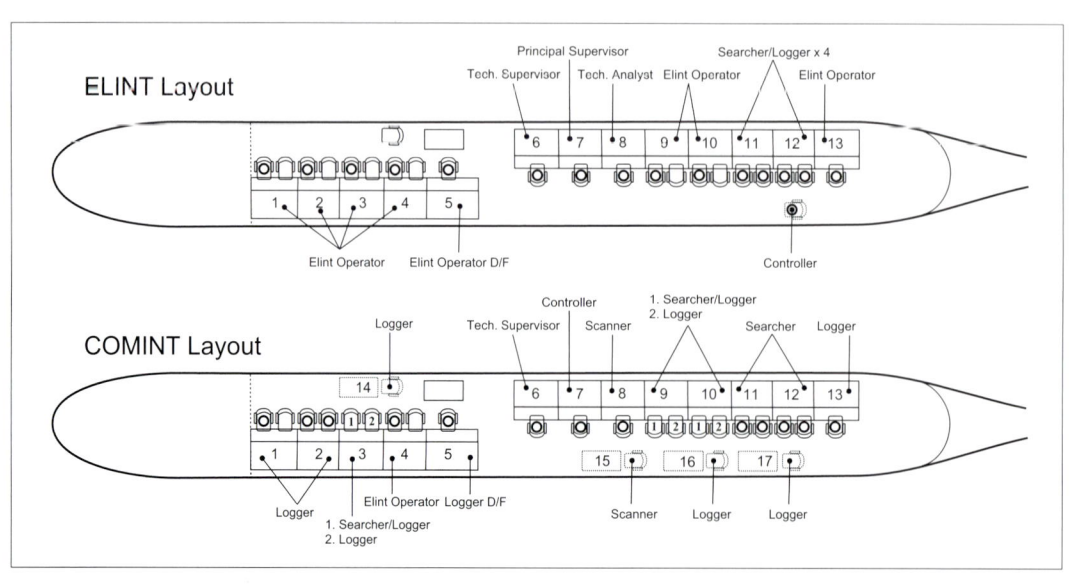

delivered in a suitable state, structurally and electrically, to take the Elint installation. The design for the Nimrod installation was evolved at the EWSU during late 1969/early 1970 and seems to have been mostly complete by the second quarter of 1970. A formal interface document, describing the physical layout, structure and electrical requirements for Elint fit was issued to Hawker Siddeley in April 1970. In October a more detailed estimate of the special fit payload was passed to HSA, this apparently being the first occasion on which the complete details of the fit had been summarised into one document.

A prototype SO equipment rack for the EW School was delivered in April 1970, followed by the delivery of a number of instructional racks in October. These were used to construct a mock-up of the rear cabin to prove the layout of various modules. The mock-up was subsequently inspected by representatives from SARL, CSDE and 51 Squadron in November to approve the layout of various racks.

In the January 1971 the EWSU (engineering and installation) and SARL (R&D) were merged into a single unit, forming the Electronic Warfare Engineering & Training Unit (EWE&TU). At the same time the new unit moved from Watton to Wyton, the intended home of the Nimrod R.1 force. The merge and move went smoothly with little disruption to the main SRIM 3623 task.

Delivery and Modification

While the EWSU and SARL were designing the Elint suite, the three Nimrod R.1s were under construction at Hawker Siddeley Aircraft at Woodford. The R.1s followed the Nimrod MR.1s on the production line, receiving the construction numbers 8039-41.

The Nimrod R.1s were built to a revised standard compared to the MR.1, with an aerial bay and associated radomes replacing the Nimrod MR.1 weapons bay, various other aerials, and a redesigned main cabin layout, accommodating a number of SO and equipment racks. The cabin was fitted with a variety of cable ducting to facilitate the wiring-up of the SO racks, but the racks themselves were left empty: the Sigint fit was secret and the equipment would be fitted at Wyton.

The first aircraft (XW664) was completed in the second quarter of 1971 and a formal clearance conference was held at Woodford over 17th-21st May 1971. The conference was attended by various interested parties from the company, the Ministry of Technology and the Ministry of Defence(Air) and seems to have gone reasonably well, with no major issues raised.

Following clearance, the first Nimrod R.1, XW664, arrived at Wyton in July 1971 where it was officially handed over to the service. The aircraft immediately went to the EWE&TU for installation of SRIM 3623. In preparation for the installation the aircraft was inhibited and a number of components were removed 'for safe keeping and to support in-use MR aircraft'. As part of this pre-installation work all four engines were removed for modification. An initiation meeting for SRIM 3623 was held in June 1972 and work on the aircraft began shortly afterwards. The installation seems to have proceeded according to plan and was concluded in mid-1973, bang on target. The aircraft was then subject to a 'de-storage' procedure, as part of which all major systems were re-installed and connected up. Power-on tests began in July and lasted around four months, during which the aircraft's systems and the SRIM 3623 equipment were checked out on the ground.

XW664 flew its first sortie after the installation of SRIM 3623 on 31 Oct 1973: a 4 hour 20 minute flight from Wyton by a 51 Squadron crew under the supervision of the EWE&TU as an airframe test and an initial check of the role equipment. Further sorties on the Development Flight Trials Programme were flown during November (3 sorties), December (4 sorties), January (6 sorties) and February (9 sorties) and the Development Flight Trial programme was declared complete at the end of February 1974.

The remaining two Nimrods (XW665 and XW666) went straight from HSA to the A&AEE during 1972 for type flight development, aerial system and tropical trials. A full CA Release programme was carried out on the engineering and navigation and radio aspects of the Nimrod R.1 and modifications resulting from the trials were applied to the aircraft by the manufacturer before delivery to Wyton. The same modifications were later applied to XW664 following the completion its Elint installation at Wyton.

XW666 completed its part of the A&AEE trials in December 1972 and was delivered to Wyton for its Elint fit in January 1973. The installation was completed in early 1974, flight trials got under way in March 1974, and the aircraft was delivered to 51 Squadron at the end of May. Nimrod XW665 was retained on A&AEE trials until July 1973. The aircraft was then delivered to Wyton for its Elint fit. Flight trials began in September 1974 and the aircraft entered service in January 1975. All three Nimrods were delivered to 51 Squadron exactly on schedule.

13 The Nimrod in Service

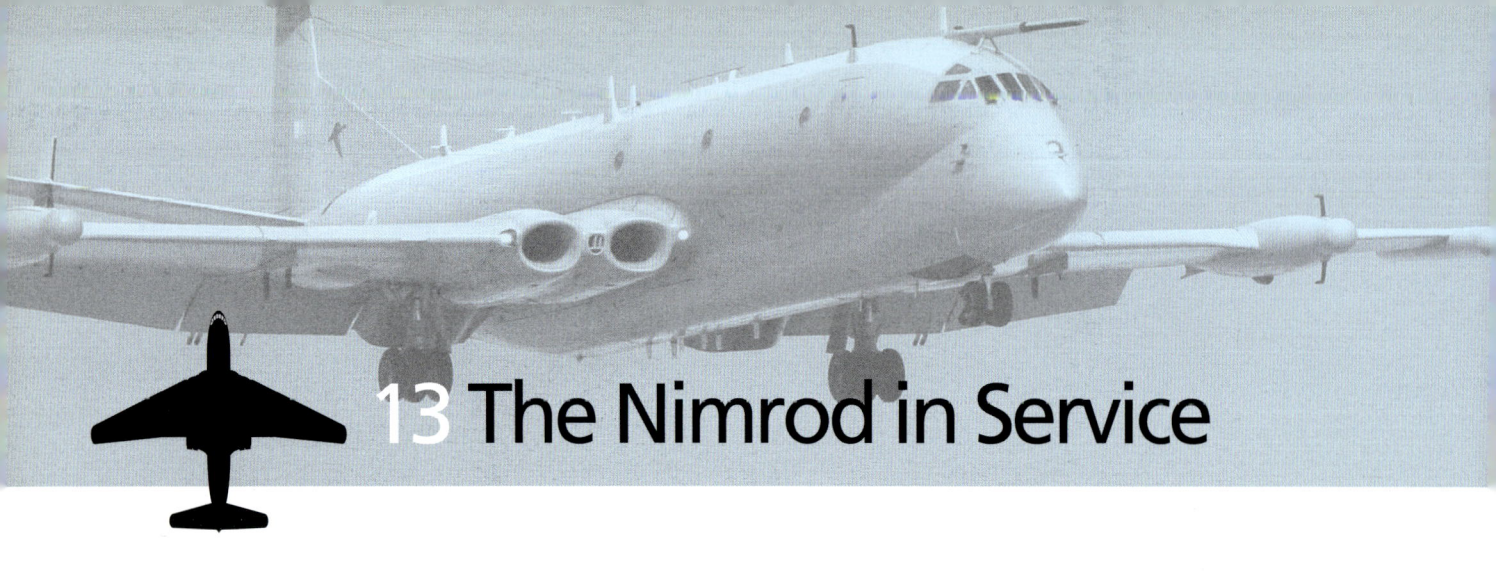

'…several modifications to the Specialist Equipment in the aircraft were also developed, designed and installed in what must be, record time'.
Summary of the preparations for Operation *Acme*, 1982

The first Nimrod (XW664) was handed over to 51 Squadron at the end of the Development Flight Trials programme in February 1974 with the second (XW666) following in May 1974. In the same period the squadron worked up with the new aircraft and prepared for operations. The first operational sortie took place on 3rd May 1974 when Nimrod XW664 carried out Operation *Mensa* over the Baltic. A further five *Mensa* sorties were flown during the month, all from Wyton. The Nimrod R.1 was formally commissioned on 10th May at a ceremony at Wyton attended by a variety of 'distinguished guests', including VCAS, Air Marshall Sir Ruthven Wade KCB, DFC.

The first Nimrod overseas detachment was to Cyprus on Operation *Sprung*, in July using XW666. The Sprung sorties were part of the longstanding programme of RPFs against Egypt and Syria. The Nimrod deployed to Akrotiri via Gibraltar on a working transit, probably taking in Libya and Algeria en-route, and flew six sorties from Cyprus along the Egyptian and Syrian coasts. During the second half of 1974 the Nimrods gradually took over operational commitments from the Comets and Canberras. The last Canberra RPFs were flown in June 1974, the third Nimrod (XW665) was handed over to 51 Squadron at the end of November 1974 and the last Comet (XK695) retired in December 1974. Thus by the end of 1974 51 Squadron had completed re-equipment with the Nimrod R.1

Operations 1975-1982

By January 1975 the Nimrods had completely taken over from the Comets and Canberras, and over the next few years the Nimrods settled into a fairly regular monthly routine. This entailed around six sorties each on two RPF programmes over Germany and the Baltic: one a Strategic programme and the other a Tactical programme. Both were flown from Wyton, as were up to four training sorties each month. In addition to the two home-based programmes, a Nimrod was also detached each month to either Akrotiri or Luqa for around six sorties over the Mediterranean against Syria, Egypt, Algeria and Libya. The regular monthly programme was interrupted three times a year for the detachment of a Nimrod to Tehran, Iran, for sorties along the southern border of the Soviet Union and over the Caspian Sea. During 1975/early 1976 a Nimrod was also detached on four occasions to Masirah for operations off the coast of Yemen.

Hawker Siddeley Nimrod R.1 XW666 at Wyton sometime in the 1970s, in the original white over grey colour scheme of the maritime Nimrods.
T Panopalis Collection

THE NIMROD IN SERVICE

Andover C.1(mod) XS644 of the EWE&TU in May 1979. Note electronic equipment pod under fuselage, similar to the tail cone radome on the Nimrod R.1, and 51 Sqn badge on fin. *A Balch via T Panopalis*

Training & Support

The Nimrod R.1s were expensive assets and training hours were at a premium. The majority of aircrew continuation training was therefore carried out at the Nimrod OCU at St Mawgan, also taking advantage of the simulator there. Travelling to St Mawgan for continuation training was hardly ideal and in April 1976 51 Squadron received its own training aircraft in the form of a standard Nimrod MR.1 (XZ283). The MR.1 also acted as a support aircraft on overseas detachments. Unfortunately the allotment only lasted for a couple of years, the aircraft departing in June 1978 for conversion to AEW.3 standard. The squadron then reverted to its previous training regime.

Technical support for 51 Squadron was initially provided by the EWE&TU, which merged with the 60 MU SRIM Squadron in March 1976 to become the Electronic Warfare Avionic Unit (EWAU). This specialist unit was responsible for modifications and upgrades to the Elint systems carried in the aircraft. As might be expected, the Nimrods were subject to a rolling programme of improvements in order to maintain their capability. The EWE&TU/EWAU provided a so-called Quick Reaction Capability (QRC), providing quick fixes and improvements to equipment as dictated by operational requirements. The first modification to the Elint fit (QRC 1/74) was applied to XW666 in June 1974 and XW664 in July, and by November, QRC 7/74 had been applied.

When the Nimrod entered service in 1974 the EWE&TU were using Varsity WJ916 as a trials aircraft, testing and proving systems in an airborne environment. In mid-1977 the Varsity was retired and replaced by Andover XS644. The Andover made its first flight with the EWAU on 1st August 1977 and continued to fly in the trials role into the 1980s.

Tactical RPFs

The Nimrods flew four to six Tactical RPFs per month from Wyton, rehearsing the squadron's war role of locating and classifying Soviet Bloc air defence systems. These operations were initially described as 'training sorties', perhaps reflecting the fact the squadron was still working up in the role. In mid-1975 the Nimrods were fitted with a secure communications installation. This was an essential part of the Tactical role fit, allowing the aircraft to relay securely tactical Elint reports to ground stations. The new equipment was exercised in October 1975 when the Nimrods carried out trials with the 'ground support facilities and associated units earmarked to support the squadron in time of war'. Sorties were also flown in conjunction with BAOR units.

In March 1981, in response to Defence cutbacks and pressures to improve fuel efficiency, the Nimrods began to fly Tactical Elint sorties from forward bases in Germany. The normal pattern was to fly the first sortie from Wyton, landing at a German base. Three further sorties would be flown from Germany, with the fourth and last sortie landing back at Wyton. This pattern of operations allowed the squadron to maximise the time the Nimrods spent 'on station' in the operational area. BAOR Elint units took advantage of the mini-detachments to Germany and sorties occasionally flew with an Army Sigint operator on-board.

Iran and Arabia

From around 1962 onwards the UK had relied on friendly relations with Iran and its ruler, Shah Reza Pahlavi, for access to the southern borders of the Soviet Union. As described earlier, RPF sorties were initially flown from the Persian Gulf over Iran but later, from 1969 onwards, the sor-

177

ties were flown from Iran itself, using Tehran as a forward operating base. The Tehran detachments continued following the introduction of the Nimrod and an aircraft was detached to Tehran approximately three times a year from 1975 onwards. Unfortunately the Iranian regime was both unpopular and repressive and by the mid-1970s the Shah increasingly relied on a brutal secret police to suppress internal dissent. Public demonstrations against the Shah began in 1977 and intensified during that year and the next, culminating in massive rallies in December 1978. In January 1979 the Shah bowed to the inevitable and left the country, ceding power to an interim government. After a short struggle the interim government was overthrown and Iran became an Islamic republic.

Following the Iranian revolution the UK was denied the use of Tehran as a forward operating base and was also refused permission to overfly the country. The last Nimrod detachment to Tehran took place in September 1978 and following this the UK lost the ability to fly Elint sorties along the southern borders of the Soviet Union.

In the period February 1975 to April 1976 a Nimrod R.1 was detached four times to Masirah in the Arabian Sea. Previous detachments to Masirah had been used for operations against the PDRY and it is possible that country was also the target of these detachments. Other possible targets include Somalia, then host to a large Soviet naval base at Berbera. The Horn of Africa was considered a strategic location, commanding the entrance to the Red Sea.

During the period December 1978 to May 1980 a Nimrod was detached three times to Taif, Saudi Arabia. The target of this detachment may have been Ethiopia, which was then being supported by the Soviet Union in its conflict with Somalia. Saudi Arabia was at that time providing military aid to Islamic groups fighting the Ethiopian government, and thus may have looked favourably on intelligence sorties against that country.

Maritime sorties

The Nimrod R.1s were not tasked with maritime sorties for the first few years of operations, possibly because the Nimrod MR.1 had a maritime Elint capability: there may also have been an absence of new Soviet naval radars requiring investigation. In any event, Nimrod R.1 participation in maritime operations did not begin until August 1978 when a single aircraft was detached to Kinloss for Operation *Dance*. The intended target did not materialise and it was not until March 1979 that a Nimrod R.1 flew a successful maritime sortie. On that occasion the Nimrod was detached to Andøya, Norway from where it flew seven sorties, presumably against a Soviet naval exercise. A similar operation, this time of nine sorties, was carried out a year later, operating from Kinloss and Bodø. The main maritime commitment came in February 1982 when, under Operation *Vicarage*, the squadron was required to keep a Nimrod at 4 hours readiness for maritime operations. A number of *Vicarage/Midcourse* sorties were flown during 1982.

First Upgrade: ASR.1851 *Astral Box*

The first major upgrade to the Nimrod R.1 came in 1980 and encompassed both a navigation refit and an Elint suite upgrade. The principal navigation system improvement was the replacement of the elderly ASV-21 radar with an EKCO 290 weather radar, allowing the radar navigator crew position to be removed. As part of the same upgrade one of the LORAN sets was replaced with a Delco AN/ASN-119 Carousel IVA INS.

Large-scale modifications to the Elint fit, to ASR.1851, were applied at the same time. The modifications were carried out by the EWAU at Wyton under SRIM 4006 with the programme apparently being known as *Astral Box*. The changes appear to have involved the installation of digital computer-controlled Elint systems, and were accompanied by the formation of a Nimrod software team at the EWAU. External evidence of the changes included the removal of all cabin windows ahead of the wing, suggesting new stations had been installed at those positions, and the installation of various new aerials includ-

The Nimrod was occasionally detached to the Middle East during the late 1970s/early 1980s for operations in the strategically-important Horn Of Africa. This area was coming under Soviet influence and the UK government wanted to keep a weather eye on it.

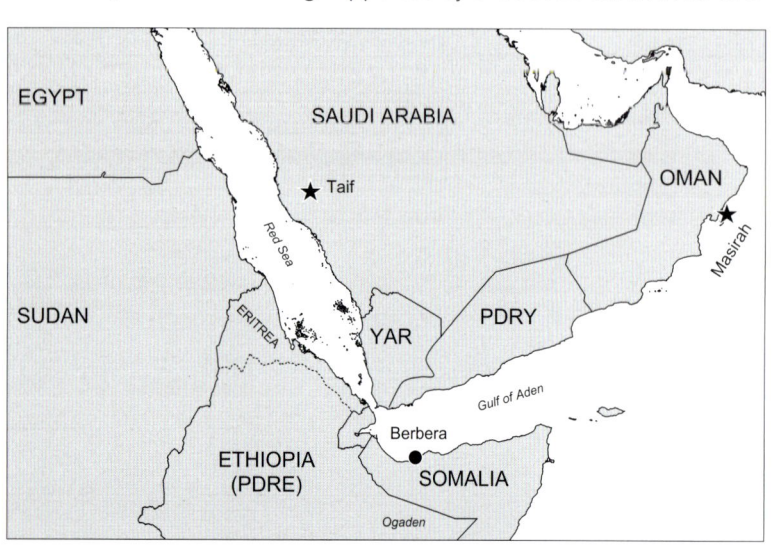

THE NIMROD IN SERVICE

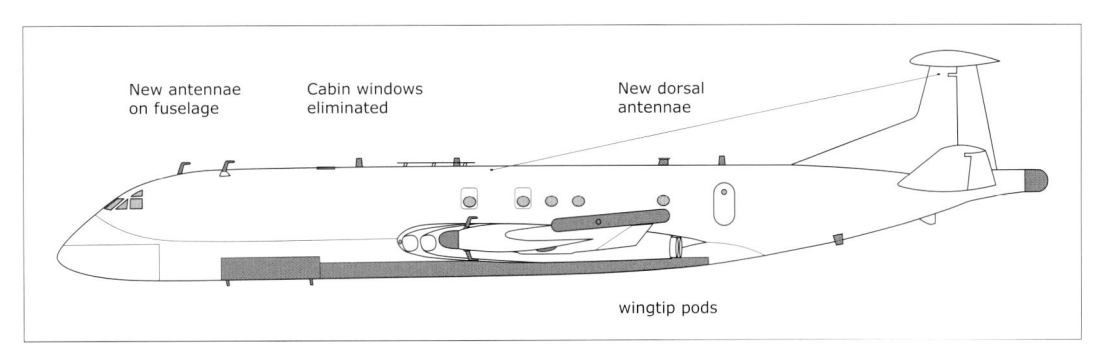

Nimrod R.1 to *Astral Box* standard. The external evidence of the refit included new aerials, the addition of wingtip pods, and the elimination of cabin windows forward of the wing.

ing three inverted 'L' aerials on the top of the fuselage above the cockpit, and similar aerials above and below the wing slipper tanks. Wingtip pods, similar in appearance to those used by the Nimrod MR.2 *Yellow Gate* installation were also fitted. The first SRIM 4006 installation was applied to XW664 during 1980-81, probably beginning in July 1980 and following a major service at Kinloss, and the Nimrod was returned to Wyton at the end of July 1981.

A series of trials, sponsored by MoD(PE), were then flown to first calibrate and then prove the new Elint systems. The calibration process required the use of radars on Aberporth range which restricted the trials to two 8-hour sorties each week. Following the completion of the calibration process in September 1981, a series of eight acceptance sorties was carried out. Designated Operation *Beat*, the intent was to work the equipment up to an operational standard and at the same time provide the Special Operators (SOs) and Supervisors with some basic instruction in its use. The acceptance phase was completed in October but, although the new system provided 'vast improvements' over the old fit, some problems remained. As the squadron CO noted: 'The first operational Strategic Reconnaissance sortie is planned for early November and it appears that much work in developing the computer programmes has to be carried out'. The Nimrod was duly flown on the planned programme (Operation *Gobble*) over the Baltic in November but so many problems were encountered with the Elint equipment that the aircraft was transferred to presumably less-demanding Tactical operations (Operation *Dubbin*) while the system was debugged. Finally, after six *Dubbin* sorties, the SRIM 4006 installation was declared fully operational. To mark the successful working-up of ASR.1851/SRIM 4006, a number of officers associated with the project were invited to a presentation at Wyton in mid-November 1981, during which XW664 was inspected.

The second SRIM 4006 modification was applied to Nimrod XW665 by the EWAU during 1981-82, with the aircraft returning to the squadron in October 1982. The third aircraft (XW666) is believed to have received a similar upgrade sometime in 1982-83. The *Astral Box* refit greatly improved the intelligence collection capability of the Nimrod but the extra equipment did not improve working conditions for the Special Operators. One observer recalled the importance of the aircraft's small galley: 'The operators had to put up with a noisy and hot environment generated by the equipment. The work required great concentration and over an 8 hour period that was difficult to achieve. In order to keep everyone going a constant supply of food and drink was provided'.

1982 Falklands: Operation *Acme*

On 2nd April 1982 Argentinian forces invaded and occupied the Falkland Islands. In response the British government despatched a Task Force to retake the Islands and by 1st May elements of the British Naval Task force were operating in vicinity of the Falklands. The sinking of the Argentinian cruiser *General Belgrano* on 2nd May largely neutralised the Argentinian Navy, leaving the Argentinian Air Force, operating from the mainland, as the main threat to the Task Force.

The introduction of computerised Elint systems, and the ability to digitise received signals, was a major step forward. This redacted output from a digital analyser shows five radars, the top graph showing signal amplitude, the bottom showing signal frequency. Correlation is provided via Time Of Intercept (TOI). *Author's collection*

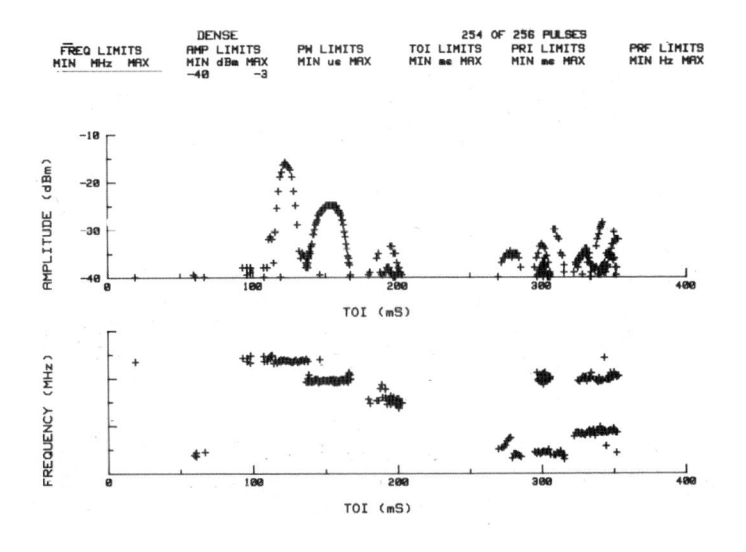

179

LISTENING IN

Nimrod R.1 after Mod.706 (Air-to-Air refuelling) and addition of tail finlets to improve handling, particularly to counter Dutch roll.

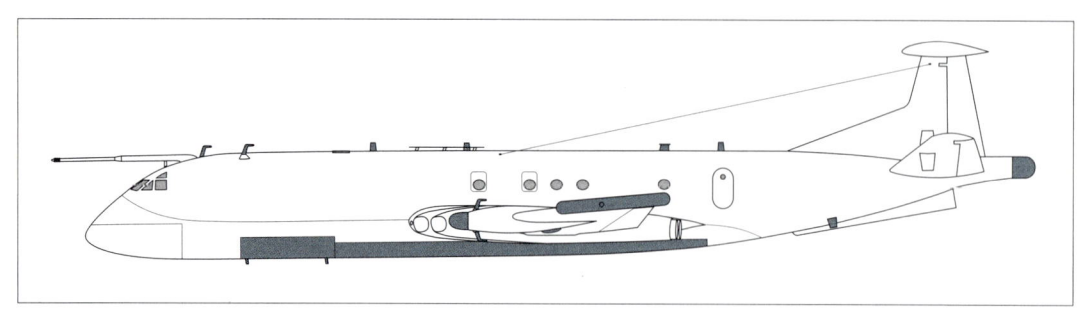

51 Squadron was placed on Standby for operations in the South Atlantic early in April. As part of the preparations for the task a detachment to Akrotiri was cancelled and 'several modifications to the Specialist Equipment in the aircraft were also developed, designed and installed in what must be, record time'. On 5th May 1982 Nimrod R.1 XW664 (the enhanced capability SRIM 4006 aircraft) and crew were detached to the South Atlantic under Operation *Acme* to support the Task Force by providing intelligence on the Argentinean threat. Exact details of the deployment are still obscure, but some sources claim that XW664, supported by a VC10, was based on San Felix Island, off the coast of Chile. The aircraft allegedly flew night sorties in Chilean airspace, intercepting Argentinean communications, refuelling at Concepción, on the Chilean mainland, before returning to San Felix each night. The first sortie reportedly took place on 9th May and further sorties were flown during the month. XW664 returned from Operation *Acme* on 22nd May, shortly after the main amphibious landings at San Carlos Water took place.

The Elint task was not one-sided and the Task Force was also subject to the attentions of an Argentinean Air Force Elint Boeing 707. The aircraft was intercepted by Sea Harriers on the voyage south, prior to the outbreak of hostilities, and escorted away. Following the British landings, the gloves came off and HMS *Cardiff* made an unsuccessful attempt to shoot the aircraft down with a Sea Dart missile on 22nd May. This seems to have had the required affect and the 707 did not reappear.

Air-to-Air Refuelling

Operations in the South Atlantic had underlined the importance of air-to-air refuelling (AAR) for long-range aircraft and consequently, in late May 1982, the MoD decided that at least one of the 51 Squadron Nimrods should have that capability. Early in June a number of aircrew attended a condensed AAR ground-school run by 232 OCU (Victor K.2) at Marham, then proceeding to Kinloss for flight training using one of the Nimrod MR.2s based there. Two pilots successfully completed the training and returned to Wyton on 10th June, fully-qualified in the AAR role.

Whilst the pilots were being trained arrangements were made to fit one Nimrod R.1 (XW664) with an air-to-air refuelling probe and associated equipment. The aircraft was flown to BAe Woodford on 2nd June for the installation work under Mod 706, returning to Wyton on the 12th June. The Nimrod AAR modification had apparently been designed by BAe/HSA some years earlier, but had never been taken up by the MoD. The modification involved the installation of a refuelling probe above the cockpit; inside a flexible pipe ran overhead and down the back of the cockpit, into the floor, connecting into the fuel system. A number of aerodynamic modifications were also applied to improve the handling of the aircraft with the probe fitted, comprising a ventral fin (designed to correct a tendency to Dutch roll), overwing vortex generators and rectangular tailplane finlets. Following an acceptance sortie, flown by a BAe crew, on 14th June, during which a fuel transfer was carried out, the aircraft was cleared as a receiver for AAR. A calibration sortie was then flown to check that the installation had no adverse impact on the Elint installation and the first 51 Squadron-crewed AAR sortie took place on 17th June when it received 5,000 lb (2,268kg) of fuel from a Victor tanker.

Elint VC10 proposal

In 1983 the prospect of an Elint VC10 re-emerged when British Aerospace proposed an Elint version of either the VC10 or Super VC10 as a possible replacement for the Nimrod R.1. The rationale was that moving the Elint role to the VC10 would allow the three Nimrod R.1s to be transferred to the Nimrod AEW.3 conversion programme. An additional benefit was that the VC10's larger cabin space would allow a more extensive Elint fit than that possible in the Nimrod.

THE NIMROD IN SERVICE

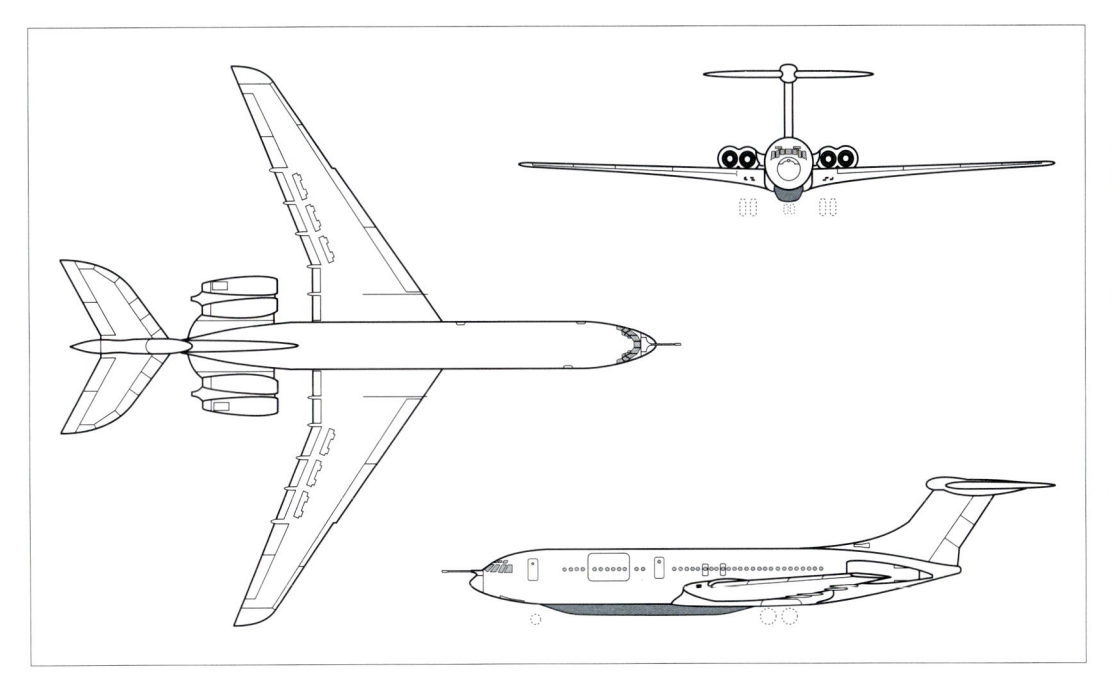

BAe proposal for an Elint VC10 circa 1983. This was the second time in twenty years the VC10 had been put forward as an Elint platform. The Elint fit was essentially that installed in the Nimrod and would have been accommodated in the under-fuselage pannier.

The core of the proposal was the provision of an aerial bay using an under-fuselage pannier, similar to that employed on the Nimrod. Two basic schemes were considered: a 'bolt on' pannier applied to the existing fuselage, and an 'integrated' pannier faired into a modified fuselage. The first scheme envisaged a 42ft (12.8m) long pannier attached to the underside of the forward fuselage via the freight-hold floor support structure. This would obviously reduce ground clearance and the proposal involved the use of an extended nosewheel. Although the pannier would increase drag it was a flexible solution, allowing re-use of Nimrod role equipment, and would have a minimum impact on the existing VC10 structure and systems.

The second scheme envisaged the modification of the VC10 forward fuselage to accommodate an aerial bay in the existing forward freight hold and systems bay. This proposal was only feasible on the Super VC10 as it required the aerial bay to be split into two sections, 26ft (7.9m) forward and 16ft (4.9m) aft. The advantage of the 'integrated' proposal was lower drag, but this was obtained at the expense of higher cost and weight. A third, hybrid scheme was also suggested, using a short 26ft (7.9m) pannier combined with an integrated aerial bay in the rear fuselage.

To make the VC10 R.1 proposal more attractive, BAe suggested the aircraft could be given a dual Elint/AAR role via the incorporation of additional internal fuel tanks. This would provide an emergency AAR capability in wartime, and would also allow an aircraft operating in the Elint role to replenish a fighter escort. In the event none of the proposals were taken up.

Operations 1983-1992

Details of 51 Squadron operations from 1982 onwards are, perhaps unsurprisingly, somewhat scarce, with publicly-available information restricted to brief descriptions of major operations and upgrades. Nonetheless it is still possible to trace the basic outline of the story.

Nimrod R.1 XW665 in 1988 in hemp over pale grey colour scheme and toned-down national markings. This scheme was adopted across much of the RAF's fleet of large jet aircraft.
T Panopalis Collection

181

LISTENING IN

The Nimrod R.1s operated in a variety of areas from 1980 up until retirement. These included (1) The Baltic (2) Germany (3) Kosovo (4) Egypt/Syria (5) Iraq (6) Afghanistan (7) Libya (8) Sierra Leone, and (9) Chile.

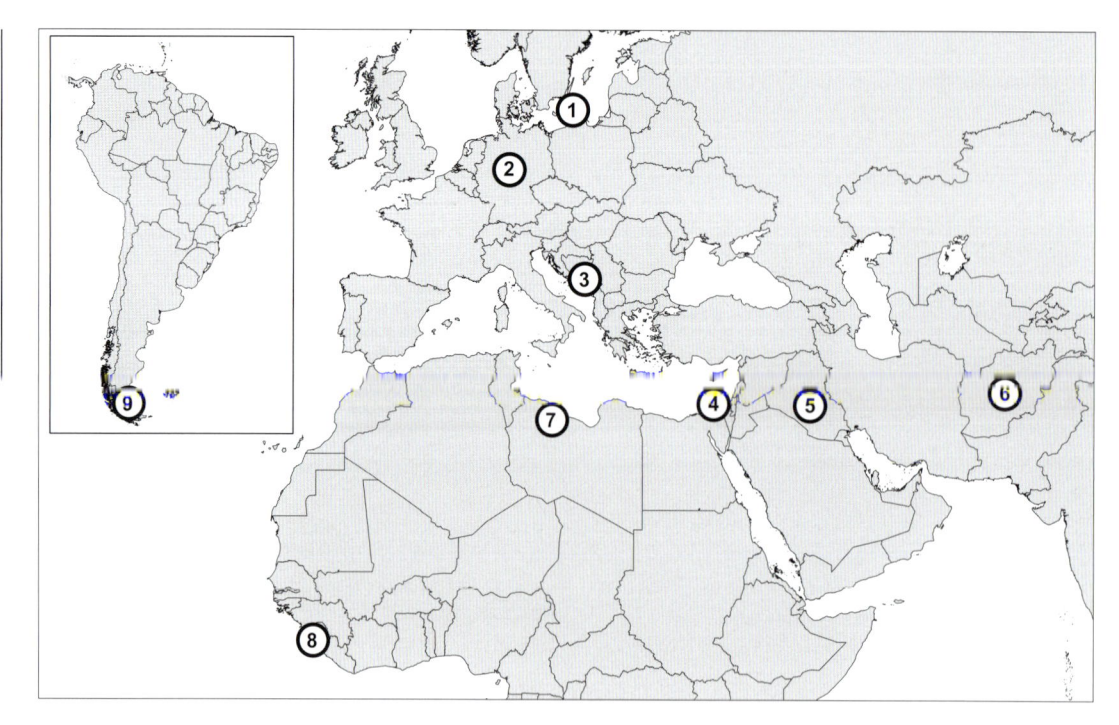

Nimrod R.1 XW666 seen at Wyton in October 1988. This aircraft was lost in an accident in May 1995, happily without loss of life. *T Panopalis Collection*

1990: Iraq (*Desert Shield/Storm*)

On 2nd August 1990, in the culmination of a long-running dispute, Iraqi armed forces invaded Kuwait. A multinational coalition, led by the US, was quickly formed to expel Iraq in an operation lasting eight months.

All three Nimrod R.1s were deployed to Akrotiri in August 1990 as part of Operation *Granby*, from where they initially flew sorties in support of *Desert Shield* during the build-up to the conflict, providing order-of-battle and communications intelligence on Iraqi forces. Two of the Nimrods also operated from a forward base in the Gulf during the conflict itself. Operations were completed on 28th February 1991 and the Nimrods returned to the UK the following month.

The Cold War finally came to an end in 1992 with the collapse of the Soviet Union. In theory this removed the primary justification for the existence of the Nimrod R.1 since the main role of the aircraft was Tactical Elint against the Soviet Bloc in the European theatre. In practice the aircraft had proved their usefulness in a variety of operations, particularly during the 1990 Gulf War, and all the indications were that an independent UK airborne Sigint capability would be required for some years to come. Hence a decision was made to retain the aircraft in service and continue to fund improvements and upgrades to the aircraft.

1993: Bosnia (*Deny Flight*)

Following the break-up of Yugoslavia in 1991 a vicious conflict broke out in Bosnia between the Serbs, Bosnians and Croats. The United Nations declared a 'no fly' zone over Bosnia in October 1992 in an attempt to limit aggression by Serbian forces and Nimrod R.1s were apparently detached to Italy in mid-1992 onwards to monitor Serbian forces. Following many breaches of the zone, NATO forces took steps in April 1993 to enforce the ban under Operation *Deny Flight* and Nimrod R.1s were regularly detached to the area to participate in the operation. In the period August-September 1995 *Deny Flight* was suspended while NATO forces took offensive action against the Serbs under Operation *Deliberate Force*. Operation *Deny Flight* was formally terminated in December 1995.

By 1995 RAF Wyton was being run down and 51 Squadron, along with its support units, moved base to RAF Waddington, home of the RAF's Boeing E-3D Sentry AEW.1 fleet. The first Nimrod flew in during April and the move apparently had little impact on operations.

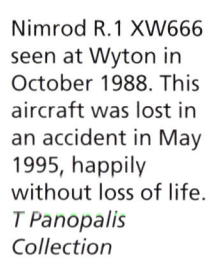

THE NIMROD IN SERVICE

51 Squadron moved to Waddington in 1995 where it joined the E-3 Sentry AEW.1s of 8 Squadron. Waddington became the RAF's main base for ISTAR (intelligence, surveillance, target acquisition and reconnaissance) operations and analysis. *MoD*

Second Upgrade: *Starwindow*

Plans for a second major upgrade of the Nimrod R.1 Elint suite were made circa 1991 under a project known as *Starwindow*. This time the work was contracted out to a specialist US contractor with experience in the airborne Sigint field, and a fixed-price contract for the conversion of three Nimrods was issued to E-Systems (later Raytheon E-Systems) in August 1992.

The *Starwindow* installation, to SRIM 6113, incorporated two high-speed search receivers, a network of 22 digital intercept receivers, a wideband digital direction-finding system and associated antennas, an improved capability against frequency-agile emitters, in-flight signal analysis equipment, the ability to generate real-time tactical data reports, and colour operator displays. The system has been described as 'computer-aided' suggesting some degree of automation and may have been derived from systems used in the USAF RC-135V/W *Rivet Joint* aircraft. The *Starwindow* installation was apparently augmented by a digital recording/playback subsystem, a multi-channel data signal demodulator and an enhanced pulsed signal processing capability introduced under a 1994 Nimrod 'Special Signals' proposal.

Starwindow is believed to have been installed in the first Nimrod R.1 during 1994, beginning flight testing in August of that year and a second aircraft was completed during 1995.

Loss of Nimrod XW666

On the 16th May 1995 the Nimrod R.1 fleet suffered its first and only loss when XW666 suffered a catastrophic engine failure during a test flight from Kinloss following a major service. The pilot (Flt Lt A Stacey) managed to ditch the aircraft safely in the Moray Firth before the wing structure burnt through. Luckily only seven crew were on board and weather conditions were good. All the crew managed to board a life raft and were quickly picked up by a RAF Sea King from nearby RAF Lossiemouth. A subsequent examination of the aircraft revealed the wing had been within 45 seconds of structural failure. Several members of the crew received awards for their actions during the ditching of XW666, with Flt Lt Stacey, the captain, receiving the Air Force Cross, Flt Sgt Hart the Queen's Commendation for Bravery in the Air while Flt Lt Hewitt, MAEOp Clay and Flt Sgt Rimmer received AOC-in-C's Commendations.

Nimrod R.1 XW666 in the 1980s, by which time the aircraft had acquired a hemp finish and Type B markings. This view shows the extent, and variety, of the antenna array along the fuselage spine. *Shaun Connor*

183

LISTENING IN

Nimrod R.1 XW666 was replaced by XV249 in November 2002 and is seen here completed to Project *Extract* standard.
T Panopalis Collection

The loss of XW666 was a major blow to the UK airborne Sigint programme since three Nimrods were just sufficient to meet tasking requirements. In many ways the situation was analogous to that encountered in 1959 following the loss of Comet XK663 in a hangar fire. This time however there appears to have been no argument regarding the need for a replacement, despite the projected £30M cost and authorisation was given within a matter of weeks. This must surely be testimony to the high regard with which the Nimrod was held by the UK defence establishment. By the end of June an aircraft (XV249) had been selected from one of four surplus Nimrod MR.2s stored at Kinloss. After a major overhaul at Kinloss, XV249 was delivered to BAe Woodford in October 1995 where it was stripped of its MR.2 role equipment and modified to R.1 standard with a new equipment bay, rear cabin layout and aerial suite. Following the structural modifications at Woodford the Nimrod was ferried to Waddington on 19th December 1996 for the Elint equipment fit. An installation to the new SRIM 6113 *Starwindow* standard was begun by the Electronic Warfare and Avionics Detachment (EWAD) at Waddington in January 1997 under Project *Anneka* (a reference to the *Challenge Anneka* TV series, in which the presenter, Anneka Rice, completed tasks to a tight deadline). Work was completed on 2nd April 1997 and XV249 was declared operational by the end of April 1997.

Yugoslavia, Kosovo and Sierra Leone

Nimrod R.1s were deployed to Italy for operations against the former Yugoslavia, enforcing the No-Fly zone. 51 Squadron were awarded the Battle Honour 'Kosovo 1999' for their participation in monitoring sorties during the Kosovo crisis. A Nimrod R.1 was also apparently detached to Sierra Leone in May 2000 to support UK operations there under Operation *Palliser*.

Third upgrade: Project *Extract*

Shortly after the completion of the *Starwindow* programme, plans were laid for a follow-on systems upgrade, codenamed Project *Extract*. The programme was apparently cancelled on cost grounds in November 1997, but then reinstated in 2000. Project *Extract* was intended to increase the level of automation of the Nimrod Elint suite and optimise the previously-fitted *Starwindow* system for RAF operating procedures. The upgrade introduced a new antenna suite, automated data collection avionics, and associated software facilities, including a central database and a data fusion capability. A £100M contract was awarded by the Defence Logistics Organisation to Raytheon Strategic Systems in 2000, covering provision of the Elint system itself, ground support equipment, rear crew trainer enhancements, simulation equipment and ten years logistic support. The three

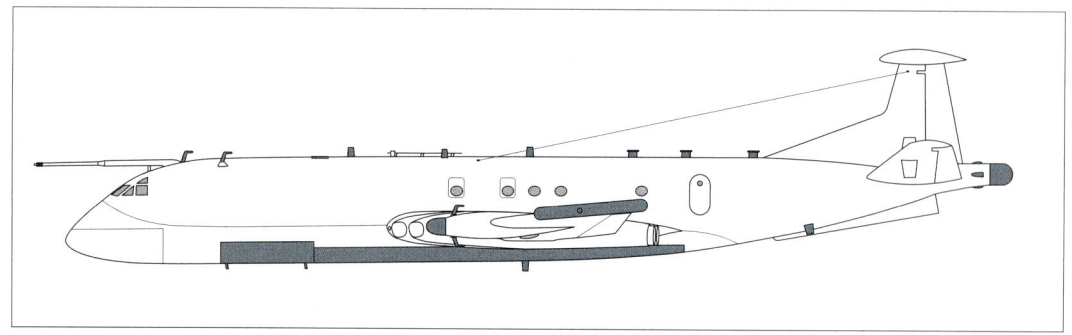

Nimrod R.1 to Project *Extract* standard. Note additional dorsal aerials on rear fuselage. These were flat disks mounted on short fins.

184

THE NIMROD IN SERVICE

Nimrod R.1 XW664 shown in February 2007 prior to a repaint. The lack of surface finish allows the various radomes and dielectric panels to be shown to advantage.
T Panopalis Collection

Nimrod R.1 XW665 at Akrotiri resplendent in an Afghanistan grey scheme.
Rob Swanson

Nimrods were upgraded sometime in the period 2001-2003 and following the aircraft installation programme, a Nimrod R.1 was detached to Patuxent River, Maryland on 17th April 2003 to test the system over the US Navy's Atlantic Test Ranges. After flying six missions, the aircraft returned to the UK on 4th May. In September 2003 the RAF announced that Project *Extract* had been successfully completed.

Operations 2003-2013

In March 2003 the UK joined with the US in Operation *Iraqi Freedom* to invade Iraq in order to overthrow the regime of Saddam Hussein. A Nimrod R.1 aircraft was deployed to the Gulf for operations against Iraq as part of Operation *Telic* (UK operations in support of *Iraqi Freedom*).

The Nimrod R.1 was also employed in Afghanistan, supporting operations there in 2001 during Operation *Enduring Freedom* following the attack on the World Trade Center. Given the nature of the Taliban and Al-Qaeda, the aircraft were almost certainly employed in the Comint role, intercepting ground communications. Further deployments were made from 2006 onwards in support of the expansion of operations in Afghanistan under Operation *Herrick*.

Comint upgrade: *Tigershark*

Operations in Afghanistan highlighted the need for improvements to the Nimrod's communications intercept capability, particularly against mobile phones, satellite phones and similar devices used by the Taliban, and resulted in the development by DRA/DERA/QinetiQ of a new Comint system, dubbed *Tigershark*. The

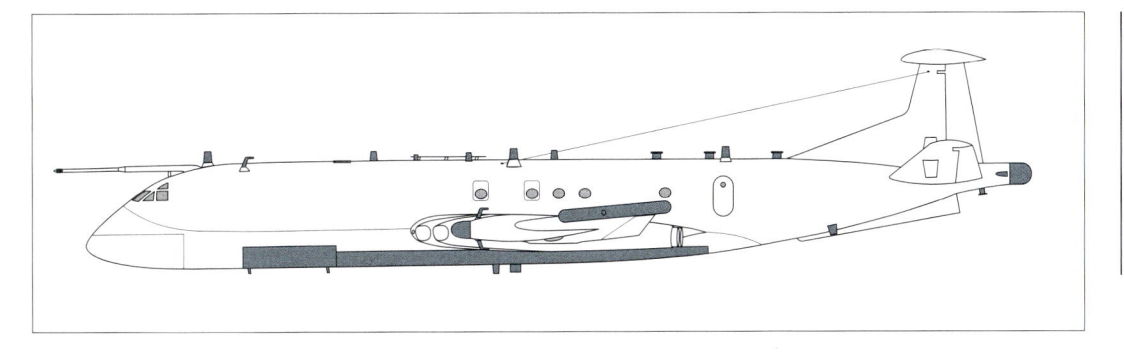

Nimrod R.1 to *Tigershark* standard. The modification added a number of communications intercept aerials offset from the fuselage centreline.

185

LISTENING IN

system, apparently developed from an earlier Army system, was installed in the three Nimrods circa 2008-2009 and involved the addition of five large blade aerials on top of the fuselage. The new system was apparently very successful, making the Nimrod the 'electronic surveillance platform of choice' for Coalition commanders in Afghanistan.

In May 2010 the Nimrod R.1 was deployed to the Arabian Gulf for its last major extended detachment prior to retirement. A 51 Squadron detachment arrived in the Gulf (Seeb, Oman) in the second week of May 2010 and the first Nimrod sortie was flown in support of Operation *Herrick* on 16th May. Approximately four sorties were flown every week during the detachment, which continued through until March 2011, making this last deployment the longest continuous deployment in the Squadron's history, clocking up 159 operational sorties in 1,177 flying hours.

Project *Helix*

Work on a fourth major upgrade of the Nimrod R.1 Sigint suite got under way in 2003 as Project *Helix*. The intention was to modernise the Nimrod aircraft systems, also upgrading ground analysis, training and support facilities. Project *Helix* was a hugely expensive undertaking, with an estimated cost of around £400M, and the programme was thus subject to a rigorous project planning and assessment process. The first phase of *Helix* began in August 2003 when eight companies were examined as possible participants in the project. By April 2004 three US companies (L-3 Communications, Lockheed

Right: Nimrod R.1 XV249 in June 2011 showing the new comint antennae arrayed along the upper fuselage. Note that the 'Flying Goose' has become less discreet as the years passed. *Jerry Gunner*

THE NIMROD IN SERVICE

A USAF Boeing RC-135V Rivet Joint in 1992. This would be the baseline model for the UK's Airseeker programme and many 51 Sqn personnel would train and fly operations on USAF RC-135Vs. T Panopalis Collection

Martin and Northrop Grumman) had been selected as likely contenders, based on their ability to meet the task. Selection of the preferred company, and validation of the project, was via a three-phase assessment process. The first phase required the companies to demonstrate an understanding of the project requirements and was completed in April 2005 with Northrop being eliminated from the process. The remaining two companies then produced detailed system proposals to meet the requirements. This second phase was completed in April 2007 when L-3 Communications (primary contractor for the US RC-135W/V *Rivet Joint* Sigint aircraft) was selected as the preferred contractor for *Helix*. The final, third phase of the programme was a risk-reduction exercise, to ensure that the proposed system was deliverable on time and on cost, and a £11.5M contract for the two-year risk-reduction phase was subsequently issued to L-3 during 2007.

The detailed examination carried out by L-3 under the risk-reduction exercise soon revealed that the cost of supporting the proposed *Helix* system in the Nimrod R.1 was likely to be much greater than originally expected. In 2008 it was decided to investigate alternatives to the Nimrod R.1 to meet the *Helix* requirement. One strong contender was the RC-135W/V *Rivet Joint*, which apparently met many, although not all, of the *Helix* requirements, and so an approach was made to the US government to explore the possibility of acquiring three of those aircraft. The US Congress was notified of the negotiations in October 2008 when the Defense Security Co-operation Agency revealed that the UK government was looking at the possibility of buying three *Rivet Joint* aircraft along with associated equipment (including ground station and mission trainer) and services. Total value of the sale was estimated at $1.068 billion. It appears that, although the US government agreed in principle to the sale in 2008, the Ministry of Defence then deferred a decision on the aircraft while it explored alternatives, including the use of the Nimrod MRA4. For a variety of reasons, the MRA4 ultimately proved unsuitable, not only for Elint, but for its maritime role as well, and a proposal to use three *Rivet Joint* aircraft to meet the *Helix* requirement was formally submitted in December 2009. The proposal was approved in March 2010 and on 1st April 2010 Project *Helix* was officially redesignated Project *Airseeker*. With the adoption of the RC-135W, plans were made to retire the Nimrod R.1s in 2011.

Libya

XW665 was retired from service on 27th October 2010 on its return from a deployment to Afghanistan, leaving 51 Squadron with two aircraft. Following the 2010 Strategic Defence & Security Review, the two remaining Nimrod R.1s were scheduled for early retirement, with a planned out-of-service date of 31st May 2011, later brought forward to 31st March. By late February 2011 the Nimrods were close to retirement, the support contracts were about to expire, personnel had been transferred, and contractors were already on site to begin dismantling the first aircraft. A retirement ceremony had been arranged for 31st March at Waddington.

However, the outbreak of the Arab Spring and the subsequent revolution in Libya led to a change of plans with a decision subsequently made to extend the service life of one aircraft by three months in order to support operations against Libya. A Nimrod R.1 Life Extension Team was formed to support the aircraft for the extra three months of service. Delaying the Nimrod out-of-service date for three months cost around £4M.

A Nimrod R.1 (XV249) was deployed to Akrotiri under Operation *Ellamy* on 4th March, (as part of 907 Expeditionary Air Wing) from where it flew sorties in support of UN Security Resolution 1973, monitoring government

Opposite page, clockwise from centre left:

Another view of Nimrod R.1 XV249 in June 2011, showing the Tigershark aerial array on top of the fuselage. The Nimrod R.1 was a much sought after asset during Operation Herrick, the British name for operations in Afghanistan. Jerry Gunner

Interior of Nimrod XW664 looking aft. The Nimrod, like the Comet it replaced, didn't have the most spacious of cabins but the installation engineers succeeded in producing a very capable set-up. Scott Hastings

Interior of Nimrod XW664, port side. The consoles on this side comprise a number of individual stations. Scott Hastings

Interior of Nimrod XW664 looking forward. By 2010 the main consoles had been arranged on the starboard side. Scott Hastings

187

LISTENING IN

A fine study of XV249 at Waddington in July 2010. The 51 Sqn had operated from Waddington since the consolidation of the RAF's ISTAR (intelligence, surveillance, target acquisition, and reconnaissance) assets at that base in 1995. *Stuart Freer/Touchdown Aviation*

Nimrod R.1 on the occasion of its retirement in June 2011, wearing a commemorative paint scheme. The 'Flying Goose' would be airborne once more on the RC-135W *Airseeker*. MoD

Nimrod R.1 XW664 at East Midlands Aeropark following decommissioning. XV665 was broken up, its nose sent to a German museum and XW249 went on display at the Cold War Museum. *Scott Hastings*

forces in Libya and the no-fly zone. XV249 was replaced by XW664 at the end of May, the latter aircraft remaining on detachment for nearly a month. Operation *Ellamy* came to an end towards the end of June and XW664 returned to Waddington on 24th June.

Retirement

The last two Nimrod R.1s were retired at the end of June 2011, shortly after XW664 returned from Cyprus, and a retirement ceremony, attended by 700 past and present RAF personnel, was held at RAF Waddington on 28th June. The last public appearance of the type in RAF service was at the Waddington Air Display in July 2011.

The Nimrod R.1, first conceived in 1966, had given the UK some 37 years of service and was by far the longest lived of all the UK's Elint aircraft. The second longest serving aircraft, the Canberra, only served 21 years. On that basis the country had got is money's-worth out of the Nimrod. The retirement of the Nimrod also marked the end of a more-or-less unbroken UK airborne Elint capability stretching back to 1946. The RAF decided to enjoy a three-year 'capability gap' while it waited for the new RC-135W/V *Airseeker* aircraft to arrive.

Following their retirement the aircraft were sold off. XW665, the first aircraft to retire, was broken up at Waddington in the summer of 2011, the nose later going to an aviation museum at Speyer in Germany. XW664 was sold to the East Midlands Aeropark, flying in to East Midlands Airport on 12th July where the aircraft underwent a decommissioning process to remove still-secret equipment and make the aircraft safe. Following this the Nimrod was towed to the adjacent Aeropark museum for static display. XW249 was allotted to the RAF Museum at Cosford. The aircraft was flown into Kemble in July 2011, then being dismantled and transported to Cosford in February 2012.

Postscript: Airseeker

Under Project *Airseeker* the UK ordered three RC-135W *Rivet Joint* aircraft as long-term replacements for the Nimrod R.1s.

In March 2011 a Memorandum of Understanding (MOU) was signed by the UK and US governments covering the future support and upgrade of the UK aircraft. Under the MOU the three RAF aircraft will be supported, maintained and upgraded by L-3 as part of a combined US/USAF fleet of 20 RC-135W/V. The aircraft will be returned to L-3 every four years for refurbishment and upgrade, ensuring the aircraft's Sigint systems remain up-to-date and relevant.

The UK aircraft were selected from low-hours KC-135R tanker aircraft (64-14833, 64-14838 and 64-14840). The first of these went into the L-3 Greenville plant for refurbishment and conversion in December 2010. The three RC-135Ws were allotted the serials ZZ664-6, matching the numbering of the previous Nimrod R.1s (XW664-6). The first aircraft (64-14833/ZZ664) was completed in April 2013 with the remaining two aircraft due to follow in 2015 (ZZ665) and 2017 (ZZ666).

Training the crews

In September 2010 the RAF signed a Co-Manning agreement with the USAF as part of the *Airseeker* project, covering the training and subsequent operational employment of UK aircrew and ground operators on US *Rivet Joint* aircraft and ground exploitation systems. The intention was to provide a cadre of trained and operationally experienced crews to man the UK *Rivet Joint* fleet. The first 22 51 Squadron personnel were assigned to the 338th Combat Training Squadron at Offutt AFB in January 2011. The training course length varied depending on specialisation, lasting between four and seven months. For the aircrew the initial two months were spent on the ground in classrooms and simulators before flying training started. The first members of the course graduated in mid-April 2011 and following completion of their training the 51 Squadron crews returned to the UK, subsequently flying operational sorties in USAF *Rivet Joint* aircraft.

By May 2013 over 100 51 Squadron aircrew had passed through the training process, and by the same date crews had logged 20,000 operational hours on the RC-135W/V during the course of 1,000 sorties over the Middle East and Mediterranean. The initial operational sorties were apparently flown in June 2011 when the first graduates of the *Airseeker* training programme participated in operations over Libya (*Unified Protector*). 51 Squadron crews subsequently took part in *Rivet Joint* sorties over Afghanistan (*Enduring Freedom*). During the course of the sorties a number of 51 Sqn pilots qualified as Aircraft Commanders, giving them overall responsibility for the aircraft, crew and mission.

The first RC135W *Airseeker* (ZZ664) was delivered ahead of schedule to the Royal Air Force at Waddington on 12th November 2013, the 'Flying Goose' emblem of 51 Sqn emblazoned on the fin.

Conclusion

The RAF pioneered Elint in the early 1940s, built up a proficiency in the task in the post-war era and, with limited resources, established a reputation in the field second to none. Despite the hiatus resulting from the retirement of the Nimrod and the introduction of the *Airseeker*, all the evidence suggests it will remain a potent force in the airborne Sigint business for years to come

Glossary of Terms

ACAS	Assistant Chief of Air Staff. Senior member of the Air Staff.
Air Ministry	Government department responsible for managing Royal Air Force. Merged into Ministry of Defence in 1964
Air Staff	Body of senior RAF officers responsible for the running of the Royal Air Force.
BAOR	British Army Of The Rhine
CAS	Chief of Air Staff. Head of the RAF.
Comint	Communications Intelligence. Signals intelligence gained from communications signals.
D/F	Direction Finding. Radio technique to determine relative bearing of a signal emitter.
DIS	Defence Intelligence Staff. Established 1964 as a clearing house for military intelligence, serving the MoD.
DV receiver	Detector Video receiver. Also known as Crystal Video receiver. Wide frequency range receiver using silicon diode detector.
Elint	Electronic Intelligence. Signals Intelligence gained from non-communications signals. Also known as Noise Listening in early post-war period.
FIR	Flight Information. Region of airspace.
GCHQ	Government Communications Headquarters. UK intelligence agency responsible for Signals Intelligence.
HF	High Frequency. Radio frequency range 3-30MHz.
Hz	Hertz. Unit of frequency. One cycle per second.
LSIB	London Signals Intelligence Board. Supervisory authority for Signals Intelligence, overseen by GCHQ.
LSIC	London Signals Intelligence Centre. Name used by GCHQ in early post-war years.
LSIC	London Signals Intelligence Committee. Sub-committee of the LSIB.
MoA	Ministry of Aviation. UK government agency. Took over responsibility for supply of military aircraft from the MoS in 1959. Superseded by MinTech in 1967.
MoD	Ministry of Defence. UK government department responsible for implementing defence policy. Assumed responsibilities of Air Ministry in 1964.
MoS	Ministry of Supply. UK Government agency responsible for equipping UK armed forces until 1959.
MinTech	Ministry of Technology. Took over functions of MoA in 1967.
RPF	Radio Proving Flight. Term given to an Elint sortie. Superseded 'Ferret Flight'.
R/T	Radio Telephone. Transmission of speech over radio.
S-band	Radio frequency range. 2-4GHz
Secretary of State for Air	UK government Cabinet member in charge of the Air Ministry. Position abolished in 1964.
Sigint	Signals Intelligence. Intelligence gained by the interception of eklectronic signals.
Superheterodyne receiver	Radio frequency receiver in which a received signal is mixed with a locally-generated signal to produce a signal at a fixed Intermediate Frequency.
VCAS	Vice Chief of Air Staff. Senior assistant to CAS.
VHF	Very High Frequency. Radio frequency range 30-300 MHz.
W/T	Wireless telegraphy. Transmission of morse over radio
X-band	Radio frequency range. 8-12GHz

Selected Bibliography

Aircraft in British Military Service: Vic Flintham; Airlife, 1998

By Any Means Necessary: William E Burroughs; Hutchinson, 2002

Electronic Warfare: AVM J P R Browne/Wg Cdr M T Thurbon; Brassey's, 1998

GCHQ: Richard J Aldrich; Harper Collins, 2011

Nimrod: The Centenarian Aircraft: Bill Gunston; Spellmount, 2009

Secretive Nimrods: Jon Lake; Air International, April 2008

The History of US Electronic Warfare Vols 1-2: Alfred Price; The Association of Old Crows, 1984

The USSR in Third World Conflicts: Bruce D Porter; Cambridge University Press, 1984

Many files (too numerous to list) at The National Archives (TNA)

Index

AIRCRAFT
727, Boeing 149
Airseeker, Boeing RC-135W 5, 187, 188, 189
Albatross, Grumman HU-16 128
Andover, HS748/780 164, 166, 167, 168, 169, 177
Anson, Avro Type 652A 10, 12, 13, 18
Badger, Tupolev Tu-16 141, 156
Beagle, Ilyushin Il-28 141
Bear, Tupolev Tu-95 6, 156
Bison, Myasishchev M-4 156
C-47, Douglas 21, 42
Camel, Tupolev Tu-104 79
Canberra, English Electric 4, 5, 6, 26, 50, 51, 52, 53, 54, 55, 57, 59, 60, 61, 62, 63, 64, 65, 66, 67, 68, 69, 70, 71, 73, 75, 76, 77, 78, 79, 80, 81, 82, 83, 84, 85, 86, 87, 89, 90, 91, 100, 101, 105, 108, 109, 110, 111, 112, 113, 114, 115, 118, 119, 120, 123, 124, 125, 126, 128, 131, 132, 133, 134, 135, 136, 137, 138, 139, 148, 150, 151, 152, 155, 156, 157, 161, 162, 173, 176, 188
Catalina, Consolidated 42
Comet, de Havilland 5, 6, 35, 50, 76, 87, 88, 90, 91, 92, 93, 94, 95, 96, 97, 98, 99, 100, 101, 102, 103, 104, 108, 110, 113, 115, 116, 117, 118, 120, 121, 122, 123, 124, 125, 126, 127, 128, 129, 131, 132, 133, 134, 135, 136, 137, 138, 139, 140, 141, 142, 143, 144, 145, 146, 147, 148, 149, 150, 151, 152, 155, 156, 158, 159, 160, 161, 162, 163, 165, 167, 168, 169, 171, 172, 173, 174, 175, 176, 184, 187
Fagot, Mikoyan-Gurevich MiG-15 42, 59, 68
Fang, Lavochkin La-11 21, 22, 30
Farmer, Mikoyan-Gurevich MiG-19 106, 141
Fin, Lavochkin La-7 21
Fishbed, Mikoyan-Gurevich MiG-21 141, 147, 148
Flashlight, Yakovlev Yak-25 80, 81, 84, 85
Frank, Yakovlev Yak-9 21
Fresco, Mikoyan-Gurevich MiG-17 66, 67, 68, 80, 81, 141
Fritz, Lavochkin La-9 21
Halifax, Handley Page 8, 10, 12, 26
Hastings, Handley Page HP.67 7, 44, 46, 50, 75, 122, 187, 188
Hercules, Lockheed C-130 132, 149, 165
Hormone, Kamov Ka-25 151
HS801(R), Hawker Siddeley 158, 161, 162, 163, 165, 166, 167, 169, 171, 173
Hurricane, Hawker 56
Hustler, Convair B-58 153
Invader, Douglas FA-26C/RB-26 35
Lancaster, Avro Type 683 8, 10, 11, 12, 13, 14, 15, 16, 17, 18, 20, 21, 22, 23, 24, 25, 27, 28, 29, 31, 50, 59
Lincoln, Avro Type 694 16, 17, 18, 22, 23, 24, 26, 27, 28, 29, 31, 32, 33, 34, 36, 37, 38, 39, 40, 41, 42, 43, 44, 48, 50, 51, 59, 88, 89, 90, 103
Mercator, Martin P4M 42
Meteor NF.14, Gloster 68, 87
Mirage 5, Dassault 149
Mosquito, de Havilland 8, 10, 12, 15, 17, 18, 21, 24, 26, 27, 29, 34, 51, 52
Neptune, Lockheed P-2/MR.1 73
Nimrod, Hawker Siddeley 4, 5, 6, 104, 121, 123, 125, 138, 151, 155, 159, 161, 162, 163, 165, 167, 168, 169, 170, 171, 172, 173, 174, 175, 176, 177, 178, 179, 180, 181, 182, 183, 184, 185, 186, 187, 188, 189, 190
Phantom, McDonnell Douglas F-4 147, 148
Privateer, Consolidated PB4Y-2 30, 31, 33, 37, 42
RB-57F, Martin 134, 137
RB-66C, Douglas 134, 153
RC-135, Boeing 5, 106, 129, 130, 131, 146, 153, 183, 187, 188, 189
Rivet Joint, Boeing RC-135W 5, 183, 187, 189
Sabre, North American F-86F 137
Sentry, Boeing E-3 182, 183
Shackleton, Avro Type 696 57, 76, 86, 92, 93, 94
Skymaster, Douglas C-54 35, 36
Skywarrior, Douglas EA-3B 150
Stratojet, Boeing RB-47 66, 67, 84, 105, 106, 107, 108, 109, 110, 114, 116, 118, 120, 129, 130, 131, 132, 133, 134, 153
Superfortress, Boeing B-29/RB-29/F-13 31, 44, 45, 46, 47, 49, 51
Thunderbolt, Republic P-47 56
U-2, Lockheed 105, 106, 107, 109, 113, 116, 133
V.1000, Vickers 50, 51, 89, 91, 92, 158
Valiant, Vickers Type 706 6, 35, 40, 41, 50, 65, 72, 133, 134
Varsity, Vickers 69, 78, 90, 97, 102, 103, 104, 105, 122, 123, 154, 177
VC10, Vickers 103, 117, 158, 159, 160, 161, 180, 181
Victor, Handley Page HP.80 6, 7, 50, 72, 101, 113, 114, 132, 136, 180
Vulcan, Avro 698 6, 50, 72, 168
Washington 5, 6, 40, 43, 44, 45, 46, 47, 48, 49, 50, 51, 52, 54, 55, 56, 57, 58, 59, 61, 62, 63, 66, 68, 69, 70, 71, 72, 73, 74, 75, 76, 77, 78, 79, 80, 82, 83, 84, 85, 86, 87, 91, 92, 94, 95, 97, 98, 99, 100, 109, 152
Wellington, Vickers 6, 7, 8
York, Avro Type 685 101

ELINT PROJECTS AND REQUIREMENTS
Airseeker 5, 187, 188, 189
Anneka 184
ASR.1851 178, 179
ASR.389 163
ASR.817 155
ASR.871 174
Astral Box 178, 179
Extract 41, 184, 185
Helix 186, 187
OR.3614 192
Starwindow 183, 184
Tigershark 185, 187
Zabra 125, 156, 157

EQUIPMENT
AI.21 radar 68
Air Position Indicator (API) 65, 95
ALA-6 D/F system 101, 102, 104, 106, 117, 118, 121
ALA-74 D/F system 106, 118
ALD-4 automatic Elint 118, 153
ALR-8 receiver 120, 121, 172
AMES Type 6 radar 21
ANQ-1A magnetic wire recorder 46, 48, 61, 64, 65, 66, 73, 77, 97, 104
APA-11 signal analyser 46, 48, 54, 55, 64, 65, 76, 97, 101, 104, 117
APA-11A signal analyser 54, 104
APA-17 direction finder 45, 47, 52, 73, 86, 90, 96, 97, 98, 101, 104
APA-24 direction finder 27, 28, 29, 33, 37, 38, 39, 45
APD-4 automatic Elint 106, 118, 153
APQ-13 radar 47, 62, 80, 84
APR-17 receiver 106, 120
APR-4 receiver 10, 12, 14, 16, 21, 22, 27, 32, 37, 43, 45, 99
APR-4Y receiver 99
APR-5 receiver 15, 22, 32, 37, 43, 45, 52, 53
APR-9 receiver 46, 48, 52, 54, 55, 61, 62, 64, 65, 76, 101, 112, 121
APS-20 radar 73
ARAR ESM suite 167
ARAX ESM suite 167
ARI 18021 receiver 40, 45, 47, 52, 54, 73, 76, 86, 96, 111, 112
ARI 18050 receiver 101, 117, 172
ARI 18058 receiver 86, 89, 101
ARI 18147 receiver 111
ARI 5428 radio compass 54, 89
ARR-5 comms receiver 12, 14, 16, 37, 63, 64, 65, 66, 104
ARR-8 receiver 37
ASD-1 automatic Elint 129, 130, 146, 153
Automatic Y Elint system 24, 25, 26
Bagful (R.1622) receiver 10, 25, 26
BC-348 receiver 63
Big Mesh 141
Blonde (R.1645) receiver 10, 16, 25, 26, 64, 65, 77, 98
Blue Shadow SLAR 52, 53, 54, 55, 60, 61, 62, 64, 65, 70, 76, 86, 90, 95, 96, 112, 120, 128, 131, 137
Breton Elint system 86, 88, 89, 90, 95, 98, 101, 102, 104, 110, 111, 112, 116, 119, 120, 121, 152
Collins 618/T3 radio 120
Complex Signal Recognition Unit (CSRU) 172
CPS-6 radar 59
CRT-2 analyser 101, 102, 104
Decca navigation system 17, 38, 40, 62, 63, 95, 136, 141, 143, 155
DLD-1 Elint system 153, 154
Dumbo, P-3 radar 21, 28, 59
Ekco E160 radar 120
Fire Shield radar 22
Flange, ARI 18021 receiver 39, 40, 43, 45, 46, 48, 52, 53, 54, 55, 61, 64, 65, 73, 76, 88
Flat Face, P-15 radar 147, 170, 171
Gee navigation sytsem 26, 52, 53, 54, 76, 89, 95
Goa, SA-3 missile 147, 170, 171
Green Satin, ARI.5871 Doppler navigaor 64, 65, 68, 69, 70, 73, 75, 76, 77, 82, 84, 90, 95, 112
Ground Position Indicator (GPI) 65, 70, 95
GSQ-17 Finder playback system 153
Guideline, SA-2 missile 105, 141, 147, 149, 164, 170, 171
H2S ground mapping radar 14, 17, 18, 26, 29, 32, 33, 36, 37, 38, 40, 132
H2S Mk.9A radar 132
Instantaneous Frequency Indicator (IFI) 112, 172, 173
Integrated Display and Measurement System (IDMS) 104, 172, 173
Kalki, SLIS 155
Kite Hawk radar 68
Knife Rest, P-8 radar 59, 171
Nysa radar 141
Passive ECM Tactical Aid (PETA) 164, 165
Pegmatit, radar 21
Periscopic Sextant 63, 64, 65, 76, 95
PTR 175 radio 120
QRC-7 receiver 73, 76, 77
R.216 receiver 104, 117, 121
RAK-2 D/F system 167, 172
RUS-2, radar 21, 22, 28, 59
Scan Fix, SRD-1 radar 67
Scan Odd, RP-1 Emerald radar 67, 68, 81
Scan Three, RP-6 radar 80, 81, 84
SCR-527 radar 21
SCR-584 radar 59
Sea Dart, HSD missile 180
Sideways Looking Intercept System (SLIS) 153, 154, 155, 172, 173
Taildog missile 155, 156
Tall King, P-14 radar 131, 171
Thorium, radar 67, 68
Token, P-20 radar 59, 62
TR1934 radio 76
Watton Box D/F system 27, 28, 29, 33, 37, 38, 39
Wurzeburg, radar 192
X.390 tail-warning receiver 111, 112
X.449 guard receiver 172
X.450 guard receiver 172
X.451 guard receiver 172

LOCATIONS
Aberporth 179
Agra 142, 143
Akrotiri 85, 113, 117, 118, 123, 124, 133, 136, 139, 141, 143, 144, 145, 147, 148, 149, 151, 170, 176, 180, 182, 185, 187
Albania 20, 23, 28, 77, 146
Andøya 126, 127, 128, 129, 150, 151, 178
Baghdad 20, 21, 61
Bahrain 113, 135, 139
Baltiysk 57
Barents Sea 58, 77, 78, 79, 83, 84, 87, 100, 101, 103, 106, 108, 109, 113, 114, 123, 126, 127, 128, 129, 130, 131, 159
Berbera 178
Berlin 13, 15, 19, 20, 62, 105, 170
Black Sea 19, 20, 21, 40, 41, 42, 43, 47, 48, 55, 56, 57, 61, 64, 65, 69, 70, 72, 74, 75, 77, 78, 79, 81, 84, 85, 100, 101, 103, 109, 113, 118, 132, 133, 134, 135, 136, 137, 150, 151
Bodø 77, 78, 84, 87, 100, 103, 113, 126, 127, 128, 178
Borneo 140, 141, 142
Boscombe Down 98
Bremerhaven 30
Cape Gata 146, 148
Cape St Vincent 72
Cape Svyatoy Nos 106
Caspian Sea 28, 42, 43, 56, 57, 61, 62, 64, 65, 66, 68, 79, 80, 84, 85, 103, 113, 124, 135, 136, 137, 176
Changi 136, 141, 142, 143
China 4, 66, 139, 142, 143
Cocos Islands 140, 141
Concepción 180
Czechoslovakia 19, 124
Darwin 140, 141
Defford 54, 55, 65, 68, 70, 76, 90
Denmark 58, 84, 170
Dhofar 139
Djakarta 141, 142
Düsseldorf 17
Egypt 23, 43, 68, 80, 82, 83, 85, 86, 100, 123, 124, 139, 143, 144, 145, 146, 147, 148, 149, 165, 176, 182
Eielson AFB 129, 130
Eindhoven 13, 15, 17
El Adem 98, 118, 135, 149, 151
English Channel 72
Falkland Islands 179
Fayid 23, 43
Frankfurt 17, 30
Fürstenfeldbruck 21
Gan 141
Germany 8, 9, 13, 14, 15, 18, 19, 20, 21, 23, 28, 29, 30, 31, 34, 35, 40, 41, 43, 57, 59, 60, 62, 66, 73, 77, 100, 113, 122, 123, 124, 131, 132, 134, 164, 170, 176, 177, 182, 188
Germany, East (DDR) 43, 122, 131, 132
Germany, West (BRD) 30, 34, 59, 60, 73, 100, 124, 131, 164, 170
Henlow 40, 73
Hong Kong 142, 143
Idris 70
Incirlik 133, 134
Iran 43, 56, 61, 79, 85, 87, 113, 123, 124, 135, 136, 137, 139, 176, 178
Iraq 20, 27, 28, 43, 48, 59, 61, 69, 77, 101, 113, 135, 136, 137, 138, 139, 165, 182, 185
Israel 68, 69, 85, 100, 143, 144, 145, 147, 149
Kassel 17
Khormaksar 135, 139, 140, 141
Kiel 43
Kola Peninsula 66, 77, 78, 79, 81, 83, 84, 106, 127
Kuwait 138, 182
Latvia 30, 43
Libya 70, 98, 141, 149, 151, 176, 182, 187, 188, 189
Lithuania 30, 43
Lofoten Islands 73, 128
Lübeck 18, 34
Malaysia 113, 140, 141
Malta 20, 22, 23, 28, 43, 74, 77, 101, 136, 145, 151
Masirah 125, 140, 176, 178
Mehrabad 137, 138
Nicosia 43, 61, 68, 69, 75, 77, 78, 79, 80, 82, 83, 84, 85, 86, 87, 99, 100, 101, 105, 133, 148
Nordholz 157
North Borneo 140, 141
North Cape 73, 78, 83, 106
North Sea 17, 57, 58, 70, 72, 73, 75, 101
Norway 41, 58, 78, 83, 106, 108, 113, 114, 117, 126, 127, 128, 129, 150, 170, 178

191

LISTENING IN

Novaya Zemlya 77, 78, 79, 81, 84, 126, 127, 131
Obernkirchen 32, 60
Offutt AFB 129, 130, 146, 189
Oman 139, 186
Paris 13, 15, 17, 105
Patuxent River 185
Pechora Sea 131
Pershore 156, 157
Persian Gulf 113, 134, 135, 137, 177
Poland 28, 30, 36, 43, 61, 66, 81, 131, 132
Portugal 72
Putlos 60
Red Sea 139, 178
Sabah 141
San Felix Island 180
Sarawak 140, 141
Saudi Arabia 139, 178
Seeb 186
Shaibah 43, 61
Sharjah 113, 117, 123, 135, 136, 137, 138, 139, 141, 142, 143, 144, 145
Silifke 132, 133
Singapore 113, 117, 140, 141
Skagerrak 72, 73
Soviet Union 6, 7, 19, 20, 21, 22, 23, 27, 28, 30, 42, 56, 57, 59, 61, 66, 67, 80, 83, 105, 106, 108, 109, 113, 114, 125, 126, 127, 128, 130, 131, 134, 137, 139, 144, 149, 150, 161, 165, 176, 177, 178, 182
Syria 68, 69, 80, 82, 83, 85, 86, 100, 123, 142, 143, 144, 145, 146, 147, 148, 149, 176, 182
Taif 178
Tengah 140, 141
Tiflis 81
Turkey 43, 48, 56, 61, 70, 79, 83, 85, 106, 113, 118, 132, 133, 134, 136, 137, 139, 143, 170
Turkmenistan 137
United Arab Republic, UAR 144, 145, 147, 148
Vietnam 164
Wiesbaden 30, 35, 134
Woodford, BAe 175, 180, 184
Yemen 125, 139, 140, 176
Yugoslavia 20, 21, 23, 28, 77, 182, 184

MISCELLANEOUS
Comint 8, 53, 125, 139, 141, 165, 167, 170, 172, 173, 174, 185, 186, 190
Future Combat Aircraft Study 170
Military Defence Aid Programme (MDAP) 28, 45, 116
North Atlantic Treaty Organisation (NATO) 6, 41, 57, 68, 83, 105, 123, 126, 135, 165, 166, 169, 170, 171, 182
Polaris, Lockheed UGM-27 6
Service Radio Installation Modification (SRIM) 54, 76, 97, 98, 99, 101, 102, 111, 112, 117, 118, 119, 120, 125, 128, 174, 175, 177, 178, 179, 180, 183, 184
Sigint 8, 9, 11, 15, 21, 30, 93, 116, 117, 118, 120, 124, 126, 132, 134, 135, 139, 141, 143, 149, 151, 159, 161, 166, 167, 168, 170, 171, 173, 175, 177, 182, 183, 184, 186, 187, 189, 190
Six Day War 144, 145
Suez Crisis 82, 83, 85
Treasury, HM 91
Yom Kippur War 149

OPERATIONS AND EXERCISES
Acme 176, 179, 180
Adage, Op 148
Adjunct, Op 62
Agatha, Op 127
Amami, Op 141, 142
Angle, Op 140
Barrow, Op 62
Beat, Op 179
Blountstown, Op 128
Bonaparte, Op 60, 61
Border, Flight 33, 34, 36, 40, 41, 43, 57, 62, 63, 71, 73, 81, 82, 100, 102, 106, 123, 124, 138, 164
Boundary, Flight 92
Breach, Op 66
Brigand, Op 69
Brigham, Op 69
Brim, Op 148
Brimstone, Op 62
Brindled, Op 69
Brummel, Op 132
Bung, Op 128
Burglary, Op 157
Calomel, Op 75
Camelia, Op 61
Capella, Op 148
Careen, Op 148
Catarrh, Op 63
Chianti, Op 86
Claret, Op 58, 70, 71, 72, 73, 79, 100
Crawford, Op 142
Dachs, Op 147
Dagger, Exercise 20
Dance, Op 178
Deliberate Force, Op 182
Denver Thomas, Op (US) 114
Deny Flight, Op 182
Desert Storm, Op (US) 192
Dubbin, Op 179
Egma, Op 140
Ellamy, Op 187, 188
Encon, Op 127
Enduring Freedom, Op (US) 185, 189
Ewe, Op 147
Ferret, Flight 19, 20, 21, 22, 23, 24, 28, 29, 34, 35, 41, 51, 190
Fervency, Op 151
Fifi, Op 129
First Flight, Exercise 43, 55, 157, 177
Fitzroy, Op 150
Flame, Op 139
Fortitude, Op 101
Foxtrot, Exercise 24
Frigid, Op 77
Gamash, Op 42, 43
Gentry, Op 78
Genus, Op 139
Glidden, Op 141
Gobble, Op 179
Granby, Op 182
Grape, Op 70
Harley, Op 144
Harpoon, Op 4, 156
Herrick, Op 185, 186, 187
Hobble, Op 43
Intact, Op 144
Iraqi Freedom, Op (US) 185
Jetsam, Op 61
July Osbourne, Op (US) 118
Jungle King, Exercise 48
Lancer, Exercise 24
Lebanon, Op 66, 101
Limbo, Op 150, 151
Local, Flight 25, 63
Long, Op 129
Lovelock, Op 192
Lyric, Op 143, 149
Mainbrace, Exercise 41, 42
Mashie, Op 43
Massive, Op 114, 165, 174, 178
Mensa, Op 176
Midcourse, Op 178
Minority, Op 66
Mirimar, Op 142
Mogul, Op 66
Monitor, Flight 16, 24, 32, 33, 36, 37, 48, 50, 52, 53, 54
Moselle, Op 70, 73, 100
Mourne, Op 62
Nadir, Op 127
Olympic, Op 140, 141
One Step 23
Palliser, Op 184
Pastry, Op 151
Permit, Op 139, 142
Planetary, Op 128
Plume, Op 128
Polymenus, Op 143
Porcupine, Exercise 27, 28
Possum, Op 43, 44, 47, 48
Probate, Op 61
Radio Proving Flight (RPF) 29, 47, 91, 190
Reason, Op 57, 58, 70
Revivor, Op 66
Ripper, Op 109
Rut, Op 131
Saint, Op 139
Salom, Op 148
Shelbourne, Op 143
Sherry, Op 76, 100
Shingle, Op 77
Sprung, Op 176
Stimulation, Op 132
Tarnish, Op 140, 141
Telic, Op 185
Thorn, Op 138
Tobias, Op 127
Unified Protector, Op (US) 189
Vicarage, Op 178
Wild Duck, Exercise 73, 75
Wild Goose, Exercise 73

ORGANISATIONS
51 Squadron 4, 5, 101, 102, 103, 105, 106, 109, 112, 113, 115, 116, 117, 118, 119, 120, 121, 122, 123, 124, 125, 126, 128, 129, 131, 133, 135, 137, 139, 140, 141, 142, 143, 147, 148, 156, 157, 158, 159, 162, 163, 164, 171, 174, 175, 176, 177, 180, 181, 182, 183, 184, 186, 187, 189
58 Squadron 113
90 (Signals) Group 14, 15, 88
54 Signals Unit 141, 173
100 Group 8, 9, 36
192 Squadron 8, 9, 10, 12, 14, 26, 36, 40, 41, 43, 46, 48, 49, 53, 54, 57, 58, 59, 61, 63, 65, 68, 70, 71, 72, 73, 74, 76, 77, 78, 82, 83, 84, 86, 87, 88, 90, 94, 97, 98, 101, 110, 135
216 Squadron 98, 115, 121
543 Squadron 113, 133
ADI (Science) 23
Admiralty, The 23, 57, 150
Air Council 23, 92
Air Ministry 8, 10, 11, 13, 14, 15, 16, 17, 18, 19, 23, 24, 25, 26, 27, 28, 29, 31, 32, 33, 34, 35, 36, 37, 40, 41, 42, 43, 44, 45, 46, 48, 50, 51, 52, 53, 54, 55, 56, 57, 58, 61, 62, 63, 64, 65, 66, 68, 69, 70, 71, 72, 74, 76, 81, 82, 83, 84, 85, 87, 90, 91, 92, 93, 94, 95, 96, 97, 100, 107, 109, 112, 113, 116, 133, 135, 146, 160, 161, 164, 168, 190
Air Staff, The 4, 24, 50, 53, 55, 78, 94, 95, 96, 154, 163, 165, 168, 169, 170, 190
Aircraft and Armament Experimental Establishment (A&AEE) 157, 175
Baltic Fleet 150
Bletchley Park 11
Bomber Command 8, 10, 16, 17, 18, 23, 24, 31, 32, 40, 41, 44, 45, 46, 48, 50, 64, 69, 70, 71, 74, 78, 81, 83, 87, 94, 105, 131, 132, 137
Bomber Support Development Unit (BSDU) 9, 10
Boulton-Paul 192
British Naval Intelligence 57
British Overseas Airways Corporation (BOAC) 50, 93, 94, 95, 96, 103
Central Reconnaissance Establishment (CRE) 104, 105, 109, 110, 111, 112, 113, 114, 115, 117, 118, 119, 120, 123, 125, 137, 152, 153, 154
Coastal Command 9, 10, 11, 13, 14, 15, 16, 17, 18, 19, 20, 21, 22, 23, 24, 25, 26, 27, 28, 29, 31, 32, 33, 34, 35, 36, 37, 39, 40, 41, 43, 44, 45, 46, 48, 49, 50, 51, 52, 53, 54, 55, 57, 58, 59, 60, 61, 62, 63, 64, 65, 68, 69, 70, 71, 73, 84, 86, 87, 88, 89, 90, 91, 92, 93, 94, 95, 96, 97, 98, 99, 100, 101, 102, 103, 104, 162, 167, 173
Defence Intelligence Staff 40, 128, 134, 145, 146, 148, 151, 156, 173, 190
Electronic Warfare Avionic Unit (EWAU) 177, 178, 179
Electronic Warfare Engineering & Training Unit (EWE&TU) 123, 175, 177
Electronic Warfare Support Unit (EWSU) 118, 123, 156, 174, 175
Electronic Warfare Support Wing (EWSW) 123, 174
Foreign & Commonwealth Office (FCO) 124, 125, 135, 137, 146, 147, 149
Foreign Office Research & Development Establishment (FORDE) 11, 14
Government Code & Cypher School (GC&CS) 10
Government Communications Headquarters (GCHQ) 10, 21, 36, 40, 41, 80, 103, 112, 116, 119, 134, 148, 151, 161, 162, 168, 170, 173, 190
Imperial Iranian Air Force 56, 137, 139
Installation Flight 33, 37, 46, 48, 64, 68, 90, 98, 99, 101
Joint Anti-Submarine School 23
Joint Intelligence Committee (JIC) 93, 101, 106, 107, 108
London Signals Intelligence Board (LSIB) 10, 11, 14, 31, 36, 41, 63, 93, 101, 161, 163, 167, 190
London Signals Intelligence Centre (LSIC) 10, 11, 14, 21, 153, 154, 155, 158, 161, 190
Marshalls of Cambridge 95, 116
Middle East Air Force (MEAF) 68, 83, 86
Ministry of Aviation 116, 117, 120, 152, 155, 156, 190
Northern Fleet 58, 150, 151, 159
Protivo Vozdushnaya Oborona (PVO) Strany 21
Radio Engineering Unit 9, 40, 73, 98
Radio Warfare Establishment (RWE) 9, 10, 11, 12, 13, 14, 16, 17, 24
Radio Warfare Squadron 36
Signals Command 18, 92, 103, 116, 118, 123, 170, 172
Signals Command Air Radio Laboratory (SARL) 118, 123, 172, 173, 174, 175
Strike Command 125, 148, 156
Telecommunications Research Establishment (TRE) 10, 25, 26, 27, 53, 54, 55, 65
Transport Command 44, 92, 95, 98, 101, 115, 116, 117, 121, 142, 159, 160, 161
United States Air Force (USAF) 5, 35, 36, 41, 42, 44, 47, 66, 67, 69, 70, 74, 84, 91, 106, 107, 109, 110, 113, 118, 129, 130, 131, 132, 134, 146, 147, 183, 187, 189
United States Navy (USN) 33, 37, 42, 134
Y Wing 9, 10, 11, 13, 14, 15, 18, 21

PEOPLE
Attlee, Clement 20
Bevin, Ernest 20
Bridges, Sir Edward 91, 92, 93, 94
Bufton, Air Vice-Marshal Sydney 78
Bulganin, Nikolai 79
Churchill, Sir Winston 42, 55, 58, 63
Cook, Sir William 168
Crabb, Commander Lionel 79, 80, 82, 83
Dickson, Marshal of the Royal Air Force Sir William F 56
Douglas-Home, Sir Alec 113, 114
Earl Alexander of Tunis 42
Earle, Air Chief Marshal Alfred 117
Eden, Anthony 41, 80
Ellwood, Air Marshal Sir Aubrey 24
Elworthy, Air Chief Marshal Sir Charles 137
Grandy, Air Chief Marshal Sir John 168
Guest, Air Vice-Marshal Charles E N 31
Hailsham, Lord 114
Hampshire, Professor Stuart 159
Healey, Denis 128
Henderson, Arthur 20
Hooper, Leonard 'Joe' 116
Ivelaw-Chapman, Air Chief Marshal Sir Ronald 71, 80
Khrushchev, Nikita 79, 106
Kyle, Air Chief Marshal Sir Wallace 137
Lemnitzer, General Lyman 133
Lloyd, Selwyn 31, 40, 55, 57, 83
Lord De L'Isle and Dudley 41, 42, 55, 56, 68, 69, 71, 74, 93
Lygren, Ingeborg 128
Malinovsky, Marshal 106
Mavor, Air Vice-Marshal Leslie 158, 168
Mazdon, Sqn Ldr JF 10
Mountbatten, Lord Louis 105, 133
Nicholas, Air Commodore B D 87
Powers, Gary 105
Qaddafi, Colonel Muammar 149
Sanders, Air Marshal Sir Arthur P M 4, 6, 28
Shah of Iran, The 137, 177, 178
Suharto, General Haji Muhammad 142
Sunay, General Cevdet 133
Thomson, George 142
Tidemand, Otto 128
Tordella, Dr Louis 21
Wade, Air Marshall Sir Ruthven 176
Walmsley, Air Marshal Sir Hugh S P 24
Ward, George 100, 109
Wellesley, Arthur, 1st Duke of Wellington 6
Wilson, Harold 118, 124, 169
Young, Air Commodore B P 137

RAF STATIONS
Ahlhorn 86, 100
Aldergrove 33, 87, 88
Ballykelly 23
Brize Norton 106, 146
Fairford 66
Foulsham 8, 10
Gütersloh 75, 77, 78, 84, 122, 125
Habbaniya 20, 21, 28, 43, 47, 48, 61, 62, 65, 66, 69, 77, 80, 85, 135
Jever 85, 87, 109, 112
Kinloss 73, 150, 170, 178, 179, 180, 183, 184
Labuan 141, 142, 143
Lossiemouth 126, 150, 183
Luqa 43, 65, 69, 74, 77, 79, 84, 85, 86, 101, 125, 134, 135, 136, 143, 144, 149, 151, 170, 176
Lyneham 22, 98, 115, 121
Marham 45, 58, 78, 84, 180
Medmenham 88
Mildenhall 130
Muharraq 135, 139
Schleswigland 20, 21, 28
Shepherds Grove 12, 13, 14, 17, 18, 20, 22, 29
St Eval 23, 27, 29
St Mawgan 151, 177
Watton 6, 9, 10, 11, 12, 14, 16, 17, 18, 27, 28, 29, 33, 34, 35, 37, 38, 39, 45, 47, 48, 53, 54, 58, 63, 65, 68, 69, 70, 71, 72, 73, 74, 75, 76, 77, 78, 80, 83, 84, 85, 86, 87, 88, 89, 97, 98, 101, 102, 111, 112, 113, 115, 117, 118, 119, 120, 122, 127, 131, 174, 175
Wünstorf 29, 43, 59, 60, 61, 62, 63, 66, 73, 75
Wyton 4, 6, 113, 123, 124, 125, 129, 137, 142, 156, 157, 170, 173, 175, 176, 177, 178, 179, 180, 182

SHIPS
General Belgrano 179
HMS Ambush 33
HMS Anchorite 33
HMS Cardiff 180
Leningrad 151
Moskva 150, 151
Neva 72
Ordzhonikidze 79
Sverdlov 57, 58, 73, 79, 85
USS Liberty 144